Chi-Squared Goodness of Fit Tests with Applications

Chi-Squared Goodness of Fit Tests with Applications

V. Voinov
KIMEP University;
Institute for Mathematics and Mathematical Modeling
of the Ministry of Education and Science,
Almaty, Kazakhstan

M. Nikulin
University Bordeaux-2,
Bordeaux, France

N. Balakrishnan
McMaster University, Hamilton,
Ontario, Canada

ELSEVIER

AMSTERDAM • BOSTON • HEIDELBERG • LONDON
NEW YORK • OXFORD • PARIS • SAN DIEGO
SAN FRANCISCO • SINGAPORE • SYDNEY • TOKYO

Academic Press is an Imprint of Elsevier

AP

Academic press is an imprint of Elsevier
225 Wyman Street, Waltham, MA 02451, USA
The Boulevard, Langford Lane, Kidlington, Oxford, OX5 1GB, UK

Library of Congress Cataloging-in-Publication Data
Application submitted

British Library Cataloguing-in-Publication Data
A catalogue record for this book is available from the British Library.

ISBN: 978-0-12-397194-4

For information on all BH publications visit our website at store.elsevier.com

Dedication

To the memory of my father,
VV
To the memory of Antonina, Alexandra and Nikolai,
MN
To the loving memory of my sister, Mrs. Chitra Ramachandran, who departed too soon leaving a huge void in my life!
NB

Contents

Preface

Many parametric models, possessing different characteristics, shapes, and properties, have been proposed in the literature. These models are commonly used to develop parametric inferential methods. The inference developed and conclusions drawn based on these methods, however, will critically depend on the specific parametric model assumed for the analysis of the observed data. For this reason, several model validation techniques and goodness of fit tests have been developed over the years.

The oldest and perhaps the most commonly used one among these is the chi-squared goodness of fit test proposed by Karl Pearson over a century ago. Since then, many modifications, extensions, and generalizations of this methodology have been discussed in the statistical literature. Yet, there are some misconceptions and misunderstandings in the use of this method even at the present time.

The main aim of this book is, therefore, to provide an in-depth account of the theory, methods, and applications of chi-squared goodness of fit tests. In the process, pertinent formulas for their use in testing for some specific prominent distributions, such as normal, exponential, and Weibull, are provided. The asymptotic properties of the tests are described in detail, and Monte Carlo simulations are also used to carry out some comparisons of the power of these tests for different alternatives.

To provide a clear understanding of the methodology and an appreciation for its wide-ranging application, several well-known data sets are used as illustrative examples and the results obtained are then carefully interpreted. In doing so, some of the commonly made mistakes and misconceptions with regard to the use of this test procedure are pointed out as well.

We hope this book will serve as an useful guide for this popular methodology to theoreticians and practitioners alike. As pointed out at a number of places in the book, there are still many open problems in this area, and it is our sincere hope that the publication of this book will rejuvenate research activity, both theoretical and applied, in this important topic of research.

Preparation of a book of this nature naturally requires the help and cooperation of many individuals. We acknowledge the overwhelming support we received from numerous researchers who willingly shared their research publications and ideas with us. The editors of Academic Press/Elsevier were greatly supportive of this project from the start, and their production department were patient and efficient while working on the final production stages of the book. Our sincere thanks also go to our respective families for their emotional support and patience during the course of this project, and to Ms. Debbie Iscoe for her diligent work on the typesetting of the entire manuscript.

Vassilly Voinov, Kazakhstan
Mikhail Nikulin, France
Narayanaswamy Balakrishnan, Canada

Chapter 1

A Historical Account

The famous chi-squared goodness of fit test was proposed by Pearson (1900). If simple observations are grouped over r disjoint intervals Δ_j and $N_j^{(n)}$ denote observed frequencies corresponding to a multinomial scheme with $np_j(\boldsymbol{\theta})$ as the expected frequencies, for $j = 1,2,\ldots,r$, the Pearson's sum is given by

$$\chi^2 = \sum_{j=1}^{r} \frac{(N_j^{(n)} - np_j(\boldsymbol{\theta}))^2}{np_j(\boldsymbol{\theta})} = \mathbf{V}^{(n)T}(\boldsymbol{\theta})\mathbf{V}^{(n)}(\boldsymbol{\theta}), \tag{1.1}$$

where $\mathbf{V}^{(n)}(\boldsymbol{\theta})$ is the vector of standardized frequencies with components

$$v_j^{(n)}(\boldsymbol{\theta}) = (N_j^{(n)} - np_j(\boldsymbol{\theta}))/(np_j(\boldsymbol{\theta}))^{1/2}, \quad j = 1,\ldots,r, \ \boldsymbol{\theta} \in \boldsymbol{\Theta} \subset R^s.$$

If the number of sample observations $n \to \infty$, the statistic in (1.1) will follow the chi-squared probability distribution with $r - 1$ degrees of freedom. We know that this remarkable result is true only for a simple null hypothesis when a hypothetical distribution is specified uniquely (i.e. the parameter $\boldsymbol{\theta}$ is considered to be known). Until 1934, Pearson believed that the limiting distribution of the statistic in (1.1) will be the same if the unknown parameters of the null hypothesis are replaced by their estimates based on a sample; see, for example, Baird (1983), Plackett (1983, p. 63), Lindley (1996), Rao (2002), and Stigler (2008, p. 266). In this regard, it is important to reproduce the words of Plackett (1983, p. 69) concerning E.S. Pearson's opinion: "I knew long ago that KP (meaning Karl Pearson) used the 'correct' degrees of freedom for (a) difference between two samples and (b) multiple contingency tables. But he could not see that χ^2 in curve fitting should be got asymptotically into the

Chi-Squared Goodness of Fit Tests with Applications. http://dx.doi.org/10.1016/B978-0-12-397194-4.00001-6

same category." Plackett explained that this crucial mistake of Pearson arose from Karl Pearson's assumption "that individual normality implies joint normality." Stigler (2008) noted that this error of Pearson "has left a positive and lasting negative impression upon the statistical world." Fisher (1924) clearly showed that the number of degrees of freedom of Pearson's test must be reduced by the number of parameters estimated from the sample. To this point, it must be added that Fisher's result is true if and only if the parameters are estimated from the vector of frequencies minimizing Pearson's chi-squared sum, using multinomial maximum likelihood estimates (MLEs), or by any other asymptotically equivalent procedure (Greenwood and Nikulin, 1996, p. 74). Such estimates based on a vector of frequencies, which is not in general the vector of sufficient statistics, are not asymptotically efficient, however, due to which the Pearson-Fisher test is not powerful in many cases. For a review on using minimum chi-squared estimators, one may refer to Harris and Kanji (1983). Nowadays, Pearson's test with unknown parameters replaced by estimates $\hat{\boldsymbol{\theta}}_n$ based on the vector of frequencies is referred to as Pearson-Fisher (PF) test given by

$$X_n^2(\hat{\boldsymbol{\theta}}_n) = \sum_{j=1}^{r} \frac{(N_j^{(n)} - np_j(\hat{\boldsymbol{\theta}}_n))^2}{np_j(\hat{\boldsymbol{\theta}}_n)} = \mathbf{V}^{(n)T}(\hat{\boldsymbol{\theta}}_n)\mathbf{V}^{(n)}(\hat{\boldsymbol{\theta}}_n). \tag{1.2}$$

Dzhaparidze and Nikulin (1974) proposed a modification of the standard Pearson statistic (DN test), valid for any $\sqrt{n}$-consistent estimator $\tilde{\boldsymbol{\theta}}_n$ of an unknown parameter, given by

$$U_n^2(\tilde{\boldsymbol{\theta}}_n) = \mathbf{V}^{(n)T}(\tilde{\boldsymbol{\theta}}_n)(\mathbf{I} - \mathbf{B}_n(\mathbf{B}_n^T\mathbf{B}_n)^{-1}\mathbf{B}_n^T)\mathbf{V}^{(n)}(\tilde{\boldsymbol{\theta}}_n), \tag{1.3}$$

where $\mathbf{B}_n$ is an estimate of the matrix $\mathbf{B}$ with elements

$$b_{jk} = \frac{1}{\sqrt{p_j(\boldsymbol{\theta})}} \int_{\Delta_j} \frac{\partial f(x,\boldsymbol{\theta})}{\partial \theta_k} dx, \quad j = 1, \ldots, r, \ k = 1, \ldots, s.$$

This test, being asymptotically equivalent to the Pearson-Fisher statistic in many cases, is not powerful for equiprobable cells (Voinov et al., 2009) but is rather powerful if an alternative hypothesis is specified and one uses the Neyman-Pearson classes for constructing the vector of frequencies.

Several authors, such as Cochran (1952), Yarnold (1970), Larntz (1978), Hutchinson (1979), and Lawal (1980), considered the problem of approximating the discrete distribution of Pearson's sum if some expected frequencies become too small. Baglivo et al. (1992) elaborated methods for calculating the exact distributions and significance levels of goodness of fit statistics that can be evaluated in polynomial time. Asymptotically normal approximation of the chi-squared test valid for very large number of observations such that $n \to \infty$, $n/r \to \alpha$ was considered by Tumanyan (1956) and Holst (1972). Haberman (1988) noted that if some expected frequencies become too small and one does not use equiprobable cells, then Pearson's test can be biased. Mann and Wald (1942) and Cohen and Sackrowitz (1975) proved that Pearson's chi-squared test will be unbiased if one uses equiprobable cells. Other tests, including modified chi-squared tests, can be biased as well. Concerning selecting category boundaries and the number of classes in chi-squared goodness of fit tests, one may refer to Williams (1950), the review of Kallenberg et al. (1985) and the references cited therein, Bajgier and Aggarwal (1987) and Lemeshko and Chimitova (2003). Ritchey (1986) showed that an application of the chi-squared goodness of fit test with equiprobable cells to daily discrete common stock returns fails, and so suggested a test based on a set of intervals defined by centered approach.

Even after Fisher's clarification, many statisticians thought that while using Pearson's test one may use estimators (such as MLEs) based on non-grouped (raw) data. Chernoff and Lehmann (1954) showed that replacing the unknown parameters in (1.1) by their MLEs based on non-grouped data would dramatically change the limiting distribution of Pearson's sum. In this case, it will not follow a chi-squared distribution and that, in general, it may depend on the unknown parameters and consequently cannot be used for testing. In our opinion, what is difficult to understand for those who use chi-squared tests is that an estimate is a realization of a random variable with its own probability distribution and that a particular estimate can be quite far from the actual unknown value of a parameter or parameters. This misunderstanding is rather typical for those who apply both parametric and nonparametric tests for compound hypotheses (Orlov, 1997). Erroneous use of Pearson's test under such settings is reproduced even in some recent textbooks; see, for example, Clark (1997, p. 273) and Weiers (1991, p. 602). While Chernoff and Lehmann (1954) derived their result considering grouping cells to be fixed, Roy (1956) and Watson (1958, 1959) extended their result to the case of random grouping intervals. Molinari (1977) derived the limiting distribution of Pearson's sum if moment-type estimators (MMEs) based on raw data are used, and like in the case of MLEs, it depends on the unknown parameters. Thus, the problem of deriving a test statistic whose limiting distribution will not depend on the parameters becomes of interest. Roy (1956) and Watson (1958) (also see Drost, 1989) suggested using Pearson's sum for random cells. Dahiya and Gurland (1972a) showed that, for location and scale families with properly chosen random cells, the limiting distribution of Pearson's sum will not depend on the unknown parameters, but only on the null hypothesis. Being distribution-free, such tests can be used in practice, but the problem is that for each specific null distribution, one has to evaluate the corresponding critical values. Therefore, two different ways of constructing distribution-free Pearson-type tests are: (i) to use proper estimates of the unknown parameters (e.g. based on grouped data) and (ii) to use specially constructed grouping intervals. Yet another way is to modify Pearson's sum such that its limiting distribution would not depend on the unknown parameters. Roy (1956), Moore (1971), and Chibisov (1971) obtained a very important result which showed that the limiting distribution of a vector of standardized frequencies with any efficient estimator (such as the MLE or the best asymptotically normal (BAN) estimator) instead of the unknown parameter would be multivariate normal and will not depend on whether the boundaries of cells are fixed or random. Nikulin (1973c), by using this result and a very general theoretical approach (nowadays known as Wald's method; see Moore (1977)) solved the problem completely for any continuous or discrete probability distribution if one uses grouping intervals based on predetermined probabilities for the cells (a detailed derivation of this result is given in Greenwood and Nikulin (1996, Sections 12 and 13)). A year later, Rao and Robson (1974), by using much less general heuristic approach, obtained the same result for a particular case of the exponential family of distributions. Formally, their result is that

$$Y1_n^2(\hat{\boldsymbol{\theta}}_n) = X_n^2(\hat{\boldsymbol{\theta}}_n) + \mathbf{V}^{(n)T}(\hat{\boldsymbol{\theta}}_n)\mathbf{B}_n(\mathbf{J}_n - \mathbf{J}_{gn})^{-1}\mathbf{B}_n^T\mathbf{V}^{(n)}(\hat{\boldsymbol{\theta}}_n), \tag{1.4}$$

where $\mathbf{J}_n$ and $\mathbf{J}_{gn} = \mathbf{B}_n^T\mathbf{B}_n$ are estimators of Fisher information matrices for non-grouped and grouped data, respectively. Incidentally, this result is Rao and Robson (1974) and Nikulin (1973c). The statistic in (1.4) can also be presented as (see Nikulin, 1973b,c; Moore and Spruill, 1975; Greenwood and Nikulin, 1996)

$$Y1_n^2(\hat{\boldsymbol{\theta}}_n) = \mathbf{V}^{(n)T}(\hat{\boldsymbol{\theta}}_n)(\mathbf{I} - \mathbf{B}_n\mathbf{J}_n^{-1}\mathbf{B}_n^T)^{-1}\mathbf{V}^{(n)}(\hat{\boldsymbol{\theta}}_n). \tag{1.5}$$

The statistic in (1.4) or (1.5), suggested first by Nikulin (1973a) for testing the normality, will be referred to in the sequel as Nikulin-Rao-Robson (NRR) test (Voinov and Nikulin, 2011). Nikulin (1973a,b,c) assumed that only efficient estimates of the unknown parameters (such as the MLEs based on non-grouped data or BAN estimates) are used for testing. Spruill (1976) showed that in the sense of approximate Bahadur slopes, the NRR test is uniformly at least as efficient as Roy (1956) and Watson (1958) tests. Singh (1987) showed that the NRR test is asymptotically optimal for linear hypotheses (see Lehmann, 1959, p. 304) when explicit expressions for orthogonal projectors on linear subspaces are used. Lemeshko (1998) and Lemeshko et al. (2001) suggested an original way of taking into account the information lost due to data grouping. Their idea is to partition the sample space into intervals that maximize the determinant of Fisher information matrix for grouped data. Implementation of the idea to NRR test showed that the power of the NRR test became superior. This optimality is not surprising because the second term in (1.4) depends on the difference between the Fisher information matrices for grouped and non-grouped data that possibly takes the information lost into account (Voinov, 2006). A unified large-sample theory of general chi-squared statistics for tests of fit was developed by Moore and Spruill (1975).

Hsuan and Robson (1976) showed that a modified statistic would be quite different in case of moment-type estimators (MMEs) of unknown parameters. They succeeded in deriving the limiting covariance matrix for standardized frequencies $v_i(\bar{\theta}_n)$, where $\bar{\theta}_n$ is the MME of θ, and established that the corresponding Wald's quadratic form will follow in the limit the chi-squared distribution. They also provided the test statistic explicitly for the exponential family of distributions in which case the MMEs coincide with MLEs, thus confirming the results of Nikulin (1973b). Voinov and Pya (2004) have shown that, for the exponential family of distributions, this test is identically equal to NRR statistic. Hsuan and Robson (1976) were unable to derive the general modified test based on MMEs $\bar{\theta}_n$ explicitly, and this was achieved later by Mirvaliev (2001). To give due credit to the contributions of Hsuan and Robson (1976) and Mirvaliev (2001), we suggest calling this test as Hsuan-Robson-Mirvaliev (HRM) statistic, which is of the form

$$Y2_n^2(\bar{\boldsymbol{\theta}}_n) = X_n^2(\bar{\boldsymbol{\theta}}_n) + R_n^2(\bar{\boldsymbol{\theta}}_n) - Q_n^2(\bar{\boldsymbol{\theta}}_n). \tag{1.6}$$

Explicit expressions for the quadratic forms $R_n^2(\bar{\boldsymbol{\theta}}_n)$ and $Q_n^2(\bar{\boldsymbol{\theta}}_n)$ are presented in Section 4.1.

Moore (1977), using Wald's approach, suggested a general recipe for constructing modified chi-squared tests for any $\sqrt{n}$-consistent estimator, which is a slight generalization of Nikulin's idea, since it includes also the case of fixed grouping cells. This is not important because nobody knows a priori how to partition the sample space into fixed cells if the probability distribution to be tested is unknown. Moore has not specified those tests for a particular $\sqrt{n}$-consistent estimator, but has noted that a resulting Wald's quadratic form does not depend on how the limiting covariance matrix is inverted. A subclass of the Moore-Spruill class of tests for location-scale models, that includes as particular cases the NRR and the DN statistics, was suggested by Drost (1989). Bol'shev and Mirvaliev (1978) and Chichagov (2006) used this approach for constructing modified tests based on minimum variance unbiased estimators (MVUEs). Bol'shev and Mirvaliev (1978), by using MVUEs and Wald's approach, constructed a chi-squared type test for the Poisson, binomial, and negative binomial distributions. The same idea was Bol'shev and Mirvaliev (1978), Nikulin and Voinov (1989), and Voinov and Nikulin (1994). For an application of the best asymptotically normal (BAN) estimators for modification of chi-squared tests, one may refer to Bemis and Bhapkar (1983). After generalizing the idea of Dzhaparidze and Nikulin (1974) and Singh (1987) suggested an elegant generalization of the NRR test in (1.5), valid for any $\sqrt{n}$-consistent estimator $\tilde{\boldsymbol{\theta}}_n$ of an unknown parameter, of the form

$$Q_s^2(\tilde{\boldsymbol{\theta}}_n) = \mathbf{V}_*^{(n)T}(\tilde{\boldsymbol{\theta}}_n)(\mathbf{I} - \mathbf{B}_n\mathbf{J}_n^{-1}\mathbf{B}_n^T)^{-1}\mathbf{V}_*^{(n)}(\tilde{\boldsymbol{\theta}}_n), \tag{1.7}$$

where

$$\mathbf{V}_*^{(n)}(\tilde{\boldsymbol{\theta}}_n) = \mathbf{V}^{(n)}(\tilde{\boldsymbol{\theta}}_n) - \mathbf{B}_n\mathbf{J}_n^{-1}\mathbf{W}(\tilde{\boldsymbol{\theta}}_n) \tag{1.8}$$

is the score vector for $\boldsymbol{\theta}$ from the raw data, and

$$\mathbf{W}(\tilde{\boldsymbol{\theta}}_n) = \frac{1}{\sqrt{n}} \sum_{i=1}^{n} \left. \frac{\partial \ln f(X_i, \boldsymbol{\theta})}{\partial \boldsymbol{\theta}} \right|_{\boldsymbol{\theta}=\tilde{\boldsymbol{\theta}}_n} .$$

Dzhaparidze and Nikulin (1992) (see also Fisher, 1925b; Dzhaparidze, 1983) generalized the idea of Fisher to improve any $\sqrt{n}$-consistent estimator to make it asymptotically as efficient as the MLE. This gives an alternative way of modifying chi-squared tests: improve the estimator first and then use the NRR statistic with that improved estimator (see Section 4.4.2).

During the last 30 years, much work has been done on the classical chi-squared tests and on proposing some very original modifications (Nikulin and Voinov, 2006; Voinov and Nikulin, 2011). Bhalerao et al. (1980) noted that the limiting distribution of Wald-type modifications of the Pearson-Fisher test does not depend on the generalized minimum chi-squared procedure used, but its power may depend on it. A numerical example for the negative binomial distribution was considered for illustration. Moore and Stubblebine (1981) generalized the NRR statistic to test for the two-dimensional circular normality (see also Follmann, 1996). It is usually supposed that observations are realizations of independent and identically distributed (i.i.d) random variables. Gleser and Moore (1983) showed "that if the observations are in fact a stationary process satisfying a positive dependence condition, the test (such as chi-squared) will reject a true null hypothesis too often." Guenther (1977) and Drost (1988) considered the problem of approximation of power and sample size selection for multinomial tests. Drost (1989) also introduced a generalized chi-square goodness of fit test, which is a subclass of the Moore-Spruill class, for location-scale families when the number of equiprobable cells tends to infinity (see also Osius, 1985). He recommended a large number of classes for heavy-tailed alternatives. Heckman (1984) and Andrews (1988) discussed the theory and applications of chi-squared tests for models with covariates. Hall (1985) proposed the chi-squared test for uniformity based on overlapping cells. He showed that modified in such a manner the statistic is able to detect alternatives that are $n^{-1/2}$ distant from the null hypothesis. Loukas and Kemp (1986) studied applications of Pearson's test for bivariate discrete distributions. Kocherlakota and Kocherlakota (1986) suggested goodness of fit tests for discrete distributions based on probability generating functions. Habib and Thomas (1986) and Bagdonavičius and Nikulin (2011) suggested modified chi-squared tests for randomly censored data. Hjort (1990), by using Wald's approach for time-continuous survival data, proposed a new class of goodness of fit tests based on cumulative hazard rates, which work well even when no censoring is present. Nikulin and Solev (1999) presented a chi-squared goodness of fit test for doubly censored data. Singh (1986) proposed a modification of the Pearson-Fisher test based on collapsing some cells. Akritas (1988) (see also Hollander and Pena, 1992; Peña, 1998a,b) proposed modified chi-squared tests when data can be subject to random censoring. Cressie and Read (1984) (see also an exhaustive review of Cressie and Read (1989)) introduced the family of power divergence statistics of the form

$$2nI^{\lambda} = \frac{2}{\lambda(\lambda+1)} \sum_{i=1}^{k} X_i \left\{ \left(\frac{X_i}{np_i} \right)^{\lambda} - 1 \right\}, \quad \lambda \in R^1. \tag{1.9}$$

Interested readers may refer to the book by Pardo (2006) for an elaborate treatment on statistical inferential techniques based on divergence measures. Pearson's X^2 statistic ($\lambda = 1$), the log-likelihood ratio statistic ($\lambda \to 0$), the Freeman-Tukey statistic ($\lambda = -1/2$), the modified log-likelihood ratio statistic ($\lambda = -1$), and the Neyman modified $X^2(\lambda = -2)$ statistic are all particular cases of (1.9). As a compromising alternative to Pearson's X^2 and to likelihood ratio statistic, Cressie and Read (1984) suggested a new goodness of fit test with $\lambda = 2/3$. Read (1984) performed exact power comparisons of different tests from that family for symmetric null hypotheses under specified alternatives. Moore (1986) wrote "for general alternatives, we recommend that the Pearson X^2 statistic be employed in practice when a choice is made among the statistics $2nI^{\lambda}$." A comparative simulation study of some tests from the power divergence family in (1.9) was performed by Koehler and Gan (1990). Karagrigoriou and Mattheou (2010) (see also the references therein) suggested a generalization of measures of divergence that include as particular cases many other previously considered measures.

A chi-squared distributed modification of the score statistic of Cox and Hinkley (1974) was introduced by Cordeiro and Ferrari (1991). Lorenzen (1992), using the concept of a logarithmic mean in two arguments, reformulated the classical Pearson's test thus displaying a possibility to bridge the gap between Pearson's statistic and the log-likelihood ratio test. Li and Doss (1993) suggested a generalization of the PF test that proves to be useful in survival analysis. McLaren et al. (1994) proposed the generalized χ^2 goodness of fit test for detecting distributions containing more than a specified lack of fit due to sampling errors. The test statistic coincides with the Pearson's sum, but its null distribution turns out to be non-central χ^2. This test proves to be useful for very large samples. Boulerice and Ducharme (1995) showed that the test statistic introduced by Rayner and Best (1986) for Neyman's smooth test of goodness of fit for location-scale families does not, in general, have the anticipated chi-squared distribution. They showed that the test is inapplicable in such cases as logistic, Laplace, and Type I extreme-value distributions. Using Wald's approach, Akritas and Torbeyns (1997) developed a Pearson-type goodness of fit test for the linear regression model. Zhang (1999) developed a chi-squared goodness of fit test for logistic regression models. Jung et al. (2001) (see also Jung et al., 2003) developed an adjusted chi-squared test for observational studies of clustered binary data. Rao (2002) introduced a new test for goodness of fit in the continuous case. Graneri (2003) proposed a χ^2-type goodness of fit test based on transformed empirical processes for location and scale families; see also Cabana and Cabana (1997). Johnson (2004) suggested a Bayesian χ^2 test for goodness of fit. It is worth mentioning here the very original goodness of fit tests proposed by Henze and Meintanis (2002), Davies (2002), and Damico (2004), though they are not χ^2-type. Zhang (2005) considered approximate and asymptotic distributions of chi-squared-type mixtures that can be used to some nonparametric goodness of fit tests, especially for nonparametric regression models. Ampadu (2008) suggested four modified chi-squared type statistics for testing the null hypothesis about the discrete uniform probability distribution. Deng et al. (2009) adapted the NRR test for logistic regression models.

An important contribution to the theory of modified chi-squared goodness of fit tests is due to McCulloch (1985) and Mirvaliev (2001) who considered two types of decomposition of tests. The first is a decomposition of a test on a sum of the DN statistic and an asymptotically independent (of the DN test) additional quadratic form. Denoting $W_n^2(\boldsymbol{\theta}) = \mathbf{V}^{(n)T}(\boldsymbol{\theta})\mathbf{B}(\mathbf{B}^T\mathbf{B})^{-1}\mathbf{B}^T\mathbf{V}^{(n)}(\boldsymbol{\theta})$ and $P_n^2(\boldsymbol{\theta}) = \mathbf{V}^{(n)T}(\boldsymbol{\theta})\mathbf{B}(\mathbf{J}-\mathbf{J}_g)^{-1}\mathbf{B}^T\mathbf{V}^{(n)}(\boldsymbol{\theta})$, the decomposition of the NRR statistic in (1.4) in the case of MLEs is given by

$$Y1_n^2(\hat{\boldsymbol{\theta}}_n) = U_n^2(\hat{\boldsymbol{\theta}}_n) + S_n^2(\hat{\boldsymbol{\theta}}), \tag{1.10}$$

where the DN statistic $U_n^2(\hat{\boldsymbol{\theta}}_n)$ is asymptotically independent of $S_n^2(\hat{\boldsymbol{\theta}}) = W_n^2(\hat{\boldsymbol{\theta}}) + P_n^2(\hat{\boldsymbol{\theta}}_n)$ and of $W_n^2(\hat{\boldsymbol{\theta}})$. The decomposition of the HRM statistic in (1.6) is

$$Y2_n^2(\bar{\boldsymbol{\theta}}_n) = U_n^2(\bar{\boldsymbol{\theta}}_n) + S1_n^2(\bar{\boldsymbol{\theta}}), \tag{1.11}$$

where $U_n^2(\bar{\boldsymbol{\theta}}_n)$ is asymptotically independent of $S1_n^2(\bar{\boldsymbol{\theta}}) = W_n^2(\bar{\boldsymbol{\theta}}) + R_n^2(\bar{\boldsymbol{\theta}}_n) - Q_n^2(\bar{\boldsymbol{\theta}}_n)$, but is correlated with $W_n^2(\bar{\boldsymbol{\theta}})$. The second type decomposes a modified test on a sum of classical Pearson's test and a correction term, which makes it chi-square distributed in the limit, and independent of unknown parameters (see (1.4) and (1.6)). This representation for NRR statistic was first used by Nikulin (1973b) (see also Rao and Robson, 1974; McCulloch, 1985). The case of MMEs was first investigated by Mirvaliev (2001). The decomposition of a modified chi-squared test on a sum of the DN statistic and an additional term is of importance since the DN test based on non-grouped data is asymptotically equivalent to the Pearson-Fisher statistic for grouped data. Hence, that additional term takes into account the Fisher information lost due to grouping. It was subsequently shown (Voinov and Pya, 2004; Voinov et al., 2009) that the DN part, like the PF test, is in many cases insensitive to an alternative hypothesis for the case of equiprobable cells (fixed or random) and would be sensitive to it in the case of non-equiprobable Neyman-Pearson classes. For equiprobable cells, this suggests using the difference between a modified statistic and the DN part that will be the most powerful in case of equiprobable cells (McCulloch, 1985; Voinov et al., 2009). At this point, it is of interest to mention the elaborate works of Lemeshko and Postovalov (1997, 1998), Lemeshko (1998), Lemeshko and Chimitova (2000, 2002), Lemeshko et al. (2001, 2007, 2008, 2011), who have thoroughly investigated different tests by Monte Carlo simulations. Through these, it has become clear that the way in which the sample space is partitioned essentially influences the power of a test.

Fisher (1925a), who made great contributions to Statistics (see Efron, 1998) was the first to note that "in some cases it is possible to separate the contributions to χ^2 made by the individual degrees of freedom, and so to test the separate components of a discrepancy." Lancaster (1951) (see also Gilula and Krieger, 1983; Nair, 1987, 1988) used the partition of χ^2 to investigate the interactions of all orders in contingency tables. Cochran (1954) wrote "that the usual χ^2 tests are often insensitive, and do not indicate significant results when the null hypothesis is actually false" and recommended to "use a single degree of freedom, or a group of degrees of freedom, from the total χ^2" for obtaining more powerful and appropriate test. The problem of an implementation of the idea of Fisher and Cochran was that decompositions of Pearson's sum and modified test statistics were not known at that time. Anderson (1994) (see also Anderson, 1996; Boero et al., 2004a) was the first to decompose Pearson's χ^2 for a simple null hypothesis into a sum of independent χ_1^2 random variables in case of equiprobable grouping cells. Using Fourier analysis technique, Eubank (1997) derived a decomposition of Pearson's sum into asymptotically independent chi-square distributed components with 1 degree of freedom. New Neyman smooth-type tests based on those components were then suggested. An algorithm for parametric Pearson-Fisher's test decomposition was proposed by Rayner (2002). Unfortunately, the components of that decomposition cannot be written down explicitly. A parametric decomposition of Pearson's χ^2 sum in case of non-equiprobable cells and decompositions of the RRN and HRM statistics, based on the ideas of Mirvaliev (2001), were obtained by Voinov et al. (2007) in an explicit form. Voinov (2010) obtained explicitly a decomposition of Pearson-Fisher's and Dzhaparidze-Nikulin's statistics. A decomposition of the Chernoff-Lehmann chi-squared test and the use of the components to test for the binomial distribution was discussed by Best and Rayner (2006). Voinov and Pya (2010) introduced vector-valued goodness of fit tests (based on those components or on any combination of parametric or nonparametric statistics) that in some cases provide a gain in power for specified alternatives.

All these variations and nuantic refinements to the chi-squared test and their characteristics and properties have provided us an impetus to prepare this volume. Our aim in the subsequent chapters is to provide a thorough up-to-date review of all these developments, and also comment on their relative performance and power under different settings.

Chapter 2

Pearson's Sum and Pearson-Fisher Test

2.1 PEARSON'S CHI-SQUARED SUM

Let $X_1, \ldots, X_n$ be i.i.d. random variables (r.v.). Consider the problem of testing a simple hypothesis H_0, according to which the distribution of X_i is a member of the parametric family

$$\mathbf{P}\{X_i \leqslant x | H_0\} = F(x; \boldsymbol{\theta}), \quad \boldsymbol{\theta} = (\theta_1, \ldots, \theta_s)^T \in \Theta \subset R^s, \qquad x \in R^1,$$

where Θ is an open set. Denote by $f(x; \boldsymbol{\theta})$ the density of the probability distribution function $F(x; \boldsymbol{\theta})$ with respect to a certain σ-finite measure μ. Let

$$\begin{aligned} N_j^{(n)} &= \text{Card}\{i : X_i \in \Delta_j, \quad i = 1, \ldots, n\}, \\ p_j(\boldsymbol{\theta}) &= \int\limits_{\Delta_j} dF(x; \boldsymbol{\theta}), \quad j = 1, \ldots, r, \end{aligned} \tag{2.1}$$

where Δ_js are non-intersecting grouping intervals such that

$$\Delta_1 \cup \cdots \cup \Delta_r = R^1, \quad \Delta_i \cap \Delta_j = \emptyset \text{ for } \quad i \neq j.$$

Chi-Squared Goodness of Fit Tests with Applications. http://dx.doi.org/10.1016/B978-0-12-397194-4.00002-8

Denote by $\mathbf{V}^{(n)}$ a column vector of dimension r of a standardized grouped frequency with components

$$v_j^{(n)} = \frac{N_j^{(n)} - np_j(\boldsymbol{\theta})}{\sqrt{np_j(\boldsymbol{\theta})}}, \quad j = 1, \ldots, r. \tag{2.2}$$

Using (2.2), the standard Pearson chi-squared statistic $X_n^2(\boldsymbol{\theta})$, under a simple null hypothesis H_0 (wherein the parameter $\boldsymbol{\theta}$ is considered to be known), given by

$$X_n^2(\boldsymbol{\theta}) = \mathbf{V}^{(n)T}\mathbf{V}^{(n)} = \sum_{j=1}^{r} \frac{[N_j^{(n)} - np_j(\boldsymbol{\theta})]^2}{np_j(\boldsymbol{\theta})}, \tag{2.3}$$

possesses a limiting chi-square probability distribution (χ_{r-1}^2) with $r-1$ degrees of freedom (Pearson, 1900). For power properties of the classical Pearson test against close alternatives, one may refer to Tumanyan (1958) and Greenwood and Nikulin (1996).

2.2 DECOMPOSITIONS OF PEARSON'S CHI-SQUARED SUM

In this section, we will discuss some decompositions of Pearson's chi-squared sum. Anderson (1994) suggested a decomposition of Pearson's chi-squared sum for the case of equiprobable grouping intervals in order to secure additional information about the nature of departure from the hypothesized distribution. A theoretical derivation of that decomposition has been presented by Boero et al. (2004a). Let us describe briefly this theory. In the case of r equiprobable intervals, Pearson's sum would become

$$X_n^2(\boldsymbol{\theta}) = \sum_{j=1}^{r} \frac{(N_j^{(n)} - n/r)^2}{n/r}.$$

Let $\mathbf{x}$ be a $r \times 1$ vector with components $N_j^{(n)}$ and $\boldsymbol{\mu}$ be a $r \times 1$ vector with its components all as n/r. Then, evidently

$$X_n^2(\boldsymbol{\theta}) = (\mathbf{x} - \boldsymbol{\mu})^T[\mathbf{I} - \mathbf{p}\mathbf{p}^T](\mathbf{x} - \boldsymbol{\mu})/(n/r),$$

where $\mathbf{p} = (\sqrt{p_1}, \ldots, \sqrt{p_r})^T, p_j = 1/r$ $(j = 1, \ldots, r)$, and $\mathbf{I}$ is the $r \times r$ identity matrix. There exists a $(r-1) \times r$ matrix $\mathbf{A}$ such that

$$\mathbf{A}\mathbf{A}^T = \mathbf{I}, \quad \mathbf{A}^T\mathbf{A} = [\mathbf{I} - \mathbf{p}\mathbf{p}^T].$$

So, with the transformation $\mathbf{y} = \mathbf{A}(\mathbf{x} - \boldsymbol{\mu})$, Pearson's sum can be expressed as

$$X_n^2(\boldsymbol{\theta}) = \mathbf{y}^T\mathbf{y}/(n/r),$$

where $r-1$ components $y_i^2/(n/r)$ are independently distributed as χ_1^2 under $\mathbf{H}_0$. Boero et al. (2004a) described the following construction for the matrix $\mathbf{A}$.

Let $\mathbf{H}$ be the Hadamard matrix with orthogonal columns, i.e. $\mathbf{H}^T\mathbf{H} = r\mathbf{I}$. For r being a power of 2 and using the basic Hadamard matrix

$$\mathbf{H}_2 = \begin{pmatrix} 1 & 1 \\ 1 & -1 \end{pmatrix},$$

one can construct matrices of higher order as follows: $\mathbf{H}_4 = \mathbf{H}_2 \otimes \mathbf{H}_2, \mathbf{H}_8 = \mathbf{H}_4 \otimes \mathbf{H}_2$, and so on. The matrix $\mathbf{A}$ is then extracted from the partition

$$\mathbf{H} = \begin{pmatrix} \mathbf{e}^T \\ \sqrt{r}\mathbf{A} \end{pmatrix},$$

where $\mathbf{e}$ is $r \times 1$ vector of 1s. When $r = 4$, for example, after rearranging the rows, we have

$$\mathbf{A} = \frac{1}{2}\begin{pmatrix} 1 & 1 & -1 & -1 \\ 1 & -1 & -1 & 1 \\ 1 & -1 & 1 & -1 \end{pmatrix}.$$

With this matrix, the transformed vector $\mathbf{y}$ for four equiprobable cells will be

$$\mathbf{y} = \mathbf{A}(\mathbf{x} - \boldsymbol{\mu}),$$

or, explicitly,

$$\mathbf{y} = \frac{1}{2}\begin{pmatrix} (N_1^{(n)} - n/4) + (N_2^{(n)} - n/4) - (N_3^{(n)} - n/4) - (N_4^{(n)} - n/4) \\ (N_1^{(n)} - n/4) - (N_2^{(n)} - n/4) - (N_3^{(n)} - n/4) + (N_4^{(n)} - n/4) \\ (N_1^{(n)} - n/4) - (N_2^{(n)} - n/4) + (N_3^{(n)} - n/4) - (N_4^{(n)} - n/4) \end{pmatrix}.$$

Thus, Pearson's sum, when $r = 4$, can be decomposed as follows:

$$\begin{aligned} X_n^2(\boldsymbol{\theta}) &= \delta_1^2 + \delta_2^2 + \delta_3^2 \\ &= \frac{[(N_1^{(n)} - n/4) + (N_2^{(n)} - n/4) - (N_3^{(n)} - n/4) - (N_4^{(n)} - n/4)]^2}{n} \\ &\quad + \frac{[(N_1^{(n)} - n/4) - (N_2^{(n)} - n/4) - (N_3^{(n)} - n/4) + (N_4^{(n)} - n/4)]^2}{n} \\ &\quad + \frac{[(N_1^{(n)} - n/4) - (N_2^{(n)} - n/4) + (N_3^{(n)} - n/4) - (N_4^{(n)} - n/4)]^2}{n}. \end{aligned} \tag{2.4}$$

All terms in (2.4), which are independently distributed in the limit as χ_1^2, may be used individually. From (2.4), we observe that the first term is sensitive to location, the second one to scale, and the last one to skewness. So, depending on the alternative hypothesis, one may select a suitable component to be used for testing. The same can be done for $r = 8$. In that case, a term sensitive to kurtosis will also appear, but it is difficult to relate the remaining three components to characteristics of the distribution (Boero et al., 2004a).

Anderson's decomposition considered above is valid only for equiprobable grouping cells. Another known decomposition of Pearson's sum that is valid for non-equiprobable cells as well is as follows (Voinov et al., 2007).

Let $\mathbf{Z} = (Z_1, \ldots, Z_r)^T$ be a random vector with $\mathbf{EZ} = \mathbf{0}, \mathbf{E}(\mathbf{ZZ}^T) = \mathbf{D} = (d_{ij})$, the rank $R(\mathbf{D})$ of $\mathbf{D}$ being $k(\leqslant r)$. Denote

$$\begin{aligned}
&\mathbf{Z}_{(i)} = (Z_1, \ldots, Z_i)^T, \quad \mathbf{D}_i = \mathbf{E}(\mathbf{Z}_{(i)}\mathbf{Z}_{(i)}^T),\\
&\mathbf{d}_{i(j)} = \mathbf{Cov}(Z_i, \mathbf{Z}_{(j)}), \quad \mathbf{d}_{(i)j} = \mathbf{Cov}(\mathbf{Z}_{(i)}, Z_j), \qquad i,j = 1,2, \ldots, k.
\end{aligned}$$

Consider the vector $\boldsymbol{\delta}_{(t)} = (\delta_1, \ldots, \delta_t)^T$ with its components as

$$\begin{aligned}
\delta_i &= \frac{1}{\sqrt{|\mathbf{D}_{i-1}||\mathbf{D}_i|}} \begin{bmatrix} d_{11} & d_{12} & \cdots & d_{1(i-1)} & Z_1 \\ d_{21} & d_{22} & \cdots & d_{2(i-1)} & Z_2 \\ \cdot & \cdot & \cdots & \cdot & \cdot \\ d_{(i-1)1} & d_{(i-1)2} & \cdots & d_{(i-1)(i-1)} & Z_{i-1} \\ d_{i(1)} & d_{i(2)} & \cdots & d_{i(i-1)} & Z_i \end{bmatrix}\\
&= \frac{1}{\sqrt{|\mathbf{D}_{i-1}||\mathbf{D}_i|}} \begin{bmatrix} \mathbf{D}_{i-1} & \mathbf{Z}_{(i-1)} \\ \mathbf{d}_{i(i-1)}^T & Z_i \end{bmatrix}, \quad i = 1,2, \ldots, t. \qquad (2.5)
\end{aligned}$$

Assume $\mathbf{Z}_{(0)} = \mathbf{0}$ and $|\mathbf{D}_{(0)}| = 1$.

Theorem 2.1. *Let $\boldsymbol{\delta}_{(t)}$ be a vector with its components as in* (2.5). *Then,*

$$\mathbf{E}\boldsymbol{\delta}_{(t)} = \mathbf{0}, \quad \mathbf{E}\left\{\boldsymbol{\delta}_{(t)}\boldsymbol{\delta}_{(t)}^T\right\} = \mathbf{I}_t, \qquad (2.6)$$

where $\mathbf{I}_t$ is a $t \times t$ identity matrix, $t = 1, \ldots, R(D)$.

By using the well-known formula

$$\begin{vmatrix} \mathbf{A} & \mathbf{C} \\ \mathbf{B} & \mathbf{D} \end{vmatrix} = |\mathbf{A}| \cdot |\mathbf{D} - \mathbf{BA}^{-1}\mathbf{C}|,$$

δ_i defined in (2.5) can be expressed as

$$\delta_i = \left(\frac{|\mathbf{D}_{i-1}|}{|\mathbf{D}_i|}\right)^{1/2} \left(Z_i - \mathbf{d}_{i(i-1)}^T \mathbf{D}_{i-1}^{-1} \mathbf{Z}_{(i-1)}\right), \quad i = 1, \ldots, t. \qquad (2.7)$$

The first equality in (2.6) follows immediately from (2.7). To prove the second equality, we have to show that the components of $\boldsymbol{\delta}_{(t)}$ are all normalized. Indeed,

$$\begin{aligned}\mathbf{Var}\delta_i &= \frac{|\mathbf{D}_{i-1}|}{|\mathbf{D}_i|}\mathbf{E}\left[\left(Z_i - \mathbf{d}_{i(i-1)}^T\mathbf{D}_{i-1}^{-1}\mathbf{Z}_{(i-1)}\right)\left(Z_i - \mathbf{d}_{i(i-1)}^T\mathbf{D}_{i-1}^{-1}\mathbf{Z}_{(i-1)}\right)\right]\\ &= \frac{|\mathbf{D}_{i-1}|}{|\mathbf{D}_i|}\left(d_{ii} - \mathbf{d}_{i(i-1)}^T\mathbf{D}_{i-1}^{-1}\mathbf{Z}_{(i-1)}\right) = 1.\end{aligned}$$

Along the same lines, it can be shown that the components of $\boldsymbol{\delta}_{(t)}$ are uncorrelated as well, i.e. $\mathbf{E}\delta_i\delta_j = 0$ for $i \neq j$.

Theorem 2.2. *The following identity holds:*

$$\mathbf{Z}_{(t)}^T\mathbf{D}_t^{-1}\mathbf{Z}_{(t)} = \delta_1^2 + \cdots + \delta_t^2 = ||\boldsymbol{\delta}_{(t)}||^2, \quad t = 1, \ldots, R(\mathbf{D}). \tag{2.8}$$

Proof. If $t = 1$, the theorem holds trivially. Now, let us assume that the above factorization holds up to $t = k$ inclusive and then prove it for $t = k + 1$. Since

$$\mathbf{Z}_{(k+1)} = \begin{pmatrix} \mathbf{Z}_{(k)} \\ Z_{k+1} \end{pmatrix}, \quad \mathbf{D}_{(k+1)} = \begin{pmatrix} \mathbf{D}_k & \mathbf{d}_{(k)k+1} \\ \mathbf{d}_{k+1(k)}^T & d_{k+1k+1} \end{pmatrix},$$

then, using block matrix inversion, we obtain

$$\begin{aligned}\mathbf{Z}_{(k+1)}^T\mathbf{D}_{k+1}^{-1}\mathbf{Z}_{(k+1)} &= \mathbf{Z}_{(k)}^T\mathbf{D}_k^{-1}\mathbf{Z}_{(k)} + \left(d_{k+1k+1} - \mathbf{d}_{k+1(k)}^T\mathbf{D}_k^{-1}\mathbf{d}_{(k)k+1}\right)\\ &\quad\times\left(Z_{k+1} - \mathbf{d}_{k+1(k)}^T\mathbf{D}_k^{-1}\mathbf{d}_{(k)k+1}\right)\\ &= \delta_1^2 + \cdots + \delta_k^2 + \delta_{k+1}^2 = ||\boldsymbol{\delta}_{(k+1)}||^2,\end{aligned}$$

as required. Thus, the linear transformation in (2.5) diagonalizes the quadratic form $\mathbf{Z}_{(t)}^T\mathbf{D}_t^{-1}\mathbf{Z}_{(t)}, t = 1, \ldots, R(\mathbf{D})$.

Remark 2.1. *If a matrix* **B** *is positive definite, it can be uniquely presented in the form* $\mathbf{B} = \mathbf{U}\mathbf{U}^T$, *where* **U** *is a real-valued upper triangular matrix with positive diagonal elements. Such a factorization is known as the Cholesky decomposition. Theorems 2.1 and 2.2 permit us to modify the Cholesky decomposition as follows. Since we shall deal with limiting covariance matrices, the condition of non-negative definiteness (n.n.d.) holds. If* $R(\mathbf{D}) = k(\leqslant r)$, *we may interchange the columns of* **D** *such that the first k columns will be linearly independent. We may also use the lower triangular matrix.*

Lemma 2.1. *Let a* $r \times r$ *matrix* **D** *be n.n.d. of rank k. Then,*

$$\mathbf{R}\mathbf{D}\mathbf{R}^T = \mathbf{I}_k, \quad \mathbf{R}^T\mathbf{R} = \begin{pmatrix} \mathbf{D}_k^{-1} & \mathbf{0} \\ \mathbf{0}^T & 0 \end{pmatrix} = \mathbf{D}^-,$$

where $\mathbf{0} = (0, \ldots, 0)^T$ *is a k-dimensional vector of zeros,* $\mathbf{R} = (\mathbf{R}_k \vdots \mathbf{0})$*, and* $\mathbf{R}_k$ *is a lower triangular matrix with elements*

$$\begin{aligned} r_{ii} &= \left(\frac{|\mathbf{D}_{i-1}|}{|\mathbf{D}_i|}\right)^{1/2}, \quad i = 1, \ldots, k, \\ r_{ij} &= -r_{ii}\mathbf{d}^T_{i(i-1)}\left(\mathbf{D}^{-1}_{i-1}\right)_j, \quad j = 1, \ldots, i-1. \end{aligned} \tag{2.9}$$

Here, $(\mathbf{A}_i)_j$ *denotes the jth column of the leading sub-matrix of order* $i \times i$ *of the matrix* $\mathbf{A}$.

Proof. The proof of the lemma follows from Theorems 2.1 and 2.2, and from the representation

$$\boldsymbol{\delta}_{(k)} = \mathbf{RZ}. \tag{2.10}$$

Under the simple null hypothesis H_0, the parameter $\boldsymbol{\theta}$ in (2.1) is supposed to be known. To simplify notation, let us denote $p_j(\boldsymbol{\theta}) = p_j, j = 1, \ldots, r$. Let the components of the vector $\mathbf{Z}$ be

$$Z_j = v_j^{(n)} = \frac{N_j^{(n)} - np_j}{\sqrt{np_j}}, \quad j = 1, \ldots, r.$$

Then, under H_0, the vector $\mathbf{Z}$ will be asymptotically normally distributed with zero vector for its mean and its covariance matrix as

$$\mathbf{D} = \mathbf{I}_r - \mathbf{qq}^T, \quad \mathbf{q} = (p_1^{1/2}, \ldots, p_r^{1/2})^T, \quad R(\mathbf{D}) = r - 1.$$

It is easily verified that if $\mathbf{D}_k = \mathbf{I}_k - \mathbf{q}_k\mathbf{q}_k^T$, then

$$|\mathbf{D}_k| = 1 - \sum_{i=1}^{k} p_i, \quad \mathbf{D}_k^{-1} = \mathbf{I}_k + \left(1 - \sum_{i=1}^{k} p_i\right)^{-1} \mathbf{q}_k\mathbf{q}_k^T, \tag{2.11}$$

where $\mathbf{q}_k = (p_1^{1/2}, \ldots, p_k^{1/2})^T, k = 1, \ldots, r-1$. Substituting (2.11) into (2.9), we obtain the elements of $\mathbf{R}$ as

$$\begin{aligned} r_{ii} &= \left(\frac{|\mathbf{D}_{i-1}|}{|\mathbf{D}_i|}\right)^{1/2}, \quad i = 1, \ldots, r-1, \\ r_{ij} &= \frac{r_{ii}}{|\mathbf{D}_{i-1}|}\sqrt{p_i}\sqrt{p_j}, \quad j = 1, \ldots, i-1. \end{aligned} \tag{2.12}$$

From (2.10) and (2.12), we obtain

$$\delta_k = \left(\frac{|\mathbf{D}_{k-1}|}{|\mathbf{D}_k|}\right)^{1/2}\left[Z_k + \frac{\sqrt{p_k}}{|\mathbf{D}_{k-1}|}\sum_{i=1}^{k-1}\sqrt{p_i}Z_i\right], \quad k = 1, \ldots, r-1. \tag{2.13}$$

Corollary 2.1. *The following decomposition of Pearson's sum holds:*

$$X_n^2 = \delta_1^2 + \delta_2^2 + \cdots + \delta_{r-1}^2, \tag{2.14}$$

where $\delta_k, k = 1,2,\ldots,r-1$, *are independent random variables distributed in the limit, under* H_0, *as chi-squared with one degree of freedom. If* $r = 3$, *for example, we have*

$$\begin{cases} \delta_1 = \dfrac{Z_1}{\sqrt{1-p_1}}, \\ \delta_2 = \dfrac{\sqrt{1-p_1}}{\sqrt{1-p_1-p_2}}\left(Z_2 + \dfrac{\sqrt{p_1p_2}}{1-p_1}Z_1\right). \end{cases}$$

Taking into account that $Z_3 = -\frac{\sqrt{p_1}}{\sqrt{p_3}}Z_1 - \frac{\sqrt{p_2}}{\sqrt{p_3}}Z_2$, *we note that* $X_n^2 = Z_1^2 + Z_2^2 + Z_3^2 = \delta_1^2 + \delta_2^2$.

2.3 NEYMAN-PEARSON CLASSES AND APPLICATIONS OF DECOMPOSITIONS OF PEARSON'S SUM

Two Neyman-Pearson classes that minimize in some sense Pearson's measure of the distance between null and alternative hypotheses are defined as (Greenwood and Nikulin, 1996) $\Delta_1 = \{x : f_N(x) < f_A(x)\}$ and $\Delta_2 = \{x : f_N(x) \geqslant f_A(x)\}$, where $f_N(x)$ and $f_A(x)$ are probability density functions for a null and an alternative hypotheses, respectively. Using points of intersection of $f_N(x)$ and $f_A(x)$, it is possible to construct four or more Neyman-Pearson type classes (Voinov, 2009). Tests based on those intervals possess the limiting chi-squared distributions with usually less than $r-1$ degrees of freedom. Hence, their variance will be small, and moreover, the rate of convergence to the limit will be essentially higher.

To illustrate some applications of the above decompositions, consider the testing of the simple null hypothesis that a random variable X follows the logistic probability distribution $L(0,1)$ with density function (see Balakrishnan (1992) for elaborate details on the logistic distribution)

$$l(x,0,1) = \frac{\frac{\pi}{\sqrt{3}}\exp\left(-\frac{\pi x}{\sqrt{3}}\right)}{\left\{1+\exp\left(-\frac{\pi x}{\sqrt{3}}\right)\right\}^2}, \quad x \in R^1, \tag{2.15}$$

against the simple alternative that X follows the triangular distribution $T(0,1)$ with mean zero, variance one, and density function

$$t(x,0,1) = \frac{1}{\sqrt{6}} - \frac{|x|}{6}, \quad |x| \leqslant \sqrt{6}. \tag{2.16}$$

For the null hypothesis $H_0 : X \sim L(0,1)$, the borders of four equiprobable cells will be $-\infty = y_0 < y_1 < y_2 < y_3 < +\infty$, where $y_i = \sqrt{3}\ln(i/(4-i))/\pi$, $i = 1,2,3$. Since for the selected null and alternative hypotheses there is no difference in location and skewness, only

$$\delta_2^2 = \frac{[(N_1^{(n)} - n/4) - (N_2^{(n)} - n/4) - (N_3^{(n)} - n/4) + (N_4^{(n)} - n/4)]^2}{n}$$

of Anderson (1994) decomposition in (2.4) will possess power. This power can be measured in terms of the non-centrality parameter, because the limiting distributions of X_n^2 and of components $\delta_i^2, i = 1,2,3$, for the simple alternative hypothesis can be approximated by the non-central chi-squared distribution with the corresponding number of degrees of freedom (three for X_n^2 and one for each of the δ_i^2 (Kallenberg et al., 1985). If, for example, the sample size $n = 200$, the non-centrality parameter of δ_2^2 of (2.4) will be $\lambda = 3.548$ that for $\alpha = 0.05$ corresponds to a power of $p_{\delta_2} = 0.47$. For the test X_n^2 with 3 degrees of freedom, the power will be $p_{X_n} = 0.32$ that is less than that of δ_2^2. In this example, the power of the tests δ_1^2 and δ_3^2 equals to zero and power of δ_2^2 in some sense "averages" between the 3 degrees of freedom of X_n^2 thus reducing the power of the test.

Since Anderson's decomposition (2.4) is valid only for equiprobable intervals (Boero et al., 2004a), it is of interest to consider this example for non-equiprobable cells that can result in substantial gain in power (see Best and Rayner, 1981, Boero et al., 2004b). For $r = 4$ intervals, the decomposition in (2.13) will be $X_n^2 = \delta_1^2 + \delta_2^2 + \delta_3^2$, where

$$\delta_j = \left(\frac{|\mathbf{D}_{j-1}|}{|\mathbf{D}_j|}\right)^{1/2}\left[Z_j + \frac{\sqrt{p_j}}{|\mathbf{D}_{j-1}|}\sum_{k=1}^{j-1}\sqrt{p_k}Z_k\right], \quad j = 1,2,3, \qquad (2.17)$$

$Z_j = \frac{N_j^{(n)} - np_j}{\sqrt{np_j}}, |\mathbf{D}_j| = 1 - \sum_{k=1}^{j} p_k$, and p_j $(j = 1,2,3)$ is the probability of falling into the jth class under H_0. The decomposition in (2.17) does not relate to distributional characteristics as the decomposition (2.4) does, but investigating power of the components in (2.17) permits sometimes to find a combination of separate degrees of freedom that can give a more powerful test. Consider the following example. There are four points of intersection of densities in (2.15) and (2.16) (see Figure 2.1).

Abscissas of those points are −2.28121, −0.827723, 0.827723, and 2.28121. Consider the following set of four Neyman-Pearson type intervals:

$$\begin{aligned}
\Delta_{11} &= (-\infty; -2.28121) \cup (-0.827723; 0),\\
\Delta_{12} &= (0; 0.827723) \cup (2.28121; +\infty),\\
\Delta_{21} &= (-2.28121; -0.827723),\\
\Delta_{22} &= (0.827723; 2.28121). \qquad (2.18)
\end{aligned}$$

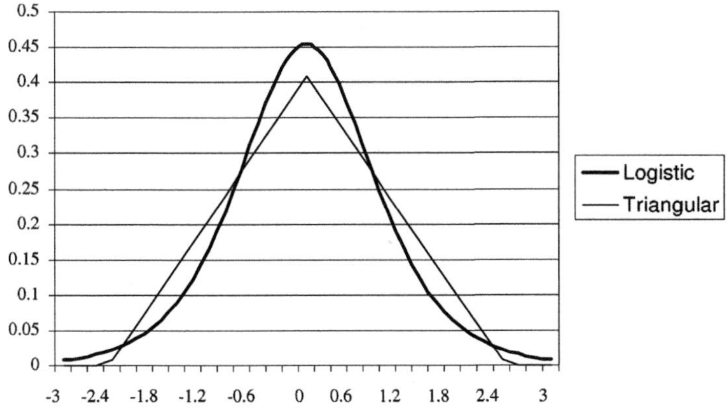

FIGURE 2.1 Graph of densities of the logistic distribution $l(x,0,1)$ in (2.15) and the triangular $t(x,0,1)$ distribution in (2.16).

Over the intervals Δ_{11} and $\Delta_{12}, t(x,0,1) < l(x,0,1)$, and over the intervals Δ_{21} and $\Delta_{22}, t(x,0,1) \geqslant l(x,0,1)$. The non-centrality parameters of the statistics $\delta_1^2, \delta_2^2, \delta_3^2$ defined in (2.17), and X_n^2 in this case are $\lambda_1 = 3.646$, $\lambda_2 = 0.910, \lambda_3 = 4.556$, and $\lambda_{X_n} = 9.112$, respectively. The power of tests based on these statistics for samples of size $n = 200$ are $p_{\delta_1} = 0.480$, $p_{\delta_2} = 0.159, p_{\delta_3} = 0.569$, and $p_{X_n} = 0.717$, respectively.

Consider the test $\delta_1^2 + \delta_3^2$ based on two independent most powerful components of X_n^2. This test possesses a power of 0.729 that is higher than that of the test X_n^2.

The above numerical example demonstrates an increase in power compared to Anderson's δ_2^2 test ($p_{\delta_2} = 0.47$) due to the following reasons: the ability to use non-equiprobable cells, and, consequently, the ability to apply Neyman-Pearson type intervals that maximize in some sense the measure of distance between hypotheses. It also illustrates the idea of Cochran (1954) to "use a single degree of freedom, or a group of degrees of freedom, from the total X_n^2, to get more powerful and appropriate tests."

2.4 PEARSON-FISHER AND DZHAPARIDZE-NIKULIN TESTS

The classical Pearson test in (2.3) will be distribution-free if and only if one is testing a simple null hypothesis. In the parametric case, it is assumed that parameters are known. During the period 1900–1924, researchers believed that Karl Pearson was of the opinion that replacing the unknown parameters in chi-squared sum in (2.3) by their estimates will not change the limiting distribution of Pearson's sum and it will still follow χ^2_{r-1} distribution (Plackett, 1983; Rao, 2002; Stigler, 2008). Fisher (1924, 1928) clearly showed that the number of degrees of freedom of Pearson's test must be reduced by the number s of parameters estimated from the sample. The result of Fisher is true if and only if the parameters are estimated by grouped data (minimizing Pearson's chi-squared sum, using multinomial maximum likelihood estimates (MLEs) for grouped data, or by any other asymptotically equivalent procedure; for details, see Greenwood and Nikulin (1996)). Nowadays, Pearson's test with unknown parameters replaced by the estimator from the grouped data, viz., $\hat{\boldsymbol{\theta}}_n$, is known as Pearson–Fisher (PF) test given by

$$X_n^2(\hat{\boldsymbol{\theta}}_n) = \mathbf{V}^{(n)T}(\hat{\boldsymbol{\theta}}_n)\mathbf{V}^{(n)}(\hat{\boldsymbol{\theta}}_n) = \sum_{j=1}^{r} \frac{[N_j^{(n)} - np_j(\hat{\boldsymbol{\theta}}_n)]^2}{np_j(\hat{\boldsymbol{\theta}}_n)}. \tag{2.19}$$

If $\hat{\boldsymbol{\theta}}_n$ is an estimator of the parameter $\boldsymbol{\theta}$ based on the grouped data, then in the limit the random vector $\mathbf{V}^{(n)}(\hat{\boldsymbol{\theta}}_n)$ will possess multivariate normal distribution with zero vector for its mean and its covariance matrix as

$$\boldsymbol{\Sigma} = \mathbf{I} - \mathbf{q}\mathbf{q}^T - \mathbf{B}(\mathbf{B}^T\mathbf{B})^{-1}\mathbf{B}^T,$$

where $\mathbf{q} = \{[p_1(\boldsymbol{\theta})]^{1/2}, \ldots, [p_r(\boldsymbol{\theta})]^{1/2}\}^T$, and $\mathbf{B} = \mathbf{B}(\boldsymbol{\theta})$ is the $r \times s$ matrix of rank s with its elements as

$$\frac{1}{\sqrt{p_j(\boldsymbol{\theta})}} \frac{\partial p_j(\boldsymbol{\theta})}{\partial \theta_k}, \quad j = 1, \ldots, r, \quad k = 1, \ldots, s. \tag{2.20}$$

In this case, the Pearson-Fisher statistic in (2.19) will be distributed in the limit as χ^2_{r-s-1}; for details, see Birch (1964) and Greenwood and Nikulin (1996).

For many years, researchers thought that instead of estimates based on grouped data one may also use other estimates such as the MLEs based on raw data. This is quite natural since it is clear that by using grouped data we lose some information about the parameters that is contained in a sample. Consider the following situation. Let in a particular interval realizations of a random variable X be grouped around, say, left or right edge, but we replace the average of those realizations by the midpoint of the interval which can be quite far from the actual average. Evidently, much information will be lost in this case.

Chernoff and Lehmann (1954) showed that replacing unknown parameters in (2.19) by estimators such as the maximum likelihood estimator $\hat{\boldsymbol{\theta}}_n$ based on non-grouped (raw) data would dramatically change the limiting distribution. In this case, it will have the distribution of

$$\chi^2_{r-s-1} + \sum_{j=r-s}^{r-1} \lambda_j Z_j^2,$$

where Z_j are standard normal variables and coefficients $0 \leqslant \lambda_j < 1$, $j = 1, \ldots, s$, may depend on $\boldsymbol{\theta}$. This result of Chernoff and Lehmann, valid for fixed grouping cells, has been generalized by Moore (1971) and Chibisov (1971) for random cells (see also Le et al., 1983). Since the statistic in (2.19) is not distribution-free, it cannot be used for hypotheses testing in general. An erroneous usage of estimates based on raw data in (2.19) is present even in some recent textbooks in statistics, as mentioned in the preceding Chapter.

So, the natural question that arises is what to do if one intends to use estimates based on raw (non-grouped) data. A possible solution to this issue was proposed by Moore (1971) and Dahiya (1972a) for continuous distributions with location and scale parameters, and in a much more general setting subsequently by Nikulin (1973b) and Dzhaparidze and Nikulin (1974).

We now consider the theory of Dzhaparidze and Nikulin (1974) since Pearson-Fisher and Dzhaparidze-Nikulin tests are asymptotically equivalent and formally possess the same decompositions, with the only difference being that the PF test uses estimators based on grouped data, while the DN test uses any $\sqrt{n}$-consistent estimator based on the raw data. Being asymptotically equivalent, these tests possess exactly the same power. In particular, they possess almost no power for equiprobable classes (see Voinov, 2009).

Let $M[\boldsymbol{\Sigma}]$ be a space spanned by the columns of the idempotent matrix

$$\boldsymbol{\Sigma} = \mathbf{I} - \mathbf{q}\mathbf{q}^T - \mathbf{B}(\mathbf{B}^T\mathbf{B})^{-1}\mathbf{B}^T.$$

Then, a vector $\mathbf{U}^{(n)}(\tilde{\boldsymbol{\theta}}_n) = \boldsymbol{\Sigma}\mathbf{V}^{(n)}(\tilde{\boldsymbol{\theta}}_n)$, where $\tilde{\boldsymbol{\theta}}_n$ is any $\sqrt{n}$-consistent estimator of $\boldsymbol{\theta}$ (not necessarily based on grouped data), will possess in the limit the multivariate normal distribution with zero vector as its mean and its covariance matrix as $\boldsymbol{\Sigma}$, same as that for the Pearson-Fisher case. Since the generalized inverse $\boldsymbol{\Sigma}^-$ of $\boldsymbol{\Sigma}$ equals $\boldsymbol{\Sigma}$, the well-known Dzhaparidze-Nikulin statistic

$$U_n^2(\tilde{\boldsymbol{\theta}}_n) = \mathbf{V}^{(n)T}(\tilde{\boldsymbol{\theta}}_n)[\mathbf{I} - \mathbf{q}_n\mathbf{q}_n^T - \mathbf{B}_n(\mathbf{B}_n^T\mathbf{B})_n^{-1}\mathbf{B}_n^T]\mathbf{V}^{(n)}(\tilde{\boldsymbol{\theta}}_n), \tag{2.21}$$

where $\mathbf{q}_n$ is the estimator of the vector $\mathbf{q}$, will possess in the limit the chi-squared probability distribution χ^2_{r-s-1}, same as that of Pearson-Fisher statistic in (2.19). Mathematical and statistical properties of (2.21) were analyzed by Dudley (1976, 1979) and Drost (1988).

Noting that $\mathbf{q}_n\mathbf{q}_n^T\mathbf{V}^{(n)}(\hat{\boldsymbol{\theta}}_n) = \mathbf{0}$, the Dzhaparidze–Nikulin test statistic in (2.21) can be represented as

$$U_n^2(\tilde{\boldsymbol{\theta}}_n) = \mathbf{V}^{(n)T}(\tilde{\boldsymbol{\theta}}_n)[\mathbf{I} - \mathbf{B}_n(\mathbf{B}_n^T\mathbf{B}_n)^{-1}\mathbf{B}_n]\mathbf{V}^{(n)}(\tilde{\boldsymbol{\theta}}_n). \tag{2.22}$$

By using the trivial orthogonal decomposition of the identity matrix

$$\mathbf{I} = \mathbf{q}\mathbf{q}^T + \mathbf{B}(\mathbf{B}^T\mathbf{B})^{-1}\mathbf{B}^T + \left[\mathbf{I} - \mathbf{q}\mathbf{q}^T - \mathbf{B}(\mathbf{B}^T\mathbf{B})^{-1}\mathbf{B}^T\right] \tag{2.23}$$

and the fact that $\mathbf{q}^T\mathbf{V}^{(n)}(\boldsymbol{\theta}) = \mathbf{0}$, Mirvaliev (2001) showed that the decomposition

$$\mathbf{V}^{(n)}(\boldsymbol{\theta}) = \mathbf{U}_n(\boldsymbol{\theta}) + \mathbf{W}_n(\boldsymbol{\theta}) \tag{2.24}$$

is the sum of orthogonal vectors

$$\mathbf{U}_n(\boldsymbol{\theta}) = \left[\mathbf{I} - \mathbf{q}\mathbf{q}^T - \mathbf{B}(\mathbf{B}^T\mathbf{B})^{-1}\mathbf{B}^T\right]\mathbf{V}^{(n)}(\boldsymbol{\theta})$$

and

$$\mathbf{W}_n(\boldsymbol{\theta}) = \mathbf{B}(\mathbf{B}^T\mathbf{B})^{-1}\mathbf{B}^T\mathbf{V}^{(n)}(\boldsymbol{\theta}). \tag{2.25}$$

From this, it follows that Pearson's sum in (2.3) is decomposed as follows:

$$X_n^2(\boldsymbol{\theta}) = U_n^2(\boldsymbol{\theta}) + W_n^2(\boldsymbol{\theta}), \tag{2.26}$$

where

$$U_n^2(\boldsymbol{\theta}) = \mathbf{V}^{(n)T}(\boldsymbol{\theta})\left[\mathbf{I} - \mathbf{q}\mathbf{q}^T - \mathbf{B}(\mathbf{B}^T\mathbf{B})^{-1}\mathbf{B}^T\right]\mathbf{V}^{(n)}(\boldsymbol{\theta}) \tag{2.27}$$

and

$$W_n^2(\boldsymbol{\theta}) = \mathbf{V}^{(n)T}(\boldsymbol{\theta})\mathbf{B}(\mathbf{B}^T\mathbf{B})^{-1}\mathbf{B}^T\mathbf{V}^{(n)}(\boldsymbol{\theta}). \tag{2.28}$$

If one replaces in (2.27) the parameter $\boldsymbol{\theta}$ by any $\sqrt{n}$-consistent estimator $\tilde{\boldsymbol{\theta}}$, the limiting distribution of the DN statistic $U_n^2(\tilde{\boldsymbol{\theta}}_n)$ would be χ^2_{r-s-1} regardless of the estimator chosen. This is precisely the reason why Mirvaliev (2001) called (2.27) the invariant part of Pearson's sum. On the contrary, the limiting distribution of the statistic $W_n^2(\tilde{\boldsymbol{\theta}}_n)$ essentially depends on the limiting properties of the estimator $\tilde{\boldsymbol{\theta}}_n$, which was called the sensitive part by Mirvaliev.

Consider a decomposition of the DN statistic (2.22) into limiting independent χ_1^2 distributed limit components (Voinov, 2010). In this case,

$$\mathbf{D}_k = \mathbf{I}_k - \mathbf{q}_k\mathbf{q}_k^T - \mathbf{B}_k(\mathbf{B}^T\mathbf{B})^{-1}\mathbf{B}_k^T, \quad k = 1, \ldots, r-s-1,$$

$$|\mathbf{D}_k| = \left(1 - \sum_{i=1}^{k} p_i\right)|\mathbf{T} - \mathbf{T}_k|/|\mathbf{T}|,$$

and

$$\mathbf{D}_k^{-1} = \mathbf{M}_k + \mathbf{M}_k\mathbf{B}_k(\mathbf{T} - \mathbf{T}_k)^{-1}\mathbf{B}_k^T\mathbf{M}_k, \tag{2.29}$$

where
$\mathbf{M}_k = \mathbf{I}_k + \mathbf{q}_k\mathbf{q}_k^T/\left(1 - \sum_{i=1}^{k} p_i\right)$, $\mathbf{T} = \mathbf{B}^T\mathbf{B}$, and $\mathbf{T}_k = \mathbf{B}_k^T(\mathbf{I}_k - \mathbf{q}_k\mathbf{q}_k^T)^{-1}\mathbf{B}_k$.
Since

$$\mathbf{d}_{i(i-1)}^T = -(\sqrt{p_i}\mathbf{q}_{i-1}^T + \mathbf{b}_i(\mathbf{B}^T\mathbf{B})^{-1}\mathbf{B}_{i-1}^T),$$

where $\mathbf{b}_i$ is the ith row of the matrix $\mathbf{B}$, then, by (2.9), we obtain

$$\begin{aligned} r_{ii} &= \sqrt{|\mathbf{D}_{i-1}|/|\mathbf{D}_i|}, \\ r_{ij} &= r_{ii}(\sqrt{p_i}\mathbf{q}_{i-1}^T + \mathbf{b}_i(\mathbf{B}^T\mathbf{B})^{-1}\mathbf{B}_{i-1}^T)(\mathbf{D}_{i-1}^{-1})_j, \end{aligned} \tag{2.30}$$

where $j = 1, \ldots, i-1, i = 1, \ldots, r-s-1$, and $(\mathbf{D}_{i-1}^{-1})_j$ is the jth column of $\mathbf{D}_{i-1}^{-1}$ defined by (2.29).

Consider the transformation

$$\boldsymbol{\delta}_{r-s-1}(\tilde{\boldsymbol{\theta}}_n) = \mathbf{R}\mathbf{U}^{(n)}(\tilde{\boldsymbol{\theta}}_n) = \mathbf{R}\boldsymbol{\Sigma}\mathbf{V}^{(n)}(\tilde{\boldsymbol{\theta}}_n), \tag{2.31}$$

where $\mathbf{R} = (\mathbf{R}_{r-s-1}\vdots\mathbf{0})$ and $\mathbf{R}_{r-s-1}$ is the lower triangular matrix with elements defined by (2.30).

Theorem 2.3. *Under the usual regularity conditions, the following decomposition of the Dzhaparidze–Nikulin statistic holds:*

$$U_n^2(\tilde{\boldsymbol{\theta}}_n) = \delta_1^2(\tilde{\boldsymbol{\theta}}_n) + \cdots + \delta_{r-s-1}^2(\tilde{\boldsymbol{\theta}}_n), \tag{2.32}$$

where components $\delta_1^2(\tilde{\boldsymbol{\theta}}_n), \ldots, \delta_{r-s-1}^2(\tilde{\boldsymbol{\theta}}_n)$ are independently distributed as χ_1^2 in the limit.

Proof. It is sufficient to prove the validity of the decomposition in (2.32). Indeed,

$$\begin{aligned}
\delta_1^2(\tilde{\boldsymbol{\theta}}_n) + \cdots + \delta_{r-s-1}^2(\tilde{\boldsymbol{\theta}}_n) &= \boldsymbol{\delta}_{(r-s-1)}^T(\tilde{\boldsymbol{\theta}}_n)\boldsymbol{\delta}_{(r-s-1)}(\tilde{\boldsymbol{\theta}}_n) \\
&= (\mathbf{R}\mathbf{U}^{(n)}(\tilde{\boldsymbol{\theta}}_n))^T(\mathbf{R}\mathbf{U}^{(n)}(\tilde{\boldsymbol{\theta}}_n)) \\
&= \mathbf{U}^{(n)T}(\tilde{\boldsymbol{\theta}}_n)\mathbf{R}^T\mathbf{R}\mathbf{U}^{(n)}(\tilde{\boldsymbol{\theta}}_n).
\end{aligned}$$

Due to Lemma 2.1, we have $\mathbf{R}^T\mathbf{R} = \boldsymbol{\Sigma}^-$ and, since $\mathbf{U}^{(n)}(\tilde{\boldsymbol{\theta}}_n) = \boldsymbol{\Sigma}\mathbf{V}^{(n)}(\tilde{\boldsymbol{\theta}}_n)$, we have

$$\begin{aligned}
\mathbf{U}^{(n)T}(\tilde{\boldsymbol{\theta}}_n)\mathbf{R}^T\mathbf{R}\mathbf{U}^{(n)}(\tilde{\boldsymbol{\theta}}_n) &= \mathbf{U}^{(n)T}(\tilde{\boldsymbol{\theta}}_n)\boldsymbol{\Sigma}^-\mathbf{U}^{(n)}(\tilde{\boldsymbol{\theta}}_n) \\
&= \mathbf{V}^{(n)T}(\tilde{\boldsymbol{\theta}}_n)\boldsymbol{\Sigma}\boldsymbol{\Sigma}^-\boldsymbol{\Sigma}\mathbf{V}^{(n)}(\tilde{\boldsymbol{\theta}}_n) \\
&= \mathbf{V}^{(n)T}(\tilde{\boldsymbol{\theta}}_n)\boldsymbol{\Sigma}\mathbf{V}^{(n)}(\tilde{\boldsymbol{\theta}}_n) = U_n^2(\tilde{\boldsymbol{\theta}}_n).
\end{aligned}$$

Corollary 2.2. *Since in the limit the vector $\mathbf{V}^{(n)}(\hat{\boldsymbol{\theta}}_n)$, where $\hat{\boldsymbol{\theta}}_n$ is an estimator based on grouped data, possesses the multivariate normal distribution with zero mean vector and covariance matrix $\boldsymbol{\Sigma} = \mathbf{I} - \mathbf{q}\mathbf{q}^T - \mathbf{B}(\mathbf{B}^T\mathbf{B})^{-1}\mathbf{B}^T$, same as for the vector $\mathbf{V}^{(n)}(\tilde{\boldsymbol{\theta}}_n)$, the following decomposition of the Pearson-Fisher statistic in (2.19) holds:*

$$X_n^2(\hat{\boldsymbol{\theta}}_n) = \delta_1^2(\hat{\boldsymbol{\theta}}_n) + \cdots + \delta_{r-s-1}^2(\hat{\boldsymbol{\theta}}_n), \tag{2.33}$$

where the components $\delta_i^2(\hat{\boldsymbol{\theta}}_n), i = 1, \ldots, r-s-1$, of the vector $\boldsymbol{\delta}_{r-s-1}(\tilde{\boldsymbol{\theta}}_n)$ are as defined in (2.31) with $\tilde{\boldsymbol{\theta}}$ replaced by $\hat{\boldsymbol{\theta}}_n$. Naturally, all matrices in (2.31) need to be estimated with the usage of $\hat{\boldsymbol{\theta}}$ instead of $\tilde{\boldsymbol{\theta}}_n$.

2.5 CHERNOFF-LEHMANN THEOREM

The famous result of Chernoff and Lehmann (1954) is indeed a milestone in the theory of chi-squared testing. Chernoff and Lehmann showed that if in (2.19) instead of the estimators based on grouped data (such as minimum χ^2 or MLEs) one will use some other estimators such as the MLEs based on original observations that are more efficient, then the limiting distribution of Pearson's sum will dramatically change. Specifically, they proved the following result.

Theorem 2.4. *Under suitable regularity conditions the asymptotic distribution of*

$$X_n^2(\hat{\boldsymbol{\theta}}_n) = \mathbf{V}^{(n)T}(\hat{\boldsymbol{\theta}}_n)\mathbf{V}^{(n)}(\hat{\boldsymbol{\theta}}_n) = \sum_{j=1}^{r} \frac{[N_j^{(n)} - np_j(\hat{\boldsymbol{\theta}}_n)]^2}{np_j(\hat{\boldsymbol{\theta}}_n)},$$

will be the same as that of

$$\sum_{j=1}^{r-s-1} Z_j^2 + \sum_{j=r-s}^{r-1} \lambda_j Z_j^2, \qquad (2.34)$$

where Z_j are standard normal random variables and λ_j are between 0 and 1, and may depend on s unknown parameters, in general.

From (2.34), we see that the limiting distribution of $X_n^2(\hat{\boldsymbol{\theta}}_n)$ is stochastically larger than that of χ^2_{r-s-1}. Chernoff and Lehmann (1954) noted that this is not a serious drawback "in case of fitting a Poisson distribution, but may be so for fitting of a normal." It is much more important to note that if λ_i's depend on the unknown parameters, then the statistic in (2.19) cannot be used in principle when testing for normality, for example. Unfortunately, as we already mentioned, this erroneous usage is reproduced even in some current textbooks in Statistics. It has to be mentioned here that Chernoff and Lehmann (1954) considered only the case of fixed grouping intervals, but their result has been generalized by Moore (1971) and Chibisov (1971) for random grouping cells.

2.6 PEARSON-FISHER TEST FOR RANDOM CLASS END POINTS

The first attempt to solve this problem was done by Roy (1956) in his unpublished report. He showed that if probability density function of X is continuous in x and differentiable in $\boldsymbol{\theta} = (\theta_1, \ldots, \theta_s)^T$, $\hat{\boldsymbol{\theta}}_n$ is a $\sqrt{n}$-consistent estimator of $\boldsymbol{\theta}$, and if the range of X, $(-\infty, +\infty)$, is partitioned into r mutually exclusive intervals depending on $\hat{\boldsymbol{\theta}}_n$, then the sum in (2.19) will be distributed in the limit as $\sum_{j=1}^{r} \lambda_j Z_j^2$, where $Z_1, \ldots, Z_r$ are independent standard normal random variables. In particular, if $\hat{\boldsymbol{\theta}}_n$ is the MLE of $\boldsymbol{\theta}$, then $r-s-1$ of the λ_j's are 1, s of λ_j's lie between 0 and 1 and may depend on the unknown parameter $\boldsymbol{\theta}$ in general, and one λ_j is zero. In this case, the sum in (2.19) will be distributed in the limit as

$$\chi^2_{r-s-1} + \lambda_1 Z_1^2 + \cdots + \lambda_s Z_s^2.$$

Roy proved that, in the case of $s = 2$, the asymptotic distribution of (2.19) will not depend on the location and scale parameters, θ_1 and $\sqrt{\theta_2}$, say, if the MLE $\hat{\boldsymbol{\theta}}_n$ is used and the class end points are of the form $\hat{\theta}_1 + c_j\sqrt{\hat{\theta}_2}, j = 1,2,\ldots,r$, where the c_j's are specified constants.

Let $X_1,\ldots,X_n$ be i.i.d. random variables and we wish to test the null hypothesis, H_0, that the probability density function of X_i is

$$f(x,\boldsymbol{\theta}) = \frac{1}{\sqrt{2\pi\theta_2}} \exp\left(-\frac{(x-\theta_1)^2}{2\theta_2}\right),$$
$$|\theta_1| < \infty, \quad |x| < \infty, \quad \theta_2 > 0, \tag{2.35}$$

where $\boldsymbol{\theta} = (\theta_1,\theta_2)^T$ and θ_1 and θ_2 are the population mean and population variance, respectively. Using the MLE $\hat{\boldsymbol{\theta}}_n = (X,S^2)^T$ of $\boldsymbol{\theta}$, define the end points of intervals as

$$g_j(\hat{\boldsymbol{\theta}}_n) = \bar{X} + c_j S, \quad j = 0,1,\ldots,r.$$

Constants c_j are chosen such that

$$\begin{aligned} p_j(\hat{\boldsymbol{\theta}}_n) &= F\{g_j(\hat{\boldsymbol{\theta}}_n)\} - F\{g_{j-1}(\hat{\boldsymbol{\theta}}_n)\} \\ &= Pr\{c_{j-1} < [(X-\theta_1)/\sqrt{\theta_2}] < c_j\} = 1/r. \end{aligned} \tag{2.36}$$

For those r equiprobable intervals, the statistic in (2.19) will be distributed in the limit as that of

$$\chi^2_{r-3} + \lambda_1 Z_1^2 + \lambda_2 Z_2^2, \tag{2.37}$$

where (Watson, 1958)

$$\lambda_1 = 1 - r\sum_{l=1}^{r} \Psi_0^2(l), \quad \lambda_2 = 1 - \frac{r}{2}\sum_{l=1}^{r} \Psi_1^2(l),$$
$$\Psi_k(l) = c_l^k \phi(c_l) - c_{l-1}^k \phi(c_{l-1}),$$

and $\phi(u) = \frac{1}{\sqrt{2\pi}} \exp(-u^2/2)$. Based on these results, Moore (1971) calculated the critical points of the limit distribution of (2.37) for odd $r = 5,7,9,11,15,21$ (see Table 1 of Moore (1971)). One year later, Dahiya (1972a) gave the table of the critical points of the limit distribution of (2.37) for both odd and even number of intervals r from 3 to 15.

Dahiya (1972a) generalized Roy's result as follows. "When $s = 2$ it is possible, under certain circumstances, for the asymptotic distribution of (2.19) to be free of the parameters θ_1 and θ_2 without assuming $\hat{\boldsymbol{\theta}}_n$ to be maximum likelihood estimator of $\boldsymbol{\theta}$." Specifically, they established the following result.

Theorem 2.5. *Suppose*

(i) *X has probability density function* $(1/\sqrt{\theta_2})P_x\{(x-\theta_1)/\sqrt{\theta_2}\}$, *where* θ_1 *and* $\sqrt{\theta_2}$ *are location and scale parameters, respectively, such that* $\mathbf{E}(X)=\theta_1$ *and* $\mathbf{Var}(X)=\theta_2$, *with* $P_x\{(x-\theta_1)/\sqrt{\theta_2}\}$ *being continuous in x and differentiable with respect to the parameters;*

(ii) $\hat{\boldsymbol{\theta}}_n=(\bar{X},S^2)^T$, *where* $\bar{X}=\frac{1}{n}\sum X_\alpha$ *and* $S^2=\frac{1}{n}\sum(X_\alpha-\bar{X})^2$ *are the sample mean and the sample variance, respectively;*

(iii) *The class end points are of the form* $\hat{\theta}_1+c_i\sqrt{\hat{\theta}_2}$, *where the* c_i*'s are specified constants,*

then the asymptotic distribution of (2.19), being independent of the parameters, is that of $\lambda_1 Z_1^2+\cdots+\lambda_r Z_r^2$.

This result of Dahiya (1972a) can, for instance, be useful when testing for the logistic null hypothesis when the sample mean and the sample variance are not efficient MLEs of the location and scale parameters.

At this point, it is worth to note that the approach considered above possesses two drawbacks. First one is that for any null hypothesis about location and scale distributions, one has to derive the limit distribution of (2.19) and needs to calculate critical points that will be different for different hypotheses. Second issue is that this test possesses not high enough power, for example, when testing for normality.

Chapter 3

Wald's Method and Nikulin-Rao-Robson Test

3.1 WALD'S METHOD

Let $X_1,\ldots,X_n$ be i.i.d. random variables following a continuous family of distributions

$$F(x;\boldsymbol{\theta}),\quad x\in R^1,\quad \boldsymbol{\theta}=(\theta_1,\ldots,\theta_s)^T\in\boldsymbol{\Theta}\subset R^s.$$

Let $f(x;\boldsymbol{\theta})$ be the corresponding pdf, $\hat{\boldsymbol{\theta}}_n$ be the MLE of $\boldsymbol{\theta}$, and $\mathbf{J}$ be the Fisher information matrix for one observation with elements $J_{ij},i,j=1,\ldots,s$.

Let $x_j(\hat{\boldsymbol{\theta}}_n) = F^{-1}(p_1+\cdots+p_j,\hat{\boldsymbol{\theta}}_n),j = 1,\ldots,r-1$. Consider partitioning the x-axis into r disjoint intervals $\Delta_j,j=1,\ldots,r$, with end points $x_0,x_1(\hat{\boldsymbol{\theta}}_n),\ldots,x_{r-1}(\hat{\boldsymbol{\theta}}_n),x_r$, where $x_r=-x_0=+\infty$. Let $\mathbf{N}^{(n)}$ be a vector with components $N_j^{(n)},j=1,\ldots,r$, that count the number of $X_1,\ldots,X_n$ falling into each interval. Let $\mathbf{p}=(p_1,\ldots,p_r)^T$. Moore (1971) then showed that the vector $\mathbf{N}^{(n)}$ is asymptotically normally distributed with the mean vector $n\mathbf{p}$ and covariance matrix $n\boldsymbol{\Sigma}$, where

$$\boldsymbol{\Sigma}=\mathbf{P}-\mathbf{p}\mathbf{p}^T-\mathbf{W}^T\mathbf{J}^{-1}\mathbf{W}, \tag{3.1}$$

with the components of the $r\times s$ matrix $\mathbf{W}$ being

$$W_{ij}=\int\limits_{\Delta_j}\frac{\partial f(x;\boldsymbol{\theta})}{\partial\theta_i}dx,\quad i=1,\ldots,s,\quad j=1,\ldots,r,$$

and $\mathbf{P}$ is the diagonal matrix with elements $p_1,\ldots,p_r$ on the diagonal. The rank $r(\boldsymbol{\Sigma})$ of the matrix $\boldsymbol{\Sigma}$ is $r-1$. What is of importance is that $\boldsymbol{\Sigma}$ does not depend on the parameter $\boldsymbol{\theta}$.

Chi-Squared Goodness of Fit Tests with Applications. http://dx.doi.org/10.1016/B978-0-12-397194-4.00003-X

Suppose we wish to test a composite null hypothesis that $X_1, \ldots, X_n$ follow a continuous family of distributions $F(x;\boldsymbol{\theta}), x \in R^1, \boldsymbol{\theta} = (\theta_1, \ldots, \theta_s)^T \in \boldsymbol{\Theta} \subset R^s$. Based on (3.1) and the theory of generalized inverses, Nikulin (1973c) proposed to use the statistic

$$Y1_n^2(\hat{\boldsymbol{\theta}}_n) = n^{-1}(\mathbf{N}^{(n)} - n\mathbf{p})^T \boldsymbol{\Sigma}^- (\mathbf{N}^{(n)} - n\mathbf{p}), \tag{3.2}$$

where $\boldsymbol{\Sigma}^-$ is the generalized inverse of $\boldsymbol{\Sigma}$ which is a matrix such that $\boldsymbol{\Sigma}\boldsymbol{\Sigma}^-\boldsymbol{\Sigma} = \boldsymbol{\Sigma}$. It is well known that it does not matter what kind of a generalized inverse we use in (3.2) (Rao, 1965). The resulting quadratic form will be the same and, since $\boldsymbol{\Sigma}$ does not depend on $\boldsymbol{\theta}$, will have chi-square distribution with $r-1$ degrees of freedom in the limit (see Lemma 9 of Khatri, 1968). To construct $\boldsymbol{\Sigma}^-$ of the matrix $\boldsymbol{\Sigma}$ of rank $r-1$, Nikulin (1973c) used the following technique due to Rao. He constructed the matrix $\widetilde{\boldsymbol{\Sigma}}$ by crossing out the last row and the last column of $\boldsymbol{\Sigma}$, found $\widetilde{\boldsymbol{\Sigma}}^{-1}$, and then added a row and a column of zeros to represent $\boldsymbol{\Sigma}^-$ as

$$\boldsymbol{\Sigma}^- = \begin{pmatrix} \widetilde{\boldsymbol{\Sigma}}^{-1} & 0 \\ \mathbf{0} & 0 \end{pmatrix}.$$

For the sake of convenience in calculations, Nikulin (1973c) represented (3.2) as

$$Y1_n^2(\hat{\boldsymbol{\theta}}_n) = X^2 + n^{-1}\boldsymbol{\alpha}^T(\hat{\boldsymbol{\theta}}_n)\boldsymbol{\Lambda}(\hat{\boldsymbol{\theta}}_n)\boldsymbol{\alpha}(\hat{\boldsymbol{\theta}}_n), \tag{3.3}$$

where

$$X^2 = \sum_{j=1}^{r} \frac{(N_j^{(n)} - np_j)^2}{np_j}, \quad \boldsymbol{\Lambda}(\hat{\boldsymbol{\theta}}_n) = \left\| J_{ij} - \sum_{l=1}^{r} \frac{W_{il}W_{jl}}{p_l} \right\|_{\boldsymbol{\theta}=\hat{\boldsymbol{\theta}}_n}^{-1}, \quad i,j = 1, \ldots, s,$$

and $\boldsymbol{\alpha}(\boldsymbol{\theta}) = (\alpha_1, \ldots, \alpha_s)^T$, with $\alpha_i = W_{i1}N_1^{(n)}/p_1 + \cdots + W_{ir}N_r^{(n)}/p_r$, $i = 1, \ldots, s$. By using the result of Moore (1971) and the theory of generalized inverses, the statistic in (3.3) follows χ_{r-1}^2 distribution in limit, and does not depend on the unknown parameter $\boldsymbol{\theta}$.

Denote by $\mathbf{B}$ the $r \times s$ matrix with elements

$$B_{jk} = \frac{1}{\sqrt{p_j}} \int_{\Delta_j} \frac{\partial f(x;\boldsymbol{\theta})}{\partial \theta_k} dx, \quad j = 1, \ldots, r, \quad k = 1, \ldots, s. \tag{3.4}$$

Let $\mathbf{N}^{(n)}/\sqrt{n\mathbf{p}}$ denote the vector with elements $N_j^{(n)}/\sqrt{np_j}, j = 1, \ldots, r$. With this notation, we have $n^{-1/2}\boldsymbol{\alpha}(\boldsymbol{\theta}) = \mathbf{B}^T\mathbf{N}^{(n)}/\sqrt{n\mathbf{p}}$. Upon noting that $\mathbf{B}^T\sqrt{n\mathbf{p}}$ is a zero s-vector, $n^{-1/2}\boldsymbol{\alpha}(\boldsymbol{\theta})$ can be presented as $\mathbf{B}^T\mathbf{V}^{(n)}$, where the vector $\mathbf{V}^{(n)}$ is a r-vector of standardized frequencies with components

$v_j^{(n)} = (N_j^{(n)} - np_j)/\sqrt{np_j}, j = 1,\dots,r$. Similarly, $n^{-1/2}\boldsymbol{\alpha}^T = \mathbf{V}^{(n)T}\mathbf{B}$ and so (3.3) can be written as

$$Y1_n^2(\hat{\boldsymbol{\theta}}_n) = X^2 + \mathbf{V}^{(n)T}\mathbf{B}_n(\mathbf{J}_n - \mathbf{J}_{gn})^{-1}\mathbf{B}_n^T\mathbf{V}^{(n)}, \tag{3.5}$$

where $\mathbf{J}_{gn} = \mathbf{B}_n^T\mathbf{B}_n$ is an estimator of the Fisher information matrix for the vector of frequencies. The formula in (3.5) is the one that has been used in statistical literature since 1975, and is known as Nikulin-Rao-Robson (NRR) test; see Habib and Thomas (1986), Drost (1988), Van der Vaart (1988), and Voinov and Nikulin (2011). Using the trivial identity

$$\mathbf{I} + \mathbf{B}(\mathbf{J} - \mathbf{B}^T\mathbf{B})^{-1} = (\mathbf{I} - \mathbf{B}\mathbf{J}^{-1}\mathbf{B}^T)^{-1}, \tag{3.6}$$

Moore and Spruill (1975) (see also Greenwood and Nikulin, 1996) presented the statistic in (3.5) as

$$Y1_n^2(\hat{\boldsymbol{\theta}}) = \mathbf{V}^{(n)T}(\mathbf{I} - \mathbf{B}_n\mathbf{J}_n^{-1}\mathbf{B}_n^T)^{-1}\mathbf{V}^{(n)}.$$

The approach in (3.2), suggested by Nikulin, is actually a generalization of Wald's (1943) idea for full rank limiting covariance matrix $\boldsymbol{\Sigma}$.

Using a heuristic approach for the exponential family of distributions, Rao and Robson (1974) obtained the test statistic which coincides with that of Nikulin based on Generalized Wald's or simply Wald's approach (see Moore, 1977, Hadi and Wells, 1990). Instead of vector of frequencies $\mathbf{N}^{(n)}$, one may use the vector of standardized frequencies with components $v_j(\hat{\boldsymbol{\theta}}_n) = (N_j^{(n)} - np_j(\hat{\boldsymbol{\theta}}_n))/\sqrt{np_j(\hat{\boldsymbol{\theta}}_n)}, j = 1,2,\dots,r$, that follows the multivariate normal distribution in limit with zero mean vector and the covariance matrix (Moore, 1971)

$$\boldsymbol{\Sigma}_1 = \mathbf{I} - \mathbf{q}\mathbf{q}^T - \mathbf{B}\mathbf{J}^{-1}\mathbf{B}^T. \tag{3.7}$$

It is easily verified that the unique Moore-Penrose matrix inverse (that satisfies the conditions $\mathbf{A}\mathbf{A}^+\mathbf{A} = \mathbf{A}, \mathbf{A}^+\mathbf{A}\mathbf{A}^+ = \mathbf{A}^+, (\mathbf{A}\mathbf{A}^+)^T = \mathbf{A}\mathbf{A}^+, (\mathbf{A}^+\mathbf{A})^T = \mathbf{A}^+\mathbf{A}$) for (3.7) is

$$\boldsymbol{\Sigma}_1^+ = \mathbf{I} - \mathbf{q}\mathbf{q}^T + \mathbf{B}(\mathbf{J} - \mathbf{B}^T\mathbf{B})^{-1}\mathbf{B}^T.$$

Then, the formula in (3.5) immediately follows from (3.6) and (3.7).

Moore and Spruill (1975) showed that, regardless of whether fixed or random intervals are used, the statistic in (3.5) can be written as

$$Y1_n^2(\hat{\boldsymbol{\theta}}_n) = X_n^2(\hat{\boldsymbol{\theta}}_n) + P_n^2(\hat{\boldsymbol{\theta}}_n), \tag{3.8}$$

where $X_n^2(\hat{\boldsymbol{\theta}}_n) = \mathbf{V}^{(n)T}(\hat{\boldsymbol{\theta}}_n)\mathbf{V}^{(n)}(\hat{\boldsymbol{\theta}}_n)$,

$$P_n^2(\hat{\boldsymbol{\theta}}_n) = \mathbf{V}^{(n)T}(\hat{\boldsymbol{\theta}}_n)\mathbf{B}_n(\mathbf{J}_n - \mathbf{J}_{gn})^{-1}\mathbf{B}_n^T\mathbf{V}^{(n)}(\hat{\boldsymbol{\theta}}_n),$$

$\mathbf{V}^{(n)}(\hat{\boldsymbol{\theta}}_n)$ is a vector with components $v_j^{(n)}(\hat{\boldsymbol{\theta}}_n) = (N_j^{(n)} - np_j(\hat{\boldsymbol{\theta}}_n))/\sqrt{np_j(\hat{\boldsymbol{\theta}}_n)}$, and $p_j(\hat{\boldsymbol{\theta}}_n) = \int\limits_{\Delta_j} f(x;\hat{\boldsymbol{\theta}}_n)dx, j = 1,2,\ldots,r$.

If one uses fixed grouping intervals, then the elements of the matrix **B** will be calculated as

$$B_{jk} = \frac{1}{\sqrt{p_j(\boldsymbol{\theta})}} \int\limits_{\Delta_j} \frac{\partial f(x;\boldsymbol{\theta})}{\partial \theta_k} dx = \frac{1}{\sqrt{p_j(\boldsymbol{\theta})}} \frac{\partial p_j(\boldsymbol{\theta})}{\partial \theta_k}, \quad j = 1,\ldots,r,\ k = 1,\ldots,s. \tag{3.9}$$

After carrying out the calculations, the unknown parameter $\boldsymbol{\theta}$ needs to be replaced by the MLE $\hat{\boldsymbol{\theta}}_n$.

For equiprobable cells $\Delta_j(\boldsymbol{\theta}) = [x_{j-1}(\boldsymbol{\theta}),x_j(\boldsymbol{\theta})), j = 1,\ldots,r$, we have

$$\frac{\partial}{\partial \theta_k} \int\limits_{x_{j-1}(\boldsymbol{\theta})}^{x_j(\boldsymbol{\theta})} f(x;\boldsymbol{\theta})dx = \frac{\partial p_j}{\partial \theta_k} = 0,$$

and, hence, due to Leibnitz formula,

$$B_{jk} = \frac{1}{\sqrt{p_j}} \int\limits_{x_{j-1}(\boldsymbol{\theta})}^{x_j(\boldsymbol{\theta})} \frac{\partial f(x;\boldsymbol{\theta})}{\partial \theta_k} dx = \frac{1}{\sqrt{p_j}} \left[f(x_{j-1}(\boldsymbol{\theta});\boldsymbol{\theta})x'_{j-1}(\boldsymbol{\theta}) - f(x_j(\boldsymbol{\theta});\boldsymbol{\theta})x'_j(\boldsymbol{\theta}) \right]. \tag{3.10}$$

Here again, after carrying out the calculations, the unknown parameter $\boldsymbol{\theta}$ needs to be replaced by the MLE $\hat{\boldsymbol{\theta}}_n$. This formula has been used by Nikulin (1973b) to test for normality, in particular.

Remark 3.1 (Test for normality). *Let $X_1,\ldots,X_n$ be i.i.d. random variables and suppose we wish to test the hypothesis, H_0, that the probability density function of X_i is*

$$f(x,\boldsymbol{\theta}) = \frac{1}{\sqrt{2\pi\theta_2}} \exp\left(-\frac{(x-\theta_1)^2}{2\theta_2} \right), \quad |\theta_1| < \infty, \quad \theta_2 > 0, \tag{3.11}$$

where $\boldsymbol{\theta} = (\theta_1,\theta_2)^T$, and θ_1 and θ_2 are the population mean and population variance, respectively. Many results for testing normality are known; see, for example, Dahiya and Gurland (1972a) and D'Agostino et al. (1990). Here, we describe in detail the idea of Nikulin. Using (3.5) and (3.10), Nikulin

(1973a,b) derived a closed-form equiprobable cells test for normality (when the parameters θ_1 and θ_2 are unknown) as follows:

$$Y^2 = X^2 + \frac{1}{n}\left(\sum_{j=1}^{r} \varepsilon_j N_j^{(n)}\right)^2 + \frac{1}{n}\left(\sum_{j=1}^{r} \omega_j N_j^{(n)}\right)^2,$$

where

$$X^2 = \frac{r}{n}\sum_{j=1}^{r} N_j^{(n)} - n,$$

$\varepsilon_j = ra_j/\sqrt{\lambda_1}, a_j = \varphi(y_j) - \varphi(y_{j-1}), \lambda_1 = 1 - r\sum_{j=1}^{r} a_j^2, \omega_j = rb_j/\sqrt{\lambda_2}, b_j = \varphi'(y_j) - \varphi'(y_{j-1}), \lambda_2 = 2 - r\sum_{j=1}^{r} b_j^2, y_j = \Phi^{-1}(j/r), j = 1,\ldots,r, y_0 = -\infty, y_r = +\infty, \Phi(x) = \frac{1}{\sqrt{2\pi}}\int_{-\infty}^{x} e^{-t^2/2}dt, \varphi(x) = \Phi'(x) = \frac{1}{\sqrt{2\pi}}e^{-x^2/2}$, *and* $N_j^{(n)}$ *are the numbers of observations falling into r intervals* $(-\infty; y_1s+\bar{x}], (y_1s+\bar{x}; y_2s+\bar{x}], \ldots, (y_{r-1}s+\bar{x}; +\infty)$, *with* $\bar{x}$ *and s being the sample mean and sample standard deviation, respectively.*

Though this formula was used and in fact necessary in the 1970s, nowadays it would be much easier to use the formula in (3.8) directly.

Let

$$\Phi(x,\theta_1,\sqrt{\theta_2}) = \frac{1}{\sqrt{2\pi\theta_2}}\int_{-\infty}^{x} \exp\left(-\frac{(t-\theta_1)^2}{2\theta_2}\right)dt \tag{3.12}$$

denote the distribution function of a normal random variable with mean θ_1 and variance θ_2, and consider r equiprobable random intervals $(x_0,x_1], (x_1,x_2], \ldots, (x_{r-1},x_r)$, where

$$x_j = \Phi^{-1}\left(j/r,\hat{\theta}_1,\sqrt{\hat{\theta}_2}\right) = \hat{\theta}_1 + \sqrt{\hat{\theta}_2}\Phi^{-1}(j/r,0,1), \quad j = 0,1,\ldots,r, \tag{3.13}$$

with $\hat{\theta}_1$ and $\hat{\theta}_2$ being the MLEs of θ_1 and θ_2, respectively. The identity in (3.13) follows easily from (3.12) by taking $x = j/r$.

Denoting $\hat{\boldsymbol{\theta}}_n = (\hat{\theta}_1,\hat{\theta}_2)^T$, the probability to fall into each such interval will be $p_j(\hat{\boldsymbol{\theta}}_n) = p_j = 1/r, j = 1,\ldots,r$. In this case, the vector $\mathbf{V}^T(\hat{\boldsymbol{\theta}}_n)$ does not depend on $\hat{\boldsymbol{\theta}}_n$, and its components are $(\nu_j - np_j)/\sqrt{np_j}$, where ν_j is the number of observations that fall into the jth interval.

For the normal distribution with density in (3.11), the Fisher information matrix $\mathbf{J}$ for one observation is

$$\mathbf{J} = \begin{pmatrix} \frac{1}{\theta_2} & 0 \\ 0 & \frac{1}{2\theta_2^2} \end{pmatrix}. \tag{3.14}$$

Considering $p_j(\boldsymbol{\theta})$ as a function of $\boldsymbol{\theta}$, we have

$$p_j(\boldsymbol{\theta}) = \frac{1}{\sqrt{2\pi\theta_2}} \int\limits_{\theta_1+\sqrt{\theta_2}\Phi^{-1}(\frac{j-1}{r},0,1)}^{\theta_1+\sqrt{\theta_2}\Phi^{-1}(\frac{j}{r},0,1)} \exp\left(-\frac{(x-\theta_1)^2}{2\theta_2}\right) dx$$
$$= \Phi\left(\theta_1 + \sqrt{\theta_2}\Phi^{-1}(\frac{j}{r},0,1),\theta_1,\sqrt{\theta_2}\right)$$
$$-\Phi\left(\theta_1 + \sqrt{\theta_2}\Phi^{-1}(\frac{j-1}{r},0,1),\theta_1,\sqrt{\theta_2}\right),$$

and the elements of the $r \times 2$ matrix $\mathbf{B}$ (see (3.10)), for $j = 1,\ldots,r$, are given by

$$B_{j1} = \frac{1}{\sqrt{2\pi\theta_2 p_j}}\left(\exp\left(-\frac{(\Delta x_{j-1})^2}{2\theta_2}\right) - \exp\left(-\frac{(\Delta x_j)^2}{2\theta_2}\right)\right), \quad (3.15)$$

$$B_{j2} = \frac{1}{2\theta_2\sqrt{2\pi\theta_2 p_j}} \times \left(\Delta x_{j-1}\exp\left(-\frac{(\Delta x_{j-1})^2}{2\theta_2}\right) - \Delta x_j \exp\left(-\frac{(\Delta x_j)^2}{2\theta_2}\right)\right), \quad (3.16)$$

where $\Delta x_{j-1} = x_{j-1} - \theta_1$ and $\Delta x_j = x_j - \theta_1$. Replacing θ_1 and θ_2 by their MLEs $\hat{\theta}_1$ and $\hat{\theta}_2$, and upon substituting (3.14), (3.15), and (3.16) into (3.8), we obtain a test statistic that is quite suitable for producing computer codes (see Section 9.6). Formulas (3.15) and (3.16) can also be used for fixed equiprobable intervals, in which case instead of the formula in (3.13) for x_j we will use

$$x_j = \Phi^{-1}(j/r,0,1), \quad j = 0,1,\ldots,r.$$

Remark 3.2 (Wald's method generalized). *Hsuan and Robson (1976) showed that the resulting modified statistic would be quite different when using moment-type estimators (MMEs) for the unknown parameters. They succeeded in deriving the limiting covariance matrix for the standardized frequencies* $v_j^{(n)}(\bar{\boldsymbol{\theta}}_n), j = 1,2,\ldots,r$, *where* $\bar{\boldsymbol{\theta}}_n$ *is the MME of* $\boldsymbol{\theta}$ *and establishing that the corresponding Wald's quadratic form will follow chi-squared distribution in the limit. They also provided the test statistic explicitly for the exponential family of distributions in which case the MMEs coincide with the MLEs, thus confirming the already known result of Nikulin (1973b). However, they were unable to derive the general modified test based on MMEs* $\bar{\boldsymbol{\theta}}_n$ *explicitly, but this was achieved later by Mirvaliev (2001) (see Chapter 4).*

Moore (1977) summarized all known results until that time pertaining to Wald's approach as follows. Assume that an estimator $\tilde{\boldsymbol{\theta}}_n$ *of* $\boldsymbol{\theta}$ *satisfies in the limit, when* $\boldsymbol{\theta} = \boldsymbol{\theta}_0$*, the condition that*

$$n^{1/2}(\tilde{\boldsymbol{\theta}}_n - \boldsymbol{\theta}_0) = n^{-1/2}\sum_{i=1}^{n} h(X_i) + o_p(1), \tag{3.17}$$

where $h(x)$ *is a s-valued function such that* $\mathbf{E}[h(\mathbf{X})] = \mathbf{0}$ *and* $\mathbf{E}[h(\mathbf{X})h(\mathbf{X})^T]$ *is a finite* $s \times s$ *matrix. The minimum chi-squared, maximum likelihood, moment type, and many other estimators possess the limiting property in (3.17). Define the* $r \times r$ *matrices*

$$\mathbf{C} = -\mathbf{BLB}^T + \mathbf{BE}[h(\mathbf{X})W(\mathbf{X})^T] + \mathbf{E}[W(\mathbf{X})h(\mathbf{X})^T]\mathbf{B}^T \tag{3.18}$$

and

$$\boldsymbol{\Sigma} = \mathbf{I} - \mathbf{qq}^T - \mathbf{C}, \tag{3.19}$$

where $W(\mathbf{X})$ *is the r-vector with components* $[\chi_j - p_j]/\sqrt{p_j}, j = 1,\ldots,r,$ *and* χ_j *is the indicator function for the jth cell. Then, under the usual regularity conditions, the vector of standardized frequencies with components* $v_j^{(n)}(\tilde{\boldsymbol{\theta}}_n) = (N_j^{(n)} - np_j(\tilde{\boldsymbol{\theta}}_n))/\sqrt{np_j(\tilde{\boldsymbol{\theta}}_n)}$ *will follow in the limit the multivariate normal distribution with zero mean vector and covariance matrix* $\boldsymbol{\Sigma}$*. If rank*$(\boldsymbol{\Sigma}) = r-1$*, and since* $\mathbf{qq}^T$ *is a projection orthogonal to* $\boldsymbol{\Sigma}$*, then rank*$(\mathbf{I} - \mathbf{C}) = r$ *and* $(\mathbf{I} - \mathbf{C})^{-1}$ *will be a generalized inverse of* $\boldsymbol{\Sigma}$ *(Moore, 1977, p. 134). From the above facts, it follows that Wald's statistic*

$$T_n(\tilde{\boldsymbol{\theta}}_n) = \mathbf{V}^{(n)T}(\tilde{\boldsymbol{\theta}}_n)\boldsymbol{\Sigma}_n^{-}\mathbf{V}^{(n)}(\tilde{\boldsymbol{\theta}}_n) \tag{3.20}$$

is invariant under the choice of $\boldsymbol{\Sigma}_n^{-}$ *and can be calculated, for instance, as*

$$T_n(\tilde{\boldsymbol{\theta}}_n) = \mathbf{V}^{(n)T}(\tilde{\boldsymbol{\theta}}_n)(\mathbf{I} - \mathbf{C}_n)^{-1}\mathbf{V}^{(n)}(\tilde{\boldsymbol{\theta}}_n), \tag{3.21}$$

where $\mathbf{C}_n$ *is an estimator of* $\mathbf{C}$*. The expression in (3.21) is very general. For a particular estimator, one has only to find the matrix* $\mathbf{C}$ *and then use (3.21). Moore and Spruill (1975) showed that Wald's approach is valid not only for continuous hypotheses, but for discrete ones as well.*

It is worth noting that all classical and modified chi-squared tests are particular cases of Wald's approach.

For a simple null hypothesis, the vector of standardized grouped frequencies with components $v_j^{(n)} = (N_j^{(n)} - np_j)/\sqrt{np_j}, j = 1,\ldots,r$, is asymptotically normally distributed with zero mean vector and covariance matrix $\boldsymbol{\Sigma} = \mathbf{I} - \mathbf{qq}^T$. This matrix is idempotent, its generalized inverse is $\boldsymbol{\Sigma}^{-} = \mathbf{I} - \mathbf{qq}^T$, and so

$$X_n^2 = \mathbf{V}^{(n)T}\boldsymbol{\Sigma}^{-}\mathbf{V}^{(n)} = \sum_{j=1}^{r} \frac{(N_j^{(n)} - np_j)^2}{np_j},$$

which is the classical Pearson's formula with X_n^2 being asymptotically distributed as χ_{r-1}^2.

If one wishes to test a composite hypothesis using the minimum chi-squared estimator $\hat{\boldsymbol{\theta}}_n$ of a parameter $\boldsymbol{\theta}$, then the limiting covariance matrix will be

$$\boldsymbol{\Sigma}_1 = \mathbf{I} - \mathbf{q}\mathbf{q}^T - \mathbf{B}(\mathbf{B}^T\mathbf{B})^{-1}\mathbf{B}^T,$$

which is symmetric and idempotent. Its Moore-Penrose inverse $\boldsymbol{\Sigma}_1^+$ identically equals $\boldsymbol{\Sigma}_1$, from which it follows that the statistic in (3.20) will become

$$T_n(\hat{\boldsymbol{\theta}}_n) = \mathbf{X}_n^2(\hat{\boldsymbol{\theta}}_n) = \mathbf{V}^{(n)T}(\hat{\boldsymbol{\theta}}_n)\mathbf{V}^{(n)}(\hat{\boldsymbol{\theta}}_n) = \sum_{j=1}^{r} \frac{[N_j^{(n)} - np_j(\hat{\boldsymbol{\theta}}_n)]^2}{np_j(\hat{\boldsymbol{\theta}}_n)},$$

which is the well-known Pearson-Fisher (PF) test discussed in the preceding Chapter.

If one wishes to test a composite hypothesis using any $\sqrt{n}$-consistent estimator $\tilde{\boldsymbol{\theta}}_n$ of a parameter $\boldsymbol{\theta}$, then the limiting covariance matrix will again be

$$\boldsymbol{\Sigma}_1 = \mathbf{I} - \mathbf{q}\mathbf{q}^T - \mathbf{B}(\mathbf{B}^T\mathbf{B})^{-1}\mathbf{B}^T,$$

using which, from (3.20), we obtain the DN statistic as

$$U_n^2(\tilde{\boldsymbol{\theta}}_n) = \mathbf{V}^{(n)T}(\tilde{\boldsymbol{\theta}}_n)[\mathbf{I} - \mathbf{q}_n\mathbf{q}_n^T - \mathbf{B}_n(\mathbf{B}_n^T\mathbf{B}_n)^{-1}\mathbf{B}_n^T]\mathbf{V}^{(n)}(\tilde{\boldsymbol{\theta}}_n).$$

The DN statistic as well as the asymptotically equivalent PF test are distributed as χ_{r-s-1}^2 in the limit.

3.2 MODIFICATIONS OF NIKULIN-RAO-ROBSON TEST

A very important decomposition of (3.8) was provided by McCulloch (1985). By using the identity

$$(\mathbf{I} - \mathbf{B}\mathbf{J}^{-1}\mathbf{B}^T)^{-1} = \mathbf{I} - \mathbf{B}(\mathbf{B}^T\mathbf{B})^{-1}\mathbf{B}^T + \mathbf{B}\left((\mathbf{J} - \mathbf{B}^T\mathbf{B})^{-1} + (\mathbf{B}^T\mathbf{B})^{-1}\right)\mathbf{B}^T,$$

McCulloch presented (3.8) as

$$Y1_n^2(\hat{\boldsymbol{\theta}}_n) = U_n^2(\hat{\boldsymbol{\theta}}_n) + S_n^2(\hat{\boldsymbol{\theta}}_n), \tag{3.22}$$

where

$$U_n^2(\hat{\boldsymbol{\theta}}_n) = \mathbf{V}^{(n)T}(\hat{\boldsymbol{\theta}}_n)[\mathbf{I} - \mathbf{q}_n\mathbf{q}_n^T - \mathbf{B}_n(\mathbf{B}_n^T\mathbf{B}_n)^{-1}\mathbf{B}_n^T]\mathbf{V}^{(n)}(\hat{\boldsymbol{\theta}}) \tag{3.23}$$

is the DN statistic (note that $\mathbf{q}_n\mathbf{q}_n^T\mathbf{V}^{(n)}(\hat{\boldsymbol{\theta}}_n) = \mathbf{0}$) and

$$\begin{aligned} S_n^2(\hat{\boldsymbol{\theta}}_n) &= W_n^2(\hat{\boldsymbol{\theta}}_n) + \mathbf{V}^{(n)T}(\hat{\boldsymbol{\theta}})(\mathbf{J}_n - \mathbf{J}_{gn})^{-1}\mathbf{B}_n^T\mathbf{V}^{(n)}(\hat{\boldsymbol{\theta}}_n) \\ &= \mathbf{V}^{(n)T}(\hat{\boldsymbol{\theta}}_n)\mathbf{B}_n\left[(\mathbf{J}_n - \mathbf{J}_{gn})^{-1} + (\mathbf{B}_n^T\mathbf{B}_n)^{-1}\right]\mathbf{B}_n^T\mathbf{V}^{(n)}(\hat{\boldsymbol{\theta}}_n). \end{aligned} \tag{3.24}$$

The test statistic in (3.24) is referred to as McCulloch test (McCu test). McCulloch (1985) proved that the statistics $U_n^2(\hat{\boldsymbol{\theta}}_n)$ and $S_n^2(\hat{\boldsymbol{\theta}}_n)$ in (3.23) and (3.24) are asymptotically independent. The DN test $U_n^2(\hat{\boldsymbol{\theta}}_n)$ in (3.23) is distributed asymptotically as χ^2_{r-s-1} while McCu statistic $S_n^2(\hat{\boldsymbol{\theta}}_n)$ in (3.24) is distributed asymptotically as χ^2_s. McCulloch (1985) showed that, under the rather mild regularity conditions of Moore and Spruill (1975), the DN test based on non-grouped data behaves locally like the Pearson-Fisher test in (2.19) based on frequencies. Monte Carlo simulations by McCulloch and the results of Voinov et al. (2009) demonstrate an essential increase in power for the NRR test over the PF and DN tests in case of equiprobable fixed or random cells. The decomposition in (3.22), the lack of power and the asymptotic equivalence of the PF and DN tests in this case, all suggest to interpret the McCu statistic as a term of the NRR test that recovers the Fisher information lost due to data grouping.

The asymptotic independence of the McCu statistic and the DN test permits to use it on its own (McCulloch, 1985). For equiprobable fixed or random cells, with respect to some alternatives, it can give a significant increase in power. In addition, this statistic possesses the lowest variance and the highest rate of convergence to the limit.

Consider the powers of $Y1_n^2(\hat{\boldsymbol{\theta}}_n), S_n^2(\hat{\boldsymbol{\theta}}_n)$, and $U_n^2(\hat{\boldsymbol{\theta}}_n)$ when testing normality against the logistic distribution which is very close to normal; see Voinov et al. (2009). For different sample sizes (from $n = 40$ to 300) and two different expected cell frequencies ($np = 5$ and $np = 10$) using equiprobable cells, powers of $Y1_n^2(\hat{\boldsymbol{\theta}}_n), S_n^2(\hat{\boldsymbol{\theta}}_n)$, and $U_n^2(\hat{\boldsymbol{\theta}}_n)$ against the logistic distribution were all simulated for $N = 10{,}000$ runs. From Figure 3.1, we see that in the range of sample sizes 40–300 the ratio of the power of $S_n^2(\hat{\boldsymbol{\theta}}_n)$ to $Y1_n^2(\hat{\boldsymbol{\theta}}_n)$ increases attaining the value of 1.76 for $np = 5$, and 1.59 for $np = 10$.

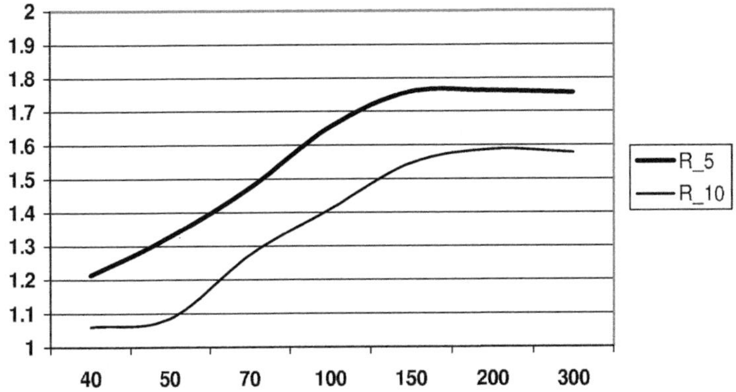

FIGURE 3.1 Ratios R_{10} and R_5 of the power of $S_n^2(\hat{\boldsymbol{\theta}}_n)$ to $Y1_n^2(\hat{\boldsymbol{\theta}}_n)$ for $np = 10$ and $np = 5$, respectively, as functions of the sample size n.

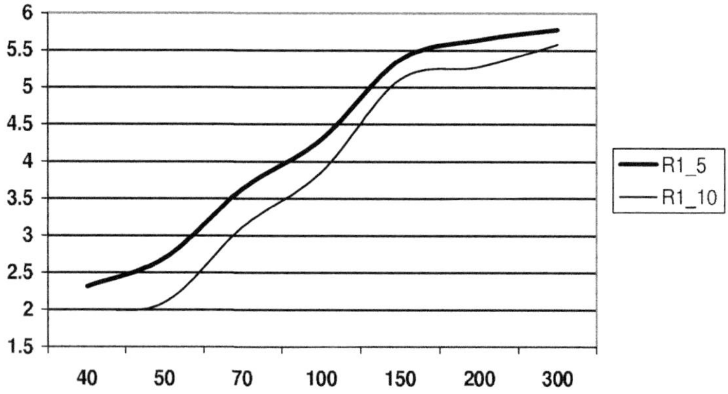

FIGURE 3.2 Ratios $R1_{10}$ and $R1_5$ of the power of $S_n^2(\hat{\boldsymbol{\theta}}_n)$ to $U_n^2(\hat{\boldsymbol{\theta}}_n)$ for $np = 10$ and $np = 5$, respectively, as functions of the sample size n.

It is useful to note that a maximal gain in power is attained if $n \sim 200$ for $np = 5$ which means that the number of equiprobable classes equals 40 (with the corresponding simulated power of $Y1_n^2(\hat{\boldsymbol{\theta}}_n)$ and $S_n^2(\hat{\boldsymbol{\theta}}_n)$ being 0.263 and 0.463).

Figure 3.2 shows the ratio of the power of $S_n^2(\hat{\boldsymbol{\theta}}_n)$ to the DN statistic $U_n^2(\hat{\boldsymbol{\theta}}_n)$ for the same expected cell frequencies ($np = 5$ and $np = 10$). We see that these ratios are monotone increasing attaining values in the range of 5.6–5.8. In this case, the power of the DN test if $n = 40, np = 5$ is only 0.082 which is 5.6 times less than that of $S_n^2(\hat{\boldsymbol{\theta}}_n)$. The fact that the power of Dzhaparidze-Nikulin test is very low for equiprobable cells has also been mentioned by Voinov et al. (2009).

Based on the idea of Dzhaparidze and Nikulin (1974), and Singh (1987) suggested the following modification to the NRR statistic valid for any $\sqrt{n}$-consistent estimator $\tilde{\boldsymbol{\theta}}_n$ of $\boldsymbol{\theta}$:

$$Q_s^2(\tilde{\boldsymbol{\theta}}_n) = \mathbf{V}_*^{(n)T}(\tilde{\boldsymbol{\theta}}_n)(\mathbf{I} - \mathbf{B}_n\mathbf{J}_n^{-1}\mathbf{B}_n^T)^{-1}\mathbf{V}_*^{(n)}(\tilde{\boldsymbol{\theta}}_n), \tag{3.25}$$

where $\mathbf{V}_*^{(n)}(\tilde{\boldsymbol{\theta}}_n) = \mathbf{V}^{(n)}(\tilde{\boldsymbol{\theta}}_n) - \mathbf{B}_n\mathbf{J}_n^{-1}\mathbf{W}(\tilde{\boldsymbol{\theta}}_n)$ and

$$\mathbf{W}(\tilde{\boldsymbol{\theta}}_n) = \frac{1}{\sqrt{n}}\sum_{i=1}^{n} \left.\frac{\partial \ln f(X_i,\boldsymbol{\theta})}{\partial \theta_j}\right|_{\boldsymbol{\theta}=\tilde{\boldsymbol{\theta}}_n}, \quad j = 1,\ldots,s.$$

The statistic in (3.25) follows the χ^2_{r-1} distribution in limit, and in fact coincides with the NRR test if the MLEs are used, since in this case the term $\mathbf{B}_n\mathbf{J}_n^{-1}\mathbf{W}(\tilde{\boldsymbol{\theta}}_n)$ vanishes (see also Thomas and Pierce, 1979). The test in (3.25) can alternatively be presented as

$$Q_s^2(\tilde{\boldsymbol{\theta}}_n) = \mathbf{V}_*^{(n)T}(\tilde{\boldsymbol{\theta}}_n)\mathbf{V}_*^{(n)}(\tilde{\boldsymbol{\theta}}_n) + (\mathbf{B}_n^T\mathbf{V}_*^{(n)}(\tilde{\boldsymbol{\theta}}_n))^T(\mathbf{J}_n - \mathbf{J}_{gn})^{-1}\mathbf{B}_n^T\mathbf{V}_*^{(n)}(\tilde{\boldsymbol{\theta}}_n). \tag{3.26}$$

However, it should be mentioned that the tests in (3.25) and (3.26) are computationally more complicated for large sample sizes than the NRR test in (3.8).

Remark 3.3. *The statistic $S_n^2(\hat{\boldsymbol{\theta}}_n)$, in addition to gaining power for many alternatives in the case of equiprobable intervals, also possesses the smallest variance if $s < r - 1$ and is, therefore, more stable. Moreover, it possesses a higher rate of convergence to the limiting distribution.*

3.3 OPTIMALITY OF NIKULIN-RAO-ROBSON TEST

Singh (1987) showed that the NRR test can be obtained through a standard method of constructing asymptotically optimal uniformly most powerful invariant tests for linear hypotheses. He derived the NRR test by using explicit expressions for orthogonal projectors on linear subspaces to solve a typical problem of testing linear hypotheses. Optimality of the NRR test immediately follows from Singh's results. He also showed the optimality of the DN statistic, though the criterion of optimality for the NRR test is different from that of the DN test (see Singh, 1987, p. 3264).

It is clear that by grouping data and using frequencies we lose information about the hypothesized distribution. Lemeshko (1998) suggested to define the end-points of the r grouping intervals, $x_0 < x_1 < \cdots < x_{r-1} < x_r$, in such a way that the determinant $|\mathbf{J}_g(\hat{\boldsymbol{\theta}}_n)|$ of the Fisher information matrix for the grouped data will be maximized. This way of data grouping was called "optimal," but the word "optimal" may not be reasonable in this case as there are many different functionals of the Fisher information matrix that could be used in this context. In particular, Lemeshko (1998) showed that if one uses the MLEs based on the raw data and the "optimal" grouping, then the large values of the PF test will be close to those of the χ^2_{r-s-1} distribution. Using Monte Carlo simulation, Lemeshko et al. (2001) showed that the power of the NRR test cannot be improved using this "optimal" way of data grouping.

Voinov (2006) provided an explanation to this by mentioning that the second term in (3.8) depends on $(\mathbf{J}-\mathbf{J}_g)$, the difference between the Fisher information for non-grouped and grouped data, respectively, and that possibly takes into account the information lost due to data grouping. This may be one reason to question the use of the term "optimal" in the procedure of Lemeshko (1998).

3.4 DECOMPOSITION OF NIKULIN-RAO-ROBSON TEST

In Section 2.2, decompositions of the Pearson-Fisher and DN statistics into asymptotically independent χ_1^2 components were presented. Here, we discuss an analogous decomposition for the NRR statistic (Voinov et al., 2007). In this case, we have

$$\mathbf{\Sigma}_{1k} = \mathbf{I}_k - \mathbf{q}_k\mathbf{q}_k^T - \mathbf{B}_k\mathbf{J}^{-1}\mathbf{B}_k^T, \quad k = 1,\dots,r-1. \tag{3.27}$$

If $rank(\mathbf{B}) = s$, then

$$|\mathbf{\Sigma}_{1k}| = \left(1 - \sum_{i=1}^{k} p_i\right) |\mathbf{J} - \mathbf{J}_k|/|\mathbf{J}|$$

and

$$\mathbf{\Sigma}_{1k}^{-1} = \mathbf{M}_k + \mathbf{M}_k\mathbf{B}_k(\mathbf{J} - \mathbf{J}_k)^{-1}\mathbf{B}_k^T\mathbf{M}_k^T, \tag{3.28}$$

where $\mathbf{M}_k = \mathbf{I}_k + \mathbf{q}_k\mathbf{q}_k^T/(1 - \sum_{i=1}^k p_i)$ and $\mathbf{J}_k = \mathbf{B}_k^T\mathbf{M}_k\mathbf{B}_k, k = 1,\dots,r-1$.
Using Lemma 2.1, we then obtain

$$r_{ii} = \sqrt{|\mathbf{\Sigma}_{1i-1}|/|\mathbf{\Sigma}_{1i}|}, \quad i = 1,\dots,r-1,$$
$$r_{ij} = r_{ii}(\sqrt{p_i}\mathbf{q}_{i-1}^T + \mathbf{b}_i\mathbf{J}^{-1}\mathbf{B}_{i-1}^T)(\mathbf{\Sigma}_{1i-1}^{-1})_j,$$

where $j = 1,\dots,i-1, i \neq j, i \geqslant 2, (\mathbf{\Sigma}_{1i-1}^{-1})_j$ is the jth column of $\mathbf{\Sigma}_{1i-1}^{-1}$ defined by (3.28), and $\mathbf{b}_i$ is the ith row of $\mathbf{B}$. With $\mathbf{R} = (\mathbf{R}_{r-1} \vdots \mathbf{0})$ and the MLEs of all the matrices involved, we finally get

$$\boldsymbol{\delta}(r-1)(\hat{\boldsymbol{\theta}}_n) = \mathbf{R}\mathbf{V}_{(r)}^{(n)}(\hat{\boldsymbol{\theta}}_n), \tag{3.29}$$

where $\hat{\boldsymbol{\theta}}_n$ is the MLE of $\boldsymbol{\theta}$. Thus, we have the following result.

Theorem 3.1. *Under proper regularity conditions (see Moore and Spruill, 1975, for example), the expansion*

$$Y1_n^2(\hat{\boldsymbol{\theta}}_n) = \delta_1^2(\hat{\boldsymbol{\theta}}_n) + \cdots + \delta_{r-1}^2(\hat{\boldsymbol{\theta}}_n)$$

of the NRR statistic holds and in the limit, under H_0, the components $\delta_i^2(\hat{\boldsymbol{\theta}}_n), i = 1,\dots,r-1$, are distributed independently as χ_1^2 and the statistic $Y1_n^2(\hat{\boldsymbol{\theta}})$ is distributed as χ_{r-1}^2.

3.5 CHI-SQUARED TESTS FOR MULTIVARIATE NORMALITY

3.5.1 Introduction

It is well known that the joint normality does not follow from the normality of marginal univariate distributions; see Kotz et al. (2000). The literature on tests of fit for the multivariate normal family is certainly not as extensive as for assessing the univariate normality, but the last three decades have seen an increased activity in this regard. Mardia (1970) and Malkovich and Afifi (1973) presented some tests based on a generalization of the univariate skewness and kurtosis measures. For some generalizations of these measures and their use in tests for multivariate normality, one may refer to Balakrishnan et al. (2007) and Balakrishnan and Scarpa (2012). For other pertinent work in this regard, see Shapiro and Wilk (1965), Royston (1983), Srivastava and Hui (1987), Tserenbat (1990),Looney (1995), Henze and Wagner (1997), and Doornik and Hansen (1994). More elaborate lists of references can be obtained from the review articles of Henze (2002) and Mecklin and Mundfrom (2004).

3.5.2 Modified chi-squared tests

Moore and Stubblebine (1981) developed a test for multivariate normality as follows. Let $\mathbf{X}_1, \cdots, \mathbf{X}_n$ be i.i.d. p-variate ($p \geqslant 2$) normal random vectors with the following joint probability density function:

$$f(\mathbf{x}|\boldsymbol{\theta}) = (2\pi)^{-p/2}|\boldsymbol{\Sigma}|^{-1/2} \exp\left[-\frac{1}{2}(\mathbf{x}-\boldsymbol{\mu})^T\boldsymbol{\Sigma}^{-1}(\mathbf{x}-\boldsymbol{\mu})\right],$$

where $\boldsymbol{\mu}$ is the p-vector of means and $\boldsymbol{\Sigma}$ is a nonsingular $p \times p$ matrix. Let a given vector $\boldsymbol{\theta} = (\boldsymbol{\mu}, \boldsymbol{\Sigma})$ of unknown parameters be

$$\boldsymbol{\theta}^T = (\mu_1, \ldots, \mu_p, \sigma_{11}, \sigma_{12}, \sigma_{22}, \ldots, \sigma_{1j}, \sigma_{2j}, \ldots, \sigma_{jj}, \ldots, \sigma_{pp});$$

that is, the elements of the matrix $\boldsymbol{\Sigma}$ are arranged column-wise by taking the elements of the upper-triangular submatrix of $\boldsymbol{\Sigma}$. The MLE $\hat{\boldsymbol{\theta}}_n$ of $\boldsymbol{\theta}$ is known to be the vector of sample means, $\bar{\mathbf{X}}$, and the matrix of sample covariances, $\mathbf{S}$. Given $0 = c_0 < c_1 < \ldots < c_M = \infty$, the M random cells can be defined as (Moore and Stubblebine, 1981)

$$E_{in}(\hat{\boldsymbol{\theta}}_n) = \left\{\mathbf{X} \in R^p : c_{i-1} \leqslant (\mathbf{X}-\bar{\mathbf{X}})^T\mathbf{S}^{-1}(\mathbf{X}-\bar{\mathbf{X}}) < c_i\right\}, \quad i = 1, \ldots, M.$$

The probability of falling into the ith cell is then

$$p_{in}(\hat{\boldsymbol{\theta}}_n) = \int_{E_{in}(\hat{\boldsymbol{\theta}}_n)} f(\mathbf{x}|\hat{\boldsymbol{\theta}}_n)d\mathbf{x}.$$

If c_i is the i/M point of the $\chi^2(p)$ distribution, then the cells are equiprobable under the estimated parameter value, $\hat{\boldsymbol{\theta}}_n$, and $p_{in}(\hat{\boldsymbol{\theta}}_n) = 1/M$.

Denote the vector of standardized cell frequencies by $\mathbf{V}_n$, with its components as

$$V_{in}(\hat{\boldsymbol{\theta}}_n) = \frac{(N_{in} - n/M)}{\sqrt{n/M}}, \quad i = 1, \ldots, M,$$

where N_{in} is the number of random vectors $\mathbf{X}_1, \cdots, \mathbf{X}_n$ falling into $E_{in}(\hat{\boldsymbol{\theta}}_n)$.

The Fisher information matrix for one observation $\mathbf{J}(\boldsymbol{\theta})$ can be evaluated as

$$\mathbf{J}(\boldsymbol{\theta}) = \begin{bmatrix} \boldsymbol{\Sigma}^{-1} & \mathbf{0} \\ \mathbf{0} & \mathbf{Q}^{-1} \end{bmatrix}.$$

The dimension of $\mathbf{J}(\boldsymbol{\theta})$ is $m \times m$, where $m = p + p(p+1)/2$ is the dimension of $\boldsymbol{\theta}$, and $\mathbf{Q}$ is the $p(p+1)/2 \times p(p+1)/2$ covariance matrix of $\mathbf{r}$, a vector of the entries of $\sqrt{n}\mathbf{S}$ (arranged column-wise by taking the upper triangular elements),

$$\mathbf{r}^T = (s_{11}, s_{12}, s_{22}, s_{13}, s_{23}, s_{33}, \ldots, s_{pp}).$$

The elements of $\mathbf{Q}$ can be expressed as (Press, 1972)

$$\mathbf{Var}(s_{ij}) = \sigma_{ij}^2 + \sigma_{ii}\sigma_{jj}, \quad i,j = 1,\ldots,p, \ i \leqslant j,$$
$$\mathbf{Cov}(s_{ij}, s_{kl}) = \sigma_{ik}\sigma_{jl} + \sigma_{il}\sigma_{jk}, \quad i,j,k,l = 1,\ldots,p, \ i \leqslant j, \ k \leqslant l,$$

where $\sigma_{ij}, i,j = 1,\ldots,p$, are the elements of $\boldsymbol{\Sigma}$.

For example, in the two-dimensional case, the matrix $\mathbf{Q}$ will be (see also McCulloch, 1980)

$$\mathbf{Q} = \begin{pmatrix} \mathbf{Var}(s_{11}) & \mathbf{Cov}(s_{11}, s_{12}) & \mathbf{Cov}(s_{11}, s_{22}) \\ \mathbf{Cov}(s_{12}, s_{11}) & \mathbf{Var}(s_{12}) & \mathbf{Cov}(s_{12}, s_{22}) \\ \mathbf{Cov}(s_{22}, s_{11}) & \mathbf{Cov}(s_{22}, s_{12}) & \mathbf{Var}(s_{22}) \end{pmatrix}$$
$$= \begin{pmatrix} 2\sigma_{11}^2 & 2\sigma_{11}\sigma_{12} & 2\sigma_{12}^2 \\ 2\sigma_{11}\sigma_{12} & \sigma_{11}\sigma_{22} + \sigma_{12}^2 & 2\sigma_{12}\sigma_{22} \\ 2\sigma_{12}^2 & 2\sigma_{12}\sigma_{22} & 2\sigma_{22}^2 \end{pmatrix}.$$

For a specified $\boldsymbol{\theta}_0 = (\boldsymbol{\mu}_0, \boldsymbol{\Sigma}_0)$, let us define

$$p_i(\boldsymbol{\theta}, \boldsymbol{\theta}_0) = \int\limits_{E_i(\boldsymbol{\theta}_0)} f(\mathbf{x}|\boldsymbol{\theta}) d\mathbf{x},$$

where

$$E_i(\boldsymbol{\theta}_0) = \left\{ \mathbf{X} \text{ in } R^p : c_{i-1} \leqslant (\mathbf{X} - \boldsymbol{\mu}_0)^T \boldsymbol{\Sigma}_0^{-1} (\mathbf{X} - \boldsymbol{\mu}_0) < c_i \right\}, \quad i = 1,\ldots,M.$$

Define $M \times m$ matrix $\mathbf{B}(\boldsymbol{\theta}, \boldsymbol{\theta}_0)$ with its elements as

$$B_{ij} = \frac{1}{\sqrt{p_i(\boldsymbol{\theta}, \boldsymbol{\theta}_0)}} \frac{\partial p_i(\boldsymbol{\theta}, \boldsymbol{\theta}_0)}{\partial \theta_j}.$$

Then, from Lemma 1 of Moore and Stubblebine (1981), it follows that, for any c_i and $\boldsymbol{\theta}_0$,

$$\left. \frac{\partial p_i(\boldsymbol{\theta}, \boldsymbol{\theta}_0)}{\partial \mu_j} \right|_{\boldsymbol{\theta}=\boldsymbol{\theta}_0} = 0, \quad 1 \leqslant i \leqslant M, \quad 1 \leqslant j \leqslant p,$$
$$\left. \frac{\partial p_i(\boldsymbol{\theta}, \boldsymbol{\theta}_0)}{\partial \sigma_{jk}} \right|_{\boldsymbol{\theta}=\boldsymbol{\theta}_0} = d_i \sigma^{jk}, \quad 1 \leqslant i \leqslant M, \quad 1 \leqslant j \leqslant k \leqslant p,$$

where σ^{jk} are the elements of $\boldsymbol{\Sigma}_0^{-1}$, and

$$d_i = \left(c_{i-1}^{p/2}\mathrm{e}^{-c_{i-1}/2} - c_i^{p/2}\mathrm{e}^{-c_i/2}\right) b_p/2,$$
$$b_p = [p(p-2)\cdots 4\cdot 2]^{-1} \quad \text{if } p \text{ is even},$$
$$b_p = (2/\pi)^{1/2}\left[p(p-2)\cdots 5\cdot 3\right]^{-1} \quad \text{if } p \text{ is odd}.$$

The Nikulin-Rao-Robson (NRR) statistic based on the MLEs can be presented as Nikulin (1973b,c), Rao and Robson (1974), and Voinov and Nikulin (2011))

$$Y_n^2 = \mathbf{V}_n^T(\hat{\boldsymbol{\theta}}_n)\mathbf{V}_n(\hat{\boldsymbol{\theta}}_n) + \mathbf{V}_n^T(\hat{\boldsymbol{\theta}}_n)\mathbf{B}_n\left[\mathbf{J}_n - \mathbf{B}_n^T\mathbf{B}_n\right]^{-1}\mathbf{B}_n^T\mathbf{V}_n(\hat{\boldsymbol{\theta}}_n), \tag{3.30}$$

where $\mathbf{J}_n = \mathbf{J}(\hat{\boldsymbol{\theta}}_n)$ and $\mathbf{B}_n = \mathbf{B}(\hat{\boldsymbol{\theta}}_n,\hat{\boldsymbol{\theta}}_n)$.

Unfortunately, in this case, the limiting covariance matrix $\mathbf{I}-\mathbf{q}\mathbf{q}^T-\mathbf{B}\mathbf{J}^{-1}\mathbf{B}^T$ of the standardized frequencies $\mathbf{V}_n$, where $\mathbf{q}$ is a M-vector with its entries as $1/\sqrt{M}$, depends on the unknown parameters. So, Lemma 9 of Khatri (1968) can not be invoked to claim that Y_n^2 in (3.30) is distributed as chi-square.

3.5.3 Testing for bivariate circular normality

Kowalski (1970) was one of the early ones to consider "some rough tests for bivariate circular normality." Gumbel (1954) pointed out some applications of the circular normal distribution in "economic statistics," geophysics, and medical studies. Following Moore and Stubblebine (1981), let us consider testing for bivariate circular normality. The hypothesized probability density function in this case is

$$f(x,y|\boldsymbol{\theta}) = (2\pi\sigma^2)^{-1}\exp\left\{-\frac{1}{2\sigma^2}[(x-\mu_1)^2 + (y-\mu_2)^2]\right\}, \tag{3.31}$$

where $\boldsymbol{\theta} = (\mu_1,\mu_2,\sigma)^T$. Using a random sample $(X_1,Y_1),\ldots,(X_n,Y_n)$, the MLEs of μ_1,μ_2, and σ^2 can be obtained as $\bar{X},\bar{Y}$, and

$$s^2 = \frac{1}{2n}\left\{\sum_{j=1}^{n}(X_j-\bar{X})^2 + \sum_{j=1}^{n}(Y_j-\bar{Y})^2\right\}.$$

If $c_i = -2\ln(1-i/M), i = 1,\ldots,M-1$, then $\hat{p}_{in} = 1/M$. Evidently, we have

$$\left.\frac{\partial p_{in}}{\partial \mu_1}\right|_{\hat{\boldsymbol{\theta}}_n} = \left.\frac{\partial p_{in}}{\partial \mu_2}\right|_{\hat{\boldsymbol{\theta}}_n} = 0,$$
$$\left.\frac{\partial p_{in}}{\partial \sigma}\right|_{\hat{\boldsymbol{\theta}}_n} = \nu_i/s,$$

where $\nu_i = 2\left[(1-\frac{i}{M})\ln(1-\frac{i}{M}) - (1-\frac{i-1}{M})\ln(1-\frac{i-1}{M})\right]$. From these, it follows that the matrix $\mathbf{B}_n$ is

$$\mathbf{B}_n = \begin{pmatrix} 0 & 0 & \sqrt{M}\nu_1/s \\ 0 & 0 & \sqrt{M}\nu_2/s \\ \cdots & \cdots & \cdots\cdots\cdots\cdots \\ 0 & 0 & \sqrt{M}\nu_M/s \end{pmatrix}. \tag{3.32}$$

Since the estimate $\mathbf{J}_n$ of the Fisher information matrix for the family in (3.31) is (Moore and Stubblebine, 1981)

$$\mathbf{J}_n = \begin{pmatrix} 1/s^2 & 0 & 0 \\ 0 & 1/s^2 & 0 \\ 0 & 0 & 4/s^2 \end{pmatrix}, \tag{3.33}$$

we obtain

$$(\mathbf{J}_n - \mathbf{B}_n^T\mathbf{B}_n)^{-1} = \begin{pmatrix} s^2 & 0 & 0 \\ 0 & s^2 & 0 \\ 0 & 0 & \frac{s^2}{4-M\sum\nu_i^2} \end{pmatrix}. \tag{3.34}$$

Denoting $\mathbf{V}_n = \left(\frac{(N_{1n}-n/M)}{\sqrt{n/M}}, \ldots, \frac{(N_{Mn}-n/M)}{\sqrt{n/M}}\right)^T = (\tilde{N}_1, \ldots, \tilde{N}_M)^T$, where N_{jn} is the number of $d_i = \frac{1}{s^2}\left[(X_i-\bar{X})^2 + (Y_i-\bar{Y})^2\right], i = 1, \ldots, n$, that fall into the interval $[c_{j-1}, c_j)$ for $j = 1, \ldots, M$, the statistic in (3.30) is easily derived to be

$$Y_n^2 = \sum_{i=1}^{M}\tilde{N}_i^2 + \frac{M}{4 - M\sum_{i=1}^{M}\nu_i^2}\left(\sum_{i=1}^{M}\tilde{N}_i\nu_i\right)^2. \tag{3.35}$$

We may note here that the formula for Y_n^2 given by Moore and Stubblebine (1981, p. 724) contains an error.

The second term of Y_n^2 in (3.30) recovers information lost due to data grouping. Another useful decomposition of Y_n^2 has been proposed by McCulloch (1985) as

$$Y_n^2 = U_n^2 + S_n^2,$$

where the Dzhaparidze-Nikulin (DN) statistic U_n^2 (Dzhaparidze and Nikulin, 1974) is

$$U_n^2 = \mathbf{V}_n^T(\hat{\boldsymbol{\theta}}_n)\left[\mathbf{I} - \mathbf{B}_n(\mathbf{B}_n^T\mathbf{B}_n)^{-1}\mathbf{B}_n^T\right]\mathbf{V}_n(\hat{\boldsymbol{\theta}}_n) \tag{3.36}$$

and

$$\begin{aligned} S_n^2 &= Y_n^2 - U_n^2 \\ &= \mathbf{V}_n^T(\hat{\boldsymbol{\theta}}_n)\mathbf{B}_n\left[(\mathbf{J}_n - \mathbf{B}_n^T\mathbf{B}_n)^{-1} + (\mathbf{B}_n^T\mathbf{B}_n)^{-1}\right]\mathbf{B}_n^T\mathbf{V}_n(\hat{\boldsymbol{\theta}}_n). \end{aligned} \tag{3.37}$$

McCulloch (1985) showed that if the rank of $\mathbf{B}$ is s, then U_n^2 and S_n^2 are asymptotically independent and distributed as χ^2_{M-s-1} and χ^2_s, respectively, in the limit.

Since the first two columns of the matrix $\mathbf{B}_n$ in our case are columns of zeros and the rest are linearly dependent, the matrix $\mathbf{B}_n$ has rank 1. From this, it readily follows that $(\mathbf{B}_n^T\mathbf{B}_n)^{-1}$ does not exist, but by using some known facts on multivariate normal distribution (Moore, 1977, p. 132), we may replace $\mathbf{A}^{-1} = (\mathbf{B}_n^T\mathbf{B}_n)^{-1}$ by $\mathbf{A}^{-} = (\mathbf{B}_n^T\mathbf{B}_n)^{-}$, where $\mathbf{A}^{-}$ is any generalized inverse of $\mathbf{A}$, which can be computed by using singular value decomposition, for instance. So, for testing two-dimensional circular normality with random cells $E_{in}(\hat{\boldsymbol{\theta}}_n)$, we may use the NRR statistic Y_n^2 defined in (3.30), the DN statistic

$$U_n^2 = \mathbf{V}_n^T(\hat{\boldsymbol{\theta}}_n)\left[\mathbf{I} - \mathbf{B}_n(\mathbf{B}_n^T\mathbf{B}_n)^{-}\mathbf{B}_n^T\right]\mathbf{V}_n(\hat{\boldsymbol{\theta}}_n), \tag{3.38}$$

where $\mathbf{I}$ is the $M \times M$ identity matrix, and

$$\begin{aligned} S_n^2 &= Y_n^2 - U_n^2 \\ &= \mathbf{V}_n^T(\hat{\boldsymbol{\theta}}_n)\mathbf{B}_n\left[(\mathbf{J}_n - \mathbf{B}_n^T\mathbf{B}_n)^{-1} + (\mathbf{B}_n^T\mathbf{B}_n)^{-}\right]\mathbf{B}_n^T\mathbf{V}_n(\hat{\boldsymbol{\theta}}_n), \end{aligned} \tag{3.39}$$

which possess asymptotically $\chi^2_{M-1}, \chi^2_{M-2}$, and χ^2_1 distributions, respectively. Note that, in this case, the limiting covariance matrix is $\mathbf{I} - \mathbf{q}\mathbf{q}^T - \mathbf{B}\mathbf{J}^{-1}\mathbf{B}^t$, where

$$\mathbf{B}\mathbf{J}^{-1}\mathbf{B}^T = \frac{M}{4}\begin{pmatrix} \nu_1^2 & \nu_1\nu_2 & \cdots & \nu_1\nu_M \\ \nu_1\nu_2 & \nu_2^2 & \cdots & \nu_2\nu_M \\ \cdot & \cdot & \cdots & \cdot \\ \nu_1\nu_M & \nu_2\nu_M & \cdots & \nu_M^2 \end{pmatrix}$$

and $\mathbf{I} - \mathbf{B}(\mathbf{B}^T\mathbf{B})^{-}\mathbf{B}^T$ do not depend on the unknown parameter σ.

From (3.32), it follows that

$$\mathbf{B}_n^T\mathbf{B}_n = \begin{pmatrix} 0 & 0 & 0 \\ 0 & 0 & 0 \\ 0 & 0 & a \end{pmatrix}, \tag{3.40}$$

where $a = \frac{M}{s^2}\sum_{i=1}^M \nu_i^2$. From the singular-value decomposition, we then have

$$(\mathbf{B}_n^T\mathbf{B}_n)^{-} = \begin{pmatrix} 0 & 0 & 0 \\ 0 & 0 & 0 \\ 0 & 0 & 1/a \end{pmatrix}. \tag{3.41}$$

After some simple matrix algebra, we obtain in this case

$$U_n^2 = \sum_{i=1}^M \widetilde{N_i^2} - \frac{\left(\sum_{i=1}^M \nu_i\tilde{N}_i\right)^2}{\sum_{i=1}^M \nu_i^2} \tag{3.42}$$

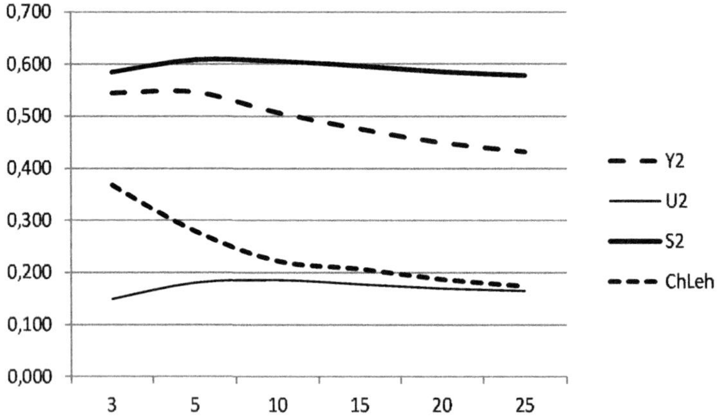

FIGURE 3.3 Powers of the tests $Y2$ in (3.35), $U2$ in (3.42), $S2$ in (3.43), and ChLeh test as functions of M.

and

$$S_n^2 = \frac{4}{\left(4 - M\sum_{i=1}^{M} \nu_i^2\right)\sum_{i=1}^{M} \nu_i^2}\left(\sum_{i=1}^{M} \tilde{N}_i \nu_i\right)^2. \tag{3.43}$$

To investigate the power of tests in (3.35), (3.42), and (3.43), we first consider their simulated distributions under the null hypothesis in (3.31).

Simulation results show that empirical values of level significance at level $\alpha = 0.05$ differ by no more than one simulated standard deviation (tested for samples of size $n = 200$ and for the number M of grouping intervals such that the expected cell frequencies are at least 5). The same is true for $n = 100$ if $M \leqslant 10$. This suggests that, while simulating the power under different alternatives, we could use rejection region by using the critical values from the corresponding chi-squared distributions.

Let the alternative be the two-dimensional standard logistic distribution with independent components distributed as

$$l(x,0,1) = \frac{\frac{\pi}{\sqrt{3}}\exp\left(-\frac{\pi x}{\sqrt{3}}\right)}{\left\{1 + \exp\left(-\frac{\pi x}{\sqrt{3}}\right)\right\}^2}, \quad x \in R^1.$$

Simulated powers of the tests in (3.35), (3.42), and (3.43), and ChLeh$=\sum_{i=1}^{M} \tilde{N}_i^2$ (see Chernoff and Lehmann, 1954; Moore and Stubblebine, 1981, p. 721) are all displayed in Figure 3.3.

First of all, we see that power of S_n^2 is the highest and that it almost does not depend on the number M of grouping intervals. Secondly, because of the decomposition $Y_n^2 = U_n^2 + S_n^2$, the power of Y_n^2 is more than that of U_n^2 and less than that of S_n^2. The same situation was observed in the univariate case (see Voinov et al., 2009, 2012). From the result of Chernoff and Lehmann (1954), it is known that the Pearson-Fisher statistic $\sum_{i=1}^{M} \tilde{N}_i^2$ does not follow the χ_{M-1}^2 distribution in the limit, and may depend on the unknown parameters. Roy (1956), Watson (1958), and Dahiya and Gurland (1972a) have shown that for a location and scale family, if one uses random intervals, then the statistic $\sum_{i=1}^{M} \tilde{N}_i^2$ will be distribution-free following the chi-squared distribution in the limit, but the number of degrees of freedom will depend on the null hypothesis. This is exactly the case considered by Moore and Stubblebine (1981) who proved that, under the two-dimensional circular normal distribution, the limiting distribution of $\sum_{i=1}^{M} \tilde{N}_i^2$ does not depend on the parameters and follows the chi-squared distribution in the limit with the number of degrees of freedom between $M-2$ and $M-1$. From Figure 3.3, we see that the power of ChLeh is indeed between that of Y_n^2 (with $M-1$ d.f.) and U_n^2 (with $M-2$ d.f.). McCulloch (1985) and Mirvaliev (2001) showed that "the DN statistic (U_n^2) behaves locally like the Pearson-Fisher statistic (ChLeh)." In the univariate case, we did not see the essential difference between the powers of the DN and Pearson-Fisher's $\sum_{i=1}^{M} \tilde{N}_i^2$ (Voinov et al., 2009), but in the two-dimensional case, the power of ChLeh is noticeably higher than that of the DN statistic. Still, the power of these two statistics is much less than that of S_n^2. This suggests us to use S_n^2 if we wish to test for the two-dimensional circular normal distribution against the two-dimensional logistic distribution with independent components using chi-squared type tests.

As another alternative, we may consider the two-dimensional normal distribution with correlated components. Let the alternative hypothesis be such that $\boldsymbol{\mu} = (0,0)^T$ and $\boldsymbol{\Sigma}_{01} = \begin{pmatrix} 2 & 1.4 \\ 1.4 & 2.5 \end{pmatrix}$. In this case, the powers of the tests in (3.35), (3.42), and (3.43), as functions of M, are displayed in Figure 3.4.

From Figure 3.4, we see that the powers of S_n^2 and Y_n^2, being high, are almost the same (still, the power of S_n^2 is slightly higher than that of Y_n^2), and that the power of U_n^2 is much lower as expected. We see also that the power of both S_n^2 and Y_n^2 is the highest and the same is when $M = 2$. The same results are also obtained if $\boldsymbol{\Sigma}_{02} = \begin{pmatrix} 2 & -1.4 \\ -1.4 & 2.5 \end{pmatrix}$.

3.5.4 Comparison of different tests

Let us test the null hypothesis in (3.31) by using the modified chi-squired test S_n^2 in (3.43), the multivariate skewness test (Sk) $b_{1,2}$ of Mardia (1970), the kurtosis test (Kur) $b_{2,2}$ of Mardia (1970), Anderson-Darling A^2, and Cramer-von Mises W^2 EDF-tests (see Henze, 2002). Considering the two-dimensional standard logistic alternative with independent components as an alternative, the powers of all these tests are as presented in Table 3.1.

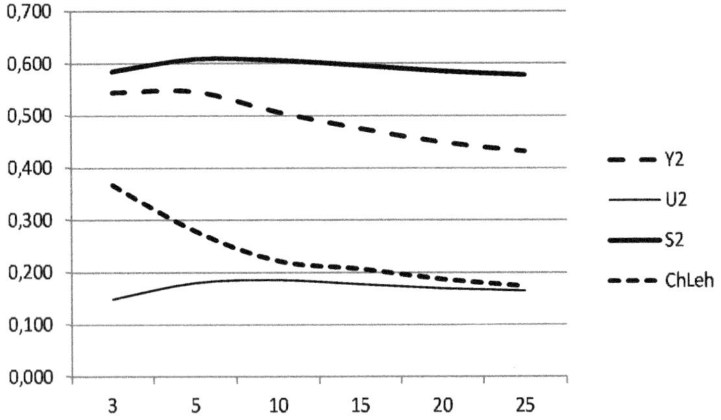

FIGURE 3.4 Powers of the tests $Y2$ in (3.35), $U2$ in (3.42) and $S2$ in (3.43), as functions of M.

TABLE 3.1 Powers of S_n^2, Sk, Kur, A^2, and W^2 tests for the circular two-dimensional normal against two-dimensional standard logistic distribution with independent components. Errors are within one simulated standard deviation.

Test	Power
S_n^2	0.609 ± 0.005
Sk	0.357 ± 0.006
Kur	0.832 ± 0.005
A^2	0.668 ± 0.009
W^2	0.671 ± 0.010

From Table 3.1, we see that Mardia's test $b_{2,2}$ is the most powerful for the particular null and alternative hypotheses. It may also be noted that the application of A^2 and W^2 is somewhat involved since one has to determine critical values of these tests by simulation.

Consider now the power of different tests for the circular two-dimensional normal null distribution against two-dimensional alternative with correlated components. Let the alternatives be two-dimensional normal distributions with zero mean vector and covariance matrices $\mathbf{\Sigma}_{01}$ and $\mathbf{\Sigma}_{02}$. The power results in this case from a simulation study are presented in Tables 3.2 and 3.3, respectively. In these tables, "Sk right" and "Sk left" denote the skewness test $b_{1,2}$ of Mardia (1970) applied with right-tailed and left-tailed rejection regions, respectively. From Tables 3.2 and 3.3, we observe that all these tests do not differentiate

TABLE 3.2 Powers of S_n^2, Sk right, Sk left, Kur, A^2, and W^2 tests for the circular two-dimensional normal against two-dimensional normal distribution with covariance matrix $\boldsymbol{\Sigma}_{01}$. Errors are within one simulated standard deviation.

Test	Power
S_n^2	0.485 ± 0.008
Sk right	0.027 ± 0.002
Sk left	0.813 ± 0.005
Kur	0.182 ± 0.005
A^2	0.622 ± 0.013
W^2	0.622 ± 0.012

TABLE 3.3 Powers of S_n^2, Sk right, Sk left, Kur, A^2, and W^2 tests for the circular two-dimensional normal against two-dimensional normal distribution with covariance matrix $\boldsymbol{\Sigma}_{02}$. Errors are within one simulated standard deviation.

Test	Power
S_n^2	0.486 ± 0.008
Sk right	0.026 ± 0.002
Sk left	0.817 ± 0.008
Kur	0.180 ± 0.006
A^2	0.617 ± 0.010
W^2	0.618 ± 0.010

between alternatives with positive ($\boldsymbol{\Sigma}_{01}$) and negative ($\boldsymbol{\Sigma}_{02}$) correlation. Next, we observe that "Sk right" is biased, in contradiction to the opinion of Mardia (1970, p. 523); see also Henze (2002). We further note that the power of "Sk left" is the highest, the power of "Kur" is considerably lower than that of the chi-squared type test S_n^2, and that the power of A^2 and W^2 are higher but close to that of S_n^2.

3.5.5 Conclusions

Among all modified chi-squared tests for the two-dimensional circular normality discussed in Section 3.5.3, the test S_n^2 possesses high enough power and has more power than that of NRR Y^2 and DN U^2 tests for the alternatives considered.

TABLE 3.4 Ratios of powers of S_n^2, Sk, Kur, A^2, and W^2 tests.

Test	Ratio
S_n^2	0.609/0.485 = 1.25
Sk	0.815/0.357 = 2.28
Kur	0.832/0.181 = 4.6
A^2	0.668/0.619 = 1.08
W^2	0.671/0.620 = 1.08

The same results were observed in the univariate case (Voinov et al., 2009, 2011), but in the two-dimensional case of the circular normal the power of S_n^2 does not seem to depend on the number of equiprobable grouping intervals. This is clearly an advantage since in the univariate case it is not easy to find the optimal number of grouping intervals; see, e.g. Greenwood and Nikulin (1996).

To examine the dependence of the power of tests to alternative hypotheses, we consider ratios of powers of tests discussed earlier in Section 3.5.4. By Ratio, we mean the ratio of power for logistic alternative to that for the two-dimensional normal distribution (we always divide the large value by a small one). From Tables 3.1, 3.2, and 3.3, we determined the Ratio values for all the tests and these are presented in Table 3.4.

From Table 3.4, we see that S_n^2, A^2, and W^2 tests are, in some sense, much less sensitive (Ratio = 1.25, 1.08, and 1.08) for the alternatives considered. We cannot conclude that S_n^2, A^2, and W^2 tests are "omnibus" tests from this result since many other alternatives need to be considered for this purpose, but with respect to the logistic and two-dimensional normal alternatives, they seem to be more preferable than other tests.

Overall, we observe that S_n^2 test is quite comparable to EDF tests and possesses some advantages. One does not have to simulate critical values to define rejection regions since for S_n^2 tests they are simply found from the chi-squared distribution with 1 d.f.. Another advantage is that there is no need to worry about the determination of the optimal number of grouping intervals. A final mention that needs to be made is that Mardia's multivariate skewness test is quite biased and therefore cannot be recommended for the considered testing problem.

3.6 MODIFIED CHI-SQUARED TESTS FOR THE EXPONENTIAL DISTRIBUTION

3.6.1 Two-parameter exponential distribution

Consider the two-parameter exponential distribution which has been used quite extensively in reliability and survival analysis; see, for example,

Balakrishnan and Basu (1995). In this case, several approaches for testing this null hypothesis are known in the literature; see, for example, Engelhardt and Bain (1975), Balakrishnan (1983), Spinelli and Stephens (1987), Ascher (1990), Ahmad and Alwasel (1999), Castillo and Puig (1999), and Gulati and Neus (2003). Here, we discuss two tests. The first one is based on the NRR statistic in (3.8). The elements of the Fisher information matrix **J** and the elements of the matrix **B** needed in this case are presented in Chapter 9, and the Excel version of the test is described in there as well.

Another way for testing H_0 has been suggested by Greenwood and Nikulin (1996, p.143) by exploiting the fact that the first-order statistic $X_{(1)}$ is a superefficient estimator of the threshold parameter μ. Their idea, as published, however contains some mistakes which we shall correct here.

Let $X_1,\dots,X_n$ be i.i.d. random variables with density

$$f(x)=\frac{1}{\theta}e^{-(x-\mu)/\theta},\quad x\geqslant\mu,\ \theta>0.$$

If H_0 is true, then

$$U_2=X_{(2)}-X_{(1)},\dots,U_n=X_{(n)}-X_{(1)},$$

where $X_{(i)},i=1,\dots,n$, are the order statistics, form a sample of i.i.d. random variables from the scale-exponential distribution with distribution function

$$F(u,\theta)=1-\exp(-u/\theta),\quad u\geqslant 0,\ \theta>0.$$

The MLE $\hat{\theta}_n$ of the parameter θ, calculated from $U_2,\dots,U_n$, is

$$\hat{\theta}_n=\frac{1}{n-1}\sum_{i=2}^{n}U_i=\frac{n}{n-1}[\bar{X}_n-X_{(1)}]=\frac{1}{n-1}\sum_{i=1}^{n}(X_i-X_{(1)}).$$

Now, construct the frequency vector $\mathbf{N}^{(n)*}=(N_1^{(n)*},\dots,N_r^{(n)*})^T$ by grouping $U_2,\dots,U_n$ over the equiprobable random intervals

$$(0,x_1\hat{\theta}_n],(x_1\hat{\theta}_n,x_2\hat{\theta}_n],\dots,(x_{r-1}\hat{\theta}_n,+\infty),$$

where $x_j=-\ln(1-j/r),j=1,\dots,r-1$. It should be mentioned that the intervals, $(X_{(1)},x_1\hat{\theta}_n],\dots,(X_{(1)}+x_{r-1}\hat{\theta}_n,+\infty)$, suggested by Greenwood and Nikulin (1996, p. 143) are not equiprobable.

An explicit expression for the test Y_n^2, as given in Greenwood and Nikulin (1996, p. 143) is

$$Y_n^2=\frac{r}{(n-1)}\sum_{j=1}^{r}N_j^{(n)*2}-(n-1)+\frac{r^2}{(n-1)\lambda_2}\left(\sum_{j=1}^{r}N_j^{(n)*}c_{2j}\right)^2,\quad(3.44)$$

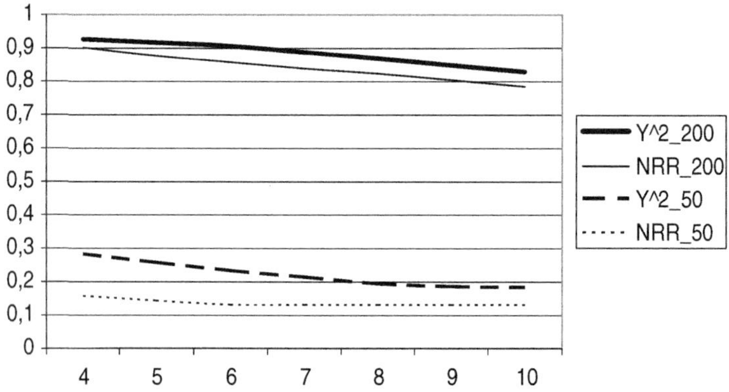

FIGURE 3.5 The power of $Y_n^2 = Y\hat{}2$ (for $n = 200$ and $n = 50, \alpha = 0.05, N = 5000$) and the power of the NRR test in (3.8), as functions of the number of cells r.

where

$$c_{2j} = \left(1 - \frac{j-1}{r}\right) \ln\left(1 - \frac{j-1}{r}\right) - \left(1 - \frac{j}{r}\right) \ln\left(1 - \frac{j}{r}\right), \quad j = 1, \ldots, r,$$

and

$$\lambda_2 = 1 - r \sum_{j=1}^{r} c_{2j}^2.$$

The formula for λ_2 given by Greenwood and Nikulin (1996, p. 141) is, however, incorrect and should read as

$$\lambda_2 = 1 - \frac{1}{r}\left[(r-1)^2 \ln^2(1 - 1/r) + \ln^2(1/r)\right]$$
$$-r \sum_{j=2}^{r-1} \left[(1 - j/r) \ln\left(\frac{r-j+1}{r-j}\right) + \frac{1}{r} \ln\left(1 - \frac{j-1}{r}\right)\right]^2.$$

A Monte Carlo simulation of the test in (3.44) shows that the limiting distribution of Y_n^2 is distribution-free and follows χ_{r-1}^2.

A comparison of the power of the NRR test in (3.8) with that of (3.44) with respect to the seminormal alternative with pdf

$$f(x; \mu, \theta) = \frac{\sqrt{2}}{\sqrt{\pi}\theta} \exp\left[-\frac{(x-\mu)^2}{2\theta^2}\right], \quad x \geqslant \mu, \ \theta > 0,$$

with parameters $\mu = 1, \theta = 1$ are presented in Figure 3.5.

From Figure 3.5, we see that, surprisingly, the power of (3.44) for the seminormal alternative is higher than that of the classical NRR test in (3.8). In addition, the statistic in (3.44) is computationally much simpler than the NRR test in (3.8), and for this reason we recommend the test in (3.44).

TABLE 3.5 Power of Y_n^2 for the seminormal distribution with parameters $\mu = 0, \theta = 1$, as a function of r.

r	2	3	4	6	8	10
Power	0.922	0.931	0.927	0.906	0.870	0.844

3.6.2 Scale-exponential distribution

Consider now the simpler situation when $X_1, \ldots, X_n$ are i.i.d. random variables from the scale-exponential distribution with distribution function

$$F(x,\theta) = 1 - \exp(-x/\theta), \quad x \geqslant 0,\ \theta > 0.$$

The MLE $\hat{\theta}_n$ of a parameter θ is $\hat{\theta}_n = \sum_{i=1}^n X_i/n$. Construct the frequency vector $\mathbf{N}^{(n)} = (N_1^{(n)}, \ldots, N_r^{(n)})^T$ by grouping $X_1, \ldots, X_n$ over the equiprobable random intervals

$$(0, x_1\hat{\theta}_n], (x_1\hat{\theta}_n, x_2\hat{\theta}_n], \ldots, (x_{r-1}\hat{\theta}_n, +\infty),$$

where $x_j = -\ln(1 - j/r), j = 1, \ldots, r-1$.

The expression for the test Y_n^2 in this case is (Greenwood and Nikulin, 1996)

$$Y_n^2 = \frac{r}{n}\sum_{j=1}^{r} N_j^{(n)2} - n + \frac{r^2}{n\lambda_2}\left(\sum_{j=1}^{r} N_j^{(n)} c_{2j}\right)^2, \tag{3.45}$$

where, as before,

$$c_{2j} = \left(1 - \frac{j-1}{r}\right)\ln\left(1 - \frac{j-1}{r}\right) - \left(1 - \frac{j}{r}\right)\ln\left(1 - \frac{j}{r}\right), \quad j = 1, \ldots, r,$$

and

$$\lambda_2 = 1 - r\sum_{j=1}^{r} c_{2j}^2.$$

The statistic in (3.45) follows asymptotically the χ_{r-1}^2 distribution and, under H_0, does not depend on the unknown parameter θ. From Table 3.5, we see that power of the statistic in (3.45) for the seminormal alternative is like that of the NRR test (see Section 4.4), being the highest for smaller number of equiprobable random cells.

A different approach for testing exponentiality has been suggested by Dahiya and Gurland (1972b).

3.7 POWER GENERALIZED WEIBULL DISTRIBUTION

In accelerated life studies, the Power Generalized Weibull (PGW) family with distribution function

$$F(t;\theta,\gamma,\nu) = 1 - \exp\left\{1 - \left[1 + \left(\frac{t}{\theta}\right)^{\nu}\right]^{1/\gamma}\right\}, \quad t,\theta,\gamma,\nu > 0, \qquad (3.46)$$

proves to be very useful (Bagdonavičius and Nikulin, 2002). The family in (3.46) has all its moments to be finite. Depending on the values of the parameters, the hazard rate function

$$\alpha(t;\theta,\gamma,\nu) = \frac{\nu}{\gamma\theta^{\nu}} t^{\nu-1}\left[1 + \left(\frac{t}{\theta}\right)^{\nu}\right]^{\frac{1}{\gamma}-1}$$

can be constant, monotone increasing or decreasing, ∩-shaped and ∪-shaped. Note also that $F(t;\theta,\nu,1) = W(t;\theta,\nu)$, the Weibull distribution, and $F(t;\theta,1,1) = E(t;\theta)$ is the exponential distribution. More details on this distribution can be found, for example, in Bagdonavičius et al. (2006).

Alloyarova et al. (2007), under the assumption that the shape parameter γ is known, developed a modified HRM chi-squared test based on the approximate MMEs of the parameters. These authors, however, erroneously concluded that the MLEs are inconsistent in this case. Here, we shall consider modified tests based on MLEs of the three parameters in (3.46) which are actually consistent (Voinov et al., 2011, 2012).

3.7.1 Estimation of parameters

Let $T_1,T_2,\ldots,T_n$ be a random sample from the distribution in (3.46). Assuming the parameters θ,γ,ν to be unknown, the log likelihood function is given by

$$\ln L = n\ln\left(\frac{e\nu}{\gamma\theta^{\nu}}\right) + (\nu-1)\sum_{i=1}^{n}\ln T_i + \left(\frac{1}{\gamma}-1\right)\sum_{i=1}^{n}\ln\left[1+\left(\frac{T_i}{\theta}\right)^{\nu}\right] - \sum_{i=1}^{n}\left[1+\left(\frac{T_i}{\theta}\right)^{\nu}\right]^{1/\gamma}. \qquad (3.47)$$

Since analytical maximization of (3.47) is not possible, we examine the maximum likelihood estimates (MLEs) $\hat{\theta}_n,\hat{\gamma}_n,\hat{\nu}_n$,of the parameters θ,γ,ν, by Monte Carlo simulation. Samples of pseudorandom numbers $T_i, i = 1,2,\ldots,n$, from the distribution in (3.46) were generated by the formula

$$T_i = \theta\left\{[1-\ln(1-\xi_i)]^{\gamma} - 1\right\}^{1/\nu},$$

where ξ_i is a pseudorandom number uniformly distributed over the interval [0,1]. MLEs from the simulated samples can be obtained by a numerical maximization

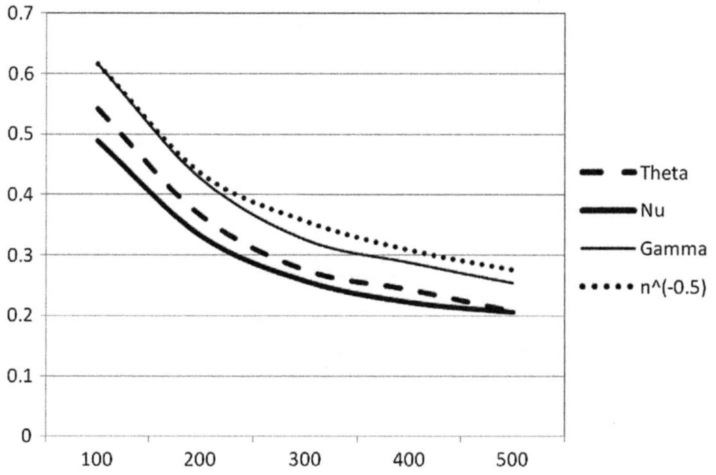

FIGURE 3.6 Simulated average absolute errors of the MLEs $\hat{\theta}_n, \hat{\gamma}_n, \hat{\nu}_n$ versus their true values as a function of sample size n. Number of runs for every n was fixed as $N = 600$. The graph for $n^{-0.5}$ is shown as the dotted line.

of the log-likelihood function in (3.47). An example of a simulation for $\theta = 3, \gamma = 2, \nu = 3$ is presented in Figure 3.6. A power function fit of the curves in Figure 3.6 gives the following rates of convergence: $\hat{\theta}_n \sim n^{-0.55}, \hat{\gamma}_n \sim n^{-0.56}$, and $\hat{\nu}_n \sim n^{-0.60}$, and so the MLEs converge faster than the rate $n^{-0.5}$ and are therefore $\sqrt{n}$-consistent.

3.7.2 Modified chi-squared test

Denote the unknown parameter by $\boldsymbol{\theta} = (\theta_1, \theta_2, \theta_3)^T$, where $\theta_1 = \theta, \theta_2 = \gamma, \theta_3 = \nu$, with the MLE $\hat{\boldsymbol{\theta}}_n$ of $\boldsymbol{\theta}$ being $\hat{\boldsymbol{\theta}}_n = (\hat{\theta}_{1n}, \hat{\theta}_{2n}, \hat{\theta}_{3n})^T$, where $\hat{\theta}_{1n} = \hat{\theta}_n, \hat{\theta}_{2n} = \hat{\gamma}_n, \hat{\theta}_{3n} = \hat{\nu}_n$. Let $\Delta_j(\hat{\boldsymbol{\theta}}_n)$ be non-intersecting random equiprobable cells with the following end points:

$$a_j(\hat{\boldsymbol{\theta}}_n) = \hat{\theta}_n \left\{ \left[1 - \ln\left(1 - \frac{j}{r}\right) \right]^{\hat{\gamma}_n} - 1 \right\}^{1/\hat{\nu}_n}, \quad j = 0,1,\ldots,r,$$

$$a_0(\hat{\boldsymbol{\theta}}_n) = 0, \quad a_r(\hat{\boldsymbol{\theta}}_n) = \infty.$$

The probability p_j of falling into jth cell is $p_j = 1/r, j = 1, \ldots, r$. The Fisher information matrix $\mathbf{J}$ can be evaluated as

$$\mathbf{J} = \mathbf{E}_{\boldsymbol{\theta}} \begin{pmatrix} \left(\frac{\partial \ln f}{\partial \theta}\right)^2 & \left(\frac{\partial \ln f}{\partial \theta}\frac{\partial \ln f}{\partial \gamma}\right) & \left(\frac{\partial \ln f}{\partial \theta}\frac{\partial \ln f}{\partial \nu}\right) \\ \left(\frac{\partial \ln f}{\partial \gamma}\frac{\partial \ln f}{\partial \theta}\right) & \left(\frac{\partial \ln f}{\partial \gamma}\right)^2 & \left(\frac{\partial \ln f}{\partial \gamma}\frac{\partial \ln f}{\partial \nu}\right) \\ \left(\frac{\partial \ln f}{\partial \nu}\frac{\partial \ln f}{\partial \theta}\right) & \left(\frac{\partial \ln f}{\partial \nu}\frac{\partial \ln f}{\partial \gamma}\right) & \left(\frac{\partial \ln f}{\partial \nu}\right)^2 \end{pmatrix}.$$

Define also the matrix **B** with elements

$$B_{jk} = \frac{1}{\sqrt{p_j}} \int_{a_{j-1}(\boldsymbol{\theta})}^{a_j(\boldsymbol{\theta})} \frac{\partial f(t;\theta,\gamma,\nu)}{\partial \theta_k}, \quad j = 1,\dots,r, \ k = 1,2,3,$$

where $f(t;\theta,\gamma,\nu)$ is the pdf of the distribution in (3.46). Explicit expressions for the elements of **J** and **B** are presented in Chapter 9.

The NRR test in (3.8) can be presented as

$$Y1_n^2(\hat{\boldsymbol{\theta}}_n) = U_n^2(\hat{\boldsymbol{\theta}}_n) + S_n^2(\hat{\boldsymbol{\theta}}_n), \tag{3.48}$$

where

$$U_n^2(\hat{\boldsymbol{\theta}}_n) = \mathbf{V}^{(n)T}(\hat{\boldsymbol{\theta}}_n)[\mathbf{I} - \mathbf{q}_n\mathbf{q}_n^T - \mathbf{B}_n(\mathbf{B}_n^T\mathbf{B}_n)^{-1}\mathbf{B}_n^T]\mathbf{V}^{(n)}(\hat{\boldsymbol{\theta}}) \tag{3.49}$$

is the DN statistic distributed under the null hypothesis as χ^2_{r-s-1}, in the limit, where s is the number of parameters under estimation (three in the present case), and

$$\begin{aligned} S_n^2(\hat{\boldsymbol{\theta}}_n) &= W_n^2(\hat{\boldsymbol{\theta}}_n) + \mathbf{V}^{(n)T}(\hat{\boldsymbol{\theta}})(\mathbf{J}_n - \mathbf{J}_{gn})^{-1}\mathbf{B}_n^T\mathbf{V}^{(n)}(\hat{\boldsymbol{\theta}}_n) \\ &= \mathbf{V}^{(n)T}(\hat{\boldsymbol{\theta}}_n)\mathbf{B}_n\left((\mathbf{J}_n - \mathbf{J}_{gn})^{-1} + (\mathbf{B}_n^T\mathbf{B}_n)^{-1}\right)\mathbf{B}_n^T\mathbf{V}^{(n)}(\hat{\boldsymbol{\theta}}_n). \end{aligned} \tag{3.50}$$

3.7.3 Evaluation of power

To assess the power of the NRR statistic $Y1_n^2(\hat{\boldsymbol{\theta}})$, the DN statistic $U_n^2(\hat{\boldsymbol{\theta}})$, and the statistic $S_n^2(\hat{\boldsymbol{\theta}})$ for the PGW null hypothesis in (3.46) against the Exponentiated Weibull (EW) distribution (Mudholkar et al., 1995)

$$F_{EW}(x) = \left\{1 - \exp\left[-\left(\frac{x}{\alpha}\right)^{\beta}\right]\right\}^{\gamma}, \quad x,\alpha,\beta,\gamma > 0. \tag{3.51}$$

Generalized Weibull (GW) distribution (Mudholkar et al., 1996)

$$F_{GW}(x) = 1 - \left[1 - \lambda\left(\frac{x}{\sigma}\right)^{1/\alpha}\right]^{1/\lambda}, \quad x,\alpha,\sigma > 0, \ \lambda \in R^1, \tag{3.52}$$

and Three-parameter Weibull (W3) distribution

$$F_{W3}(x) = 1 - \exp\left[-\left(\frac{x-\mu}{\theta}\right)^{p}\right], \quad x \geqslant \mu,\theta, \ p > 0, \tag{3.53}$$

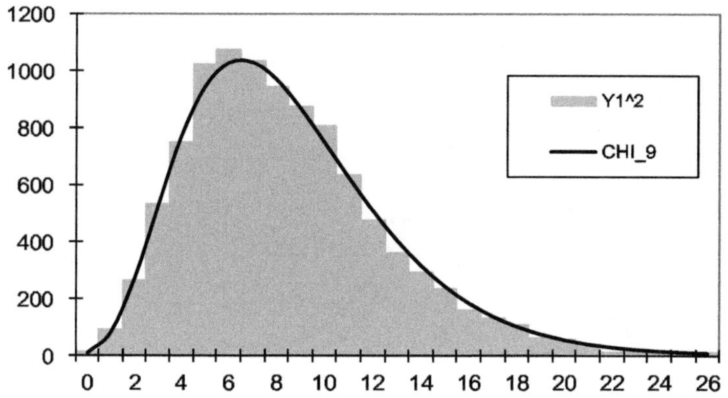

FIGURE 3.7 Simulated distribution of the NRR statistic ($Y1\hat{}2$) under the null hypothesis ($\theta = 3, \gamma = 2, \nu = 3$) against the chi-squared distribution with 9 degrees of freedom, for $n = 200, r = 10, N = 10{,}000$.

Monte Carlo simulation was conducted for different number r of equiprobable random cells.

Figure 3.7 presents the simulated distribution of the NRR statistic $Y1_n^2(\hat{\boldsymbol{\theta}})$ under the null hypothesis in (3.46) (with $\theta = 3, \gamma = 2, \nu = 3$) against the chi-squared distribution with 9 degrees of freedom (with $r = 10$).

From Figure 3.7, we see that, as expected, the NRR statistic is quite close to the limiting chi-squared distribution with $r - 1$ degrees of freedom. Same results were also obtained for different values of the parameters and different numbers of equiprobable random cells, which does mean in this case that the limiting distribution of the NRR test is distribution-free. Since the simulated levels of significance for the left-hand sided rejection region coincide with those corresponding to theoretical levels of the chi-squared distribution, we estimated the power as probability of falling into the theoretical rejection region under the alternatives.

To analyze the power of the tests for alternatives that are close to each other, we selected parameters for the alternatives in such a manner that graphs of the densities would be quite close (see Figure 3.8).

The simulation results show that the simulated probability density functions of $Y1_n^2(\hat{\boldsymbol{\theta}})$ for EW, GW, and W3 alternatives are all shifted to the left compared to the density of χ_{r-1}^2 valid for the null hypothesis. This means that the NRR test $Y1_n^2(\hat{\boldsymbol{\theta}})$ and its asymptotically independent component $S_n^2(\hat{\boldsymbol{\theta}})$ are biased if one uses the right-tailed rejection region, i.e. $S_n^2(\hat{\boldsymbol{\theta}}) > \chi_{r-1}^2(\alpha)$. This situation is analogous to that when testing for Poisson against the binomial alternative using the index of dispersion test (see Section 7.1.3, for example).

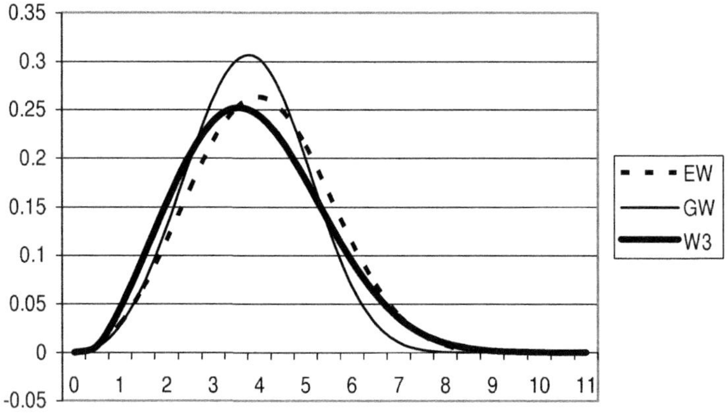

FIGURE 3.8 Probability density functions of EW (with $\alpha = 4.50, \beta = 3.00, \gamma = 1.03$), GW (with $\alpha = 0.3, \sigma = 4.2, \lambda = 0.0006$), and W3 (with $p = 2.5, \theta = 4.0, \mu = 0.3$) distributions.

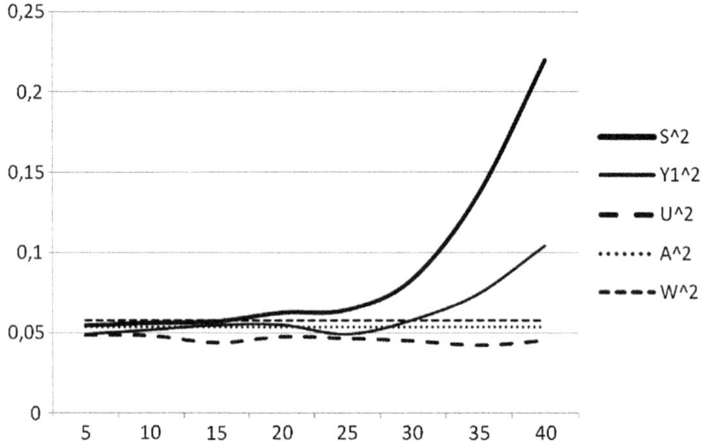

FIGURE 3.9 Power of $Y1_n^2(\hat{\boldsymbol{\theta}})(Y1\hat{\ }2), U_n^2(\hat{\boldsymbol{\theta}})(U\hat{\ }2), S_n^2(\hat{\boldsymbol{\theta}})(S\hat{\ }2)$, $A^2(\hat{\boldsymbol{\theta}})(A\hat{\ }2)$ and $W^2(\hat{\boldsymbol{\theta}})(W\hat{\ }2)$ tests for the Exponentiated Weibull alternative in (3.51) as a function of the number r of equiprobable random intervals (with $\alpha = 0.05, n = 200, N = 5000$ runs).

Results of power simulation for the chi-squared tests $Y1_n^2(\hat{\boldsymbol{\theta}}), U_n^2(\hat{\boldsymbol{\theta}}), S_n^2(\hat{\boldsymbol{\theta}})$, and, for comparison, Anderson-Darling $A^2(\hat{\boldsymbol{\theta}})$ and Cramer-von Mises $W^2(\hat{\boldsymbol{\theta}})$ tests, are all presented in Figures 3.9, 3.10, and 3.11.

From these figures, several important conclusions can be drawn. First of all, we see that, for equiprobable cells, the DN statistic and, of course, its

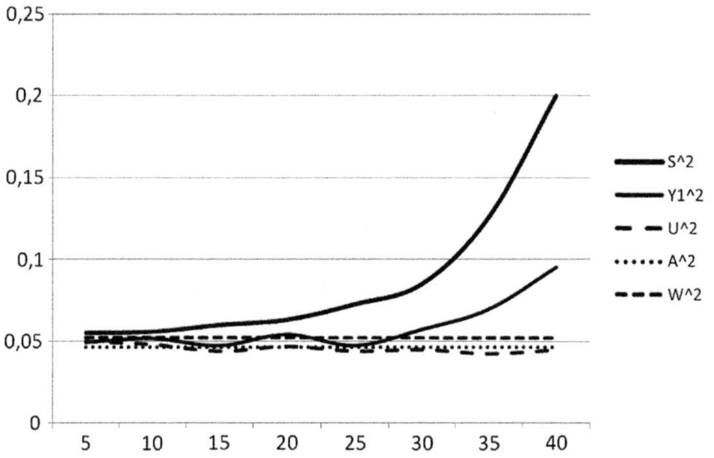

FIGURE 3.10 Power of $Y1_n^2(\hat{\boldsymbol{\theta}})(Y1\hat{}2), U_n^2(\hat{\boldsymbol{\theta}})(U\hat{}2), S_n^2(\hat{\boldsymbol{\theta}})(S\hat{}2), A^2(\hat{\boldsymbol{\theta}})(A\hat{}2)$ and $W^2(\hat{\boldsymbol{\theta}})(W\hat{}2)$ tests for the Generalized Weibull alternative in (3.52) as a function of the number r of equiprobable random intervals (with $\alpha = 0.05, n = 200, N = 5000$ runs).

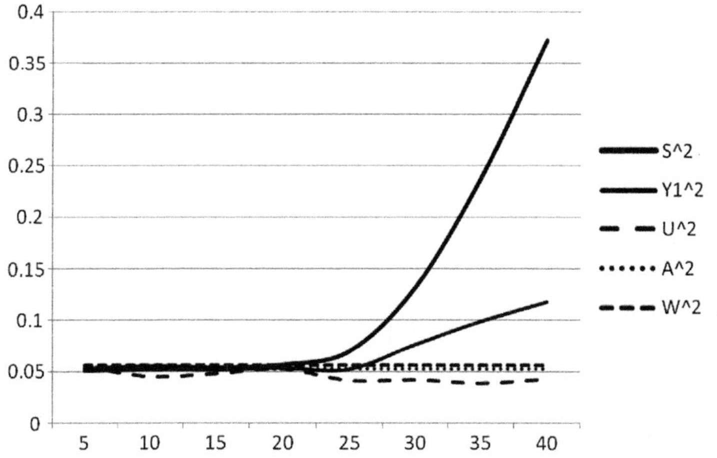

FIGURE 3.11 Power of $Y1_n^2(\hat{\boldsymbol{\theta}})(Y1\hat{}2), U_n^2(\hat{\boldsymbol{\theta}})(U\hat{}2), S_n^2(\hat{\boldsymbol{\theta}})(S\hat{}2), A^2(\hat{\boldsymbol{\theta}})(A\hat{}2)$ and $W^2(\hat{\boldsymbol{\theta}})(W\hat{}2)$ tests for the Three Parameter Weibull alternative in (3.53) as a function of the number r of equiprobable random intervals (with $\alpha = 0.05$, $n = 200, N = 5000$ runs).

asymptotically equivalent Pearson-Fisher test, possess no power and, therefore, cannot be recommended. The same is true for the Anderson-Darling and

Cramer-von Mises tests as well. Secondly, $Y1_n^2(\hat{\boldsymbol{\theta}}), U_n^2(\hat{\boldsymbol{\theta}})$, and $S_n^2(\hat{\boldsymbol{\theta}})$, even for samples of size $n = 200$, possess almost no power for small and moderate number r of cells (5–20), and they can be used only with $r = 35$–40. It may be impractical to use $r > 40$ since expected frequencies may become less than 5 in this case. We also observe that the power of $S_n^2(\hat{\boldsymbol{\theta}})$ test is approximately two times more than that of RNN test $Y1_n^2(\hat{\boldsymbol{\theta}})$. This power, in some sense, "averages" the very low power of the DN statistic and rather high power of its second asymptotically independent component $S_n^2(\hat{\boldsymbol{\theta}})$ (see Section 3.2).

Overall, for the problem of testing for the PGW distribution, we can recommend only the use of $S_n^2(\hat{\boldsymbol{\theta}})$ statistic with relatively large number of equiprobable intervals. At the same time, we note that the power of $S_n^2(\hat{\boldsymbol{\theta}})$ for the PGW null hypothesis against the EW, GW, and W3 alternatives is not high enough to discriminate between these four models with confidence. So, for the selection of one of these models for a survival analysis, one needs to have a test that will compare their hazard rate functions directly. This problem has been successfully solved for time-continuous survival data by Hjort (1990).

3.8 MODIFIED CHI-SQUARED GOODNESS OF FIT TEST FOR RANDOMLY RIGHT CENSORED DATA

3.8.1 Introduction

In this section, following the lines of Bagdonavičius and Nikulin (2011), Bagdonavičius et al. (2011a,b), Bagdonavičius et al. (2010), and Nikulin et al. (2011), we describe a chi-squared test for testing composite parametric hypotheses when data are right censored. This problem arises naturally in reliability and survival analysis. In particular, we consider the tests for the Power Generalized Weibull and Birnbaum-Saunders families of distributions, and the latter possesses some very interesting properties; see, for example Ng et al. (2003), Sanhueza et al. (2008), Leiva et al. (2008).

In the case of complete data, well-known modification of the classical chi-squared test is the NRR statistic Y_n^2 which is based on the differences between two estimators of the probabilities of falling into grouping intervals with one estimator being based on the empirical distribution function, and the other being the maximum likelihood estimators of the unknown parameters of the tested model using initial non-grouped data; see Nikulin (1973a,b,c), Rao and Robson (1974), Moore (1977), Drost (1988), LeCam et al. (1983), Van der Vaart (1988), Voinov and Nikulin (1993), Voinov et al. (2007, 2008a,b, 2009).

Habib and Thomas (1986) and Hollander and Pena (1992) considered natural modifications of the NRR statistic to the case of censored data. These tests are also based on the differences between two estimators of the probabilities of falling into grouping intervals with one being based on the Kaplan-Meier estimator of the cumulative distribution function, and the other being the maximum likelihood estimators of unknown parameters of the tested model using initial non-grouped censored data.

The idea of comparing observed and expected numbers of failures in time intervals is due to Akritas (1988), which was developed further by Hjort (1990). We discuss this here by considering the choice of random grouping intervals as data functions and then present simple formulas for computing the test statistics for some common classes of survival distributions.

3.8.2 Maximum likelihood estimation for right censored data

Suppose the failure times $T_1, \ldots, T_n$ are absolutely continuous i.i.d. random variables, and the p.d.f. of the random variable T_i belongs to a parametric family $\{f(\cdot,\boldsymbol{\theta}),\boldsymbol{\theta} \in \boldsymbol{\Theta} \subset R^m\}$, where $f(\cdot,\boldsymbol{\theta})$ is the density with respect to σ-finite measure μ, and $\boldsymbol{\Theta}$ is an open set. Denote by

$$S(t,\boldsymbol{\theta}) = \mathbf{P}_{\boldsymbol{\theta}}\{T_1 > t\}, \quad \lambda(t,\boldsymbol{\theta}) = \lim_{h \downarrow 0} \frac{1}{h}\mathbf{P}_{\boldsymbol{\theta}}\{t < T_1 \leqslant t + h | T_1 > t\}$$

the corresponding survival function and the hazard rate, respectively. For any $t > 0$, the value $S(t,\boldsymbol{\theta})$ of the survival function is the probability not to fail by time t, and the value $\lambda(t,\boldsymbol{\theta})$ of the hazard rate characterizes the failure risk just after the time t for the objects survived until this time. Suppose, instead of the sample $T_1, \ldots, T_n$, we observe the right censored simple

$$(X_1,\delta_1), \ldots, (X_n,\delta_n), \tag{3.54}$$

where

$$X_i = T_i \wedge C_i = \min\{T_i,C_i\}, \quad \delta_i = \mathbf{1}_{\{T_i \leqslant C_i\}}.$$

Denote by $\bar{G}_i$ the survival function of the censoring time C_i. Thus, for any $t > 0$, the value $\bar{G}_i(t)$ is the probability for the ith object not be censored by time t.

Let us consider the distribution of the random vector (X_i,δ_i) in the case of random censoring with absolutely continuous censoring times C_i with p.d.f. $g_i(t)$. In this case, we have the likelihood function as

$$L(\boldsymbol{\theta}) = \prod_{i=1}^{n} f^{\delta_i}(X_i,\boldsymbol{\theta})S^{1-\delta_i}(X_i,\boldsymbol{\theta})\bar{G}_i^{\delta_i}(X_i)g_i^{1-\delta_i}(X_i), \quad \boldsymbol{\theta} \in \boldsymbol{\Theta}. \tag{3.55}$$

Now, we suppose that the censoring is *non-informative*, meaning that we suppose that the function $\bar{G}_i$ does not depend on the parameter $\boldsymbol{\theta} = (\theta_1, \ldots, \theta_m)^T$.

For the estimation of the parameter $\boldsymbol{\theta}$, we can omit the multipliers which does not depend on this parameter. So, under non-informative censoring, the likelihood function has the form

$$L(\boldsymbol{\theta}) = \prod_{i=1}^{n} f^{\delta_i}(X_i,\boldsymbol{\theta})S^{1-\delta_i}(X_i,\boldsymbol{\theta}), \quad \boldsymbol{\theta} \in \boldsymbol{\Theta}. \tag{3.56}$$

Using the relationship $f(t,\boldsymbol{\theta}) = \lambda(t,\boldsymbol{\theta})S(t,\boldsymbol{\theta})$, the likelihood function in (3.56) can be written as

$$L(\boldsymbol{\theta}) = \prod_{i=1}^{n} \lambda^{\delta_i}(X_i,\boldsymbol{\theta})S(X_i,\boldsymbol{\theta}), \quad \boldsymbol{\theta} \in \boldsymbol{\Theta}. \tag{3.57}$$

The estimator $\hat{\boldsymbol{\theta}}$, maximizing the likelihood function $L(\boldsymbol{\theta}),\boldsymbol{\theta} \in \boldsymbol{\Theta}$, is the *maximum likelihood estimator*. Evidently, the log-likelihood function

$$\ell(\boldsymbol{\theta}) = \sum_{i=1}^{n} \delta_i \ln \lambda(X_i,\boldsymbol{\theta}) + \sum_{i=1}^{n} \ln S(X_i,\boldsymbol{\theta}), \quad \boldsymbol{\theta} \in \boldsymbol{\Theta}, \tag{3.58}$$

is maximized at the same point as the likelihood function.

If $\lambda(u,\boldsymbol{\theta})$ is a sufficiently smooth function of the parameter $\boldsymbol{\theta}$, then the ML estimator satisfies the equation:

$$\dot{\ell}(\hat{\boldsymbol{\theta}}) = \mathbf{0}; \tag{3.59}$$

here $\mathbf{0} = \mathbf{0}_s = (0,\ldots,0)^T \in R^m$, and $\dot{\ell}$ is the *score vector* given by

$$\dot{\ell}(\boldsymbol{\theta}) = \frac{\partial}{\partial \boldsymbol{\theta}}\ell(\boldsymbol{\theta}) = \left(\frac{\partial}{\partial \theta_1}\ell(\boldsymbol{\theta}),\ldots,\frac{\partial}{\partial \theta_m}\ell(\boldsymbol{\theta})\right)^T. \tag{3.60}$$

Equation (3.58) implies that

$$\dot{\ell}(\boldsymbol{\theta}) = \sum_{i=1}^{n} \delta_i \frac{\partial}{\partial \boldsymbol{\theta}} \ln \lambda(X_i,\boldsymbol{\theta}) - \sum_{i=1}^{n} \frac{\partial}{\partial \boldsymbol{\theta}} \Lambda(X_i,\boldsymbol{\theta}),$$

where

$$\Lambda(t,\boldsymbol{\theta}) = -\ln S(t,\boldsymbol{\theta}) = \int_0^t \lambda(u,\boldsymbol{\theta})du.$$

The *Fisher information matrix* is defined as

$$\mathbf{I}(\boldsymbol{\theta}) = -\mathbf{E}_{\boldsymbol{\theta}}\ddot{\ell}(\boldsymbol{\theta}), \tag{3.61}$$

where

$$\ddot{\ell}(\boldsymbol{\theta}) = \sum_{i=1}^{n} \delta_i \frac{\partial^2}{\partial \boldsymbol{\theta}^2} \ln \lambda(X_i,\boldsymbol{\theta}) - \sum_{i=1}^{n} \frac{\partial^2}{\partial \boldsymbol{\theta}^2} \Lambda(X_i,\boldsymbol{\theta}). \tag{3.62}$$

In the case of non-informative and random censoring, the asymptotic properties of the ML estimators result from general results on asymptotic statistics; see, for example, Van der Vaart (1998). Suppose $\boldsymbol{\theta}_0$ is the true value of $\boldsymbol{\theta}$. Then, in this case, under some regularity conditions, we have

$$\hat{\boldsymbol{\theta}} \xrightarrow{P} \boldsymbol{\theta}_0, \sqrt{n}(\hat{\boldsymbol{\theta}} - \boldsymbol{\theta}_0) = \mathbf{i}^{-1}(\boldsymbol{\theta}_0)\frac{1}{\sqrt{n}}\dot{\ell}(\boldsymbol{\theta}_0) + o_P(\mathbf{1}),$$

$$-\frac{1}{n}\ddot{\ell}(\hat{\boldsymbol{\theta}}) \xrightarrow{P} \mathbf{i}(\boldsymbol{\theta}_0), \tag{3.63}$$

$$\sqrt{n}(\hat{\boldsymbol{\theta}} - \boldsymbol{\theta}_0) \xrightarrow{d} N_m(\mathbf{0},\mathbf{i}^{-1}(\boldsymbol{\theta}_0)), \quad \frac{1}{\sqrt{n}}\dot{\ell}(\boldsymbol{\theta}_0) \xrightarrow{d} N_m(\mathbf{0},\mathbf{i}(\boldsymbol{\theta}_0)), \tag{3.64}$$

where $\hat{\boldsymbol{\theta}}$ is the solution to Eq. (3.59), and the matrix

$$\mathbf{i}(\boldsymbol{\theta}_0) = \lim_{n\to\infty} \mathbf{I}(\boldsymbol{\theta}_0)/n. \tag{3.65}$$

Let us write the log-likelihood function $\ell(\theta)$ and the score vector $\dot{\ell}(\theta)$ in terms of stochastic processes $N_i(t)$ and $Y_i(t)$. The trajectories of the counting processes N_i have the form

$$N_i(t) = \begin{cases} 0, & 0 \leqslant t < X_i \\ 1, & t \geqslant X_i \end{cases} \tag{3.66}$$

if $\delta_i = 1$, and $N_i(t) \equiv 0$ if $\delta_i = 0$. So,

$$\int_0^\infty \ln \lambda(u,\boldsymbol{\theta}) dN_i(u) = \begin{cases} \ln \lambda(X_i,\boldsymbol{\theta}), & \delta_i = 1, \\ 0, & \delta_i = 0, \end{cases}$$
$$= \delta_i \ln \lambda(X_i,\boldsymbol{\theta}). \tag{3.67}$$

The trajectories of the stochastic process Y_i have the form

$$Y_i(t) = \begin{cases} 1, & 0 \leqslant t \leqslant X_i \\ 0, & t > X_i. \end{cases} \tag{3.68}$$

So, we have

$$\int_0^\infty Y_i(u)\lambda(u,\boldsymbol{\theta})du = \int_0^{X_i} \lambda(u,\boldsymbol{\theta})du = -\ln S(X_i,\boldsymbol{\theta}), \quad \{\boldsymbol{\theta} \in \boldsymbol{\Theta}\}. \tag{3.69}$$

These relations imply that, under non-informative and random censoring, the log-likelihood function may be written in the form

$$\ell(\boldsymbol{\theta}) = \sum_{i=1}^n \int_0^\infty \{\ln \lambda(u,\boldsymbol{\theta}) dN_i(u) - Y_i(u)\lambda(u,\boldsymbol{\theta})\} du, \quad \boldsymbol{\theta} \in \boldsymbol{\Theta}, \tag{3.70}$$

from which it follows that

$$\dot{\ell}(\boldsymbol{\theta}) = \sum_{i=1}^{n} \int_{0}^{\infty} \frac{\partial}{\partial \boldsymbol{\theta}} \ln \lambda(u,\boldsymbol{\theta}) dM_i(u,\boldsymbol{\theta}), \quad \boldsymbol{\theta} \in \boldsymbol{\Theta},$$

and

$$\begin{aligned}\ddot{\ell}(\theta) = & \sum\nolimits_{i=1}^{n} \int_{0}^{\infty} \frac{\partial^2}{\partial \theta^2} \ln \lambda(u,\theta) dM_i(u,\theta) \\ & - \sum\nolimits_{i=1}^{n} \int_{0}^{\infty} \frac{\partial}{\partial \boldsymbol{\theta}} \ln \lambda(u,\boldsymbol{\theta}) \left(\frac{\partial}{\partial \boldsymbol{\theta}} \ln \lambda(u,\boldsymbol{\theta}) \right)^T \\ & \times \lambda(u,\boldsymbol{\theta}) Y_i(u) du, \quad \boldsymbol{\theta} \in \boldsymbol{\Theta},\end{aligned}$$

where

$$M_i(t,\boldsymbol{\theta}) = N_i(t) - A_i(t) = N_i(t) - \int_{0}^{t} Y_i(u)\lambda(u,\boldsymbol{\theta}) du, \quad \boldsymbol{\theta} \in \boldsymbol{\Theta}, \tag{3.71}$$

is a martingale of counting process $N_i(t), t \geqslant 0$. Let us consider the following two processes

$$N(t) = \sum_{i=1}^{n} N_i(t) \text{ and } Y(t) = \sum_{i=1}^{n} Y_i(t), \quad t \geqslant 0. \tag{3.72}$$

It is quite common in survival analysis and reliability to assume that the processes N_i and Y_i are observed for finite time $\tau > 0$, which means that at time τ, observation on all surviving objects are censored, and so instead of using censoring time C_i, we shall use censoring time $C_i \wedge \tau$. We shall denote them once again by C_i. The process $N(t)$ shows, for any $t > 0$, the number of observed failures in the interval $[0,t]$ and the process $Y(t)$ shows the number of objects which are *at risk* (not failed, not truncated, and not censored) just prior to time t, for $t < \tau$. We note that, in this case, the Fisher information matrix can be written as

$$\begin{aligned}\mathbf{I}(\boldsymbol{\theta}) &= -\mathbf{E}_{\boldsymbol{\theta}} \ddot{\ell}(\boldsymbol{\theta}) \\ &= \mathbf{E}_{\theta} \sum_{i=1}^{n} \int_{0}^{\tau} \frac{\partial}{\partial \boldsymbol{\theta}} \ln \lambda(u,\boldsymbol{\theta}) \left(\frac{\partial}{\partial \boldsymbol{\theta}} \ln \lambda(u,\boldsymbol{\theta}) \right)^T \\ &\quad \times \lambda(u,\boldsymbol{\theta}) Y_i(u) du. \end{aligned} \tag{3.73}$$

3.8.3 Chi-squared goodness of fit test

Let us consider the hypothesis

$$H_0 : F(x) \in \mathcal{F}_0 = \{F_0(x;\boldsymbol{\theta}),\boldsymbol{\theta} \in \boldsymbol{\Theta} \subset R^m\}$$

meaning that F belongs to the class $\mathcal{F}_0$ of the form $F_0(x;\boldsymbol{\theta})$; here,

$$\boldsymbol{\theta} = (\theta_1,\ldots,\theta_m)^T \in \boldsymbol{\Theta} \subset R^m$$

is an unknown m-dimensional parameter and F_0 is a known distribution function. Divide the interval $[0,\tau]$ into $k > m$ smaller intervals $I_j = (a_{j-1},a_j]$, with $a_0 = 0, a_k = \tau$, and denote by

$$U_j = N(a_j) - N(a_{j-1}) \tag{3.74}$$

the number of observed failures in the jth interval $I_j, j = 1,2,\ldots,k$. Then, what is the "expected" number of failures in the interval I_j, under the null hypothesis H_0? Taking into account the equality

$$\mathbf{E}N(t) = \mathbf{E}\int_0^t \lambda(u,\boldsymbol{\theta}_0)Y(u)du, \tag{3.75}$$

we can "expect" to have

$$e_j = \int_{a_{j-1}}^{a_j} \lambda(t,\hat{\boldsymbol{\theta}})Y(u)du \tag{3.76}$$

failures, where $\hat{\boldsymbol{\theta}}$ is the MLE of the parameter $\boldsymbol{\theta}$ and $\lambda(t,\boldsymbol{\theta})$ is the hazard function. A chi-squared test considered by Akritas (1988) is based on the vector

$$\mathbf{Z} = (Z_1,\ldots,Z_k)^T, \text{ with } Z_j = \frac{1}{\sqrt{n}}(U_j - e_j), \quad j = 1,\ldots,k. \tag{3.77}$$

To investigate the properties of the statistic $\mathbf{Z}$, we need the properties of the stochastic process

$$H_n(t) = \frac{1}{\sqrt{n}}\left(N(t) - \int_0^t \lambda(u,\hat{\boldsymbol{\theta}})Y(u)du\right) \tag{3.78}$$

in the interval $[0,\tau]$. We remind that consistency and asymptotic normality of the ML estimator $\hat{\boldsymbol{\theta}}$ hold under the following sufficient conditions.

Conditions A:

(1) There exists a neighborhood $\boldsymbol{\Theta}_0$ of $\boldsymbol{\theta}_0$ such that for all n and $\boldsymbol{\theta} \in \boldsymbol{\Theta}_0$, and almost all $t \in [0,\tau]$, the partial derivatives of $\lambda(t,\boldsymbol{\theta})$ of the first, second, and third order with respect to $\boldsymbol{\theta}$ exist and are continuous in $\boldsymbol{\theta}$ for $\boldsymbol{\theta} \in \boldsymbol{\Theta}_0$. Moreover, they are bounded in $[0,\tau] \times \boldsymbol{\Theta}_0$ and the log-likelihood function in (3.70) may be differentiated three times with respect to $\boldsymbol{\theta} \in \boldsymbol{\Theta}_0$ by interchanging the order of integration and differentiation;
(2) $\lambda(t,\boldsymbol{\theta})$ is bounded away from zero in $[0,\tau] \times \boldsymbol{\Theta}_0$;
(3) There exists a positive deterministic function $y(t)$ such that

$$\sup_{t\in[0,\tau]} |Y(t)/n - y(t)| \xrightarrow{P} 0;$$

(4) The matrix $\mathbf{i}(\boldsymbol{\theta}_0) = \lim_{n\to\infty} \mathbf{I}(\boldsymbol{\theta}_0)/n$ (the limit exists under Conditions (1)–(3)) is positive definite.

Lemma 3.1. *Under Conditions A, the following convergence holds:*

$$H_n \xrightarrow{d} V \text{ on } D[0,\tau],$$

where V is zero mean Gaussian martingale such that, for all $0 \leqslant s \leqslant t$,

$$\mathbf{Cov}(V(s),V(t)) = A(s) - \mathbf{C}^T(s)\mathbf{i}^{-1}(\boldsymbol{\theta}_0)\mathbf{C}(t), \tag{3.79}$$

with

$$A(t) = \int_0^t \lambda(u,\boldsymbol{\theta}_0)y(u)du, \quad \mathbf{C}(t) = \int_0^t \frac{\partial}{\partial\boldsymbol{\theta}} \ln \lambda(u,\boldsymbol{\theta})\lambda(u,\boldsymbol{\theta})\bigg|_{\boldsymbol{\theta}=\boldsymbol{\theta}_0} y(u)du,$$

and $\xrightarrow{d}$ *means weak convergence in the space* $D[0,\tau]$ *of cadlag functions with Skorokhod metric.*

The proof is given, for example, in Bagdonavičius and Nikulin (2011) and Bagdonavičius et al. (2010, 2011a,b).

Let us set, for $i = 1,\ldots,m$ and $j,j' = 1,\ldots,k$,

$$V_j = V(a_j) - V(a_{j-1}), \quad v_{jj'} = \mathbf{Cov}(V_j,V_{j'}),$$
$$A_j = A(a_j) - A(a_{j-1}), \quad \mathbf{C}_j = (C_{1j},\ldots,C_{sj})^T = \mathbf{C}(a_j) - \mathbf{C}(a_{j-1}),$$
$$\mathbf{V} = [v_{jj'}]_{k\times k}, \quad \mathbf{C} = [C_{ij}]_{m\times k},$$

and denote by $\mathbf{A}$ the $k \times k$ diagonal matrix with diagonal elements $A_1,\ldots,A_k$.

From Lemma 3.1, it follows that, under Conditions A, we have the following result.

Assertion 3.1.

$$\mathbf{Z} \xrightarrow{d} \mathbf{Y} \sim N_k(\mathbf{0},\mathbf{V}), \quad \text{as } n \to \infty, \tag{3.80}$$

where

$$\mathbf{V} = \mathbf{A} - \mathbf{C}^T \mathbf{i}^{-1}(\boldsymbol{\theta}_0)\mathbf{C}. \tag{3.81}$$

Remark 3.4. *Set*

$$\mathbf{G} = \mathbf{i} - \mathbf{C}\mathbf{A}^{-1}\mathbf{C}^T. \tag{3.82}$$

The generalized inverse $\mathbf{V}^-$ *of the matrix* $\mathbf{V}$ *in* (3.81) *may be written as*

$$\mathbf{V}^- = \mathbf{A}^{-1} + \mathbf{A}^{-1}\mathbf{C}^T\mathbf{G}^-\mathbf{C}\mathbf{A}^{-1}. \tag{3.83}$$

We see that we need to invert only the diagonal $k \times k$ *matrix* $\mathbf{A}$ *and then find the generalized inverse of the* $m \times m$ *matrix* $\mathbf{G} = \mathbf{i} - \mathbf{C}\mathbf{A}^{-1}\mathbf{C}^T$ *(usually* $m = 1, m = 2$, *or* $m = 3$*).*

Assertion 3.1 then implies the next assertion.

Assertion 3.2. *Under Conditions A, the following estimators of* $A_j, \mathbf{C}_j, \mathbf{i}(\boldsymbol{\theta}_0)$, *and* $\mathbf{V}$ *are consistent:*

$$\widehat{A}_j = \frac{U_j}{n}, \quad \hat{\mathbf{C}}_j = \frac{1}{n}\int_{a_{j-1}}^{a_j} \frac{\partial}{\partial\boldsymbol{\theta}} \ln\lambda(u,\hat{\boldsymbol{\theta}})dN(u),$$

and

$$\hat{\mathbf{i}} = \frac{1}{n}\int_0^{\tau} \frac{\partial}{\partial\boldsymbol{\theta}} \ln\lambda(u,\boldsymbol{\theta}) \left(\frac{\partial}{\partial\boldsymbol{\theta}} \ln\lambda(u,\boldsymbol{\theta})\right)^T \Bigg|_{\boldsymbol{\theta}=\hat{\boldsymbol{\theta}}} dN(u),$$

$$\widehat{\mathbf{V}} = \hat{\mathbf{A}} - \widehat{\mathbf{C}}^T\hat{\mathbf{i}}^{-1}\hat{\mathbf{C}}.$$

Proper demonstrations of Assertions 3.1 and 3.2 can be found, for example, in Bagdonavičius and Nikulin (2011).

Remark 3.5. *The above written estimators may be expressed in the form*

$$\widehat{A}_j = \frac{U_j}{n}, \quad U_j = \sum_{i:X_i\in I_j} \delta_i, \quad \widehat{\mathbf{C}}_j = \frac{1}{n}\sum_{i:X_i\in I_j} \delta_i \frac{\partial}{\partial\boldsymbol{\theta}} \ln\lambda(X_i,\hat{\boldsymbol{\theta}}).$$

$$e_j = \sum_{i:X_i>a_{j-1}} [\Lambda(a_j \wedge X_i;\hat{\boldsymbol{\theta}}) - \Lambda(a_{j-1};\hat{\boldsymbol{\theta}})],$$

$$\hat{\mathbf{i}} = \frac{1}{n}\sum_{i=1}^{n} \delta_i \frac{\partial \ln\lambda(X_i,\hat{\boldsymbol{\theta}})}{\partial\boldsymbol{\theta}} \left(\frac{\partial \ln\lambda(X_i,\hat{\boldsymbol{\theta}})}{\partial\boldsymbol{\theta}}\right)^T.$$

Then, Assertions 3.1 and 3.2 imply that a test for the hypothesis H_0 can be based on the statistic

$$Y^2 = \mathbf{Z}^T \widehat{\mathbf{V}}^- \mathbf{Z}, \tag{3.84}$$

where $\widehat{\mathbf{V}}^-$ is the generalized inverse of the matrix $\widehat{\mathbf{V}}$. Using the expressions

$$\widehat{\mathbf{V}}^- = \widehat{\mathbf{A}}^{-1} + \widehat{\mathbf{A}}^{-1}\widehat{\mathbf{C}}^T\widehat{\mathbf{G}}^-\widehat{\mathbf{C}}\widehat{\mathbf{A}}^{-1} \text{and} \widehat{\mathbf{G}} = \hat{\mathbf{i}} - \widehat{\mathbf{C}}\widehat{\mathbf{A}}^{-1}\widehat{\mathbf{C}}^T,$$

and the definition of $\mathbf{Z}$ in (3.77), the test statistic can be expressed as

$$Y^2 = \sum_{j=1}^{k} \frac{(U_j - e_j)^2}{U_j} + Q, \tag{3.85}$$

where

$$Q = \mathbf{W}^T \widehat{\mathbf{G}}^- \mathbf{W}, \quad \mathbf{W} = \widehat{\mathbf{C}}\widehat{\mathbf{A}}^{-1}\mathbf{Z} = (W_1, \ldots, W_m)^T,$$

$$\widehat{\mathbf{G}} = [\hat{g}_{ll'}]_{m\times m}, \quad \hat{g}_{ll'} = \hat{i}_{ll'} - \sum_{j=1}^{k} \widehat{C}_{lj}\widehat{C}_{l'j}\widehat{A}_j^{-1}, \quad W_l = \sum_{j=1}^{k} \widehat{C}_{lj}\widehat{A}_j^{-1}Z_j,$$

$$\hat{i}_{ll'} = \frac{1}{n}\sum_{i=1}^{n} \delta_i \frac{\partial \ln \lambda(X_i,\hat{\boldsymbol{\theta}})}{\partial \theta_l} \frac{\partial \ln \lambda(X_i,\hat{\boldsymbol{\theta}})}{\partial \theta_{l'}},$$

$$\widehat{C}_{lj} = \frac{1}{n}\sum_{i:X_i \in I_j} \delta_i \frac{\partial}{\partial \theta_l} \ln \lambda(X_i,\hat{\boldsymbol{\theta}}), \quad j = 1,\ldots,k; \; l,l' = 1,\ldots,m.$$

From results on the distributions of quadratic forms (see, for example, Rao, 2002 and Greenwood and Nikulin, 1996), the limiting distribution of the statistic Y_n^2 is chi-square with its degrees of freedom as

$$r = rank\mathbf{V}^- = Tr(\mathbf{V}^-\mathbf{V}).$$

A formal proof of this limiting results remains open, however. A careful study of the limiting distribution of Y^2 in (3.85) and its power is warranted. The null hypothesis is rejected with approximate significance level α if $Y_n^2 > \chi_\alpha^2(r)$ or $Y_n^2 < \chi_{1-\alpha}^2(r)$ depending on an alternative, where $\chi_\alpha^2(r)$ and $\chi_{1-\alpha}^2(r)$ are the upper and lower α percentage points of the χ_r^2 distribution, respectively.

If $\mathbf{G}$ is non-degenerate, then $r = k$ since, by using the equality $\mathbf{G} = \mathbf{i} - \mathbf{CA}^{-1}\mathbf{C}^T$, we obtain

$$\begin{aligned}\mathbf{V}^-\mathbf{V} &= (\mathbf{A}^{-1} + \mathbf{A}^{-1}\mathbf{C}^T\mathbf{G}^{-1}\mathbf{CA}^{-1})(\mathbf{A} - \mathbf{C}^T\mathbf{i}^{-1}\mathbf{C}) \\ &= \mathbf{I} - \mathbf{A}^{-1}\mathbf{C}^T(\mathbf{i}^{-1} - \mathbf{G}^{-1}\mathbf{G}\mathbf{i}^{-1})\mathbf{C} = \mathbf{I},\end{aligned}$$

so that $Tr(\mathbf{V}^-\mathbf{V}) = k$, as desired.

Remark 3.6. *The matrix* $\mathbf{G} = \mathbf{i} - \mathbf{C}\mathbf{A}^{-1}\mathbf{C}^T$ *is degenerate if and only if the vector-function*

$$\boldsymbol{\psi}(t,\boldsymbol{\theta}) = \frac{\partial}{\partial \boldsymbol{\theta}} \ln \lambda(t,\boldsymbol{\theta})$$

is such that there exists a vector $\mathbf{x} = (x_1,\ldots,x_m) \neq (0,\ldots,0)^T$ *with* $\mathbf{x}^T \boldsymbol{\psi}(t,\boldsymbol{\theta}_0)$ *being constant on each interval* I_j *(see Hjort, 1990).*

Remark 3.7. *Replacing* $\boldsymbol{\psi}(t,\boldsymbol{\theta}_0)$ *by* $\boldsymbol{\psi}(t,\hat{\boldsymbol{\theta}})$ *and* $y(u)\lambda(u,\boldsymbol{\theta}_0)du$ *by* $dN(u)$ *in the expressions of* $\mathbf{i},\mathbf{C}$ *and* $\mathbf{A}$, *we get* $\hat{\mathbf{i}},\widehat{\mathbf{C}}$ *and* $\widehat{\mathbf{A}}$, *respectively. So, the matrix* $\widehat{\mathbf{G}} = \hat{\mathbf{i}} - \widehat{\mathbf{C}}\widehat{\mathbf{A}}^{-1}\widehat{\mathbf{C}}^T$ *is degenerate if and only if the vector-function* $\boldsymbol{\psi}(t,\hat{\boldsymbol{\theta}})$ *is such that there exists a vector* $\mathbf{x} = (x_1,\ldots,x_m) \neq (0,\ldots,0)^T$ *with* $\mathbf{x}^T\boldsymbol{\psi}(X_i,\hat{\boldsymbol{\theta}})$ *being the same for all* $X_i \in I_j$ *such that* $\delta_i = 1$.

Remark 3.8. *As in the complete sample case, it is recommended to take* a_j *as random data functions. The idea is to divide the interval* $[0,\tau]$ *into k intervals with equal expected numbers of failures (which need not be integers). It seems that it is better to divide* $[0,\tau]$ *into intervals with equal estimated probabilities under the model because in most cases, most of the right censored times will be concentrated on the right side of the data, and so small number of failures or no failures may be observed in the end intervals.*

Define

$$E_k = \int_0^\tau \lambda(u,\hat{\boldsymbol{\theta}})Y(u)du = \sum_{i=1}^n [\Lambda(X_i,\hat{\boldsymbol{\theta}}) - \Lambda(D_i,\hat{\boldsymbol{\theta}})],$$

$$E_j = \frac{j}{k}E_k, \quad j = 1,\ldots,k. \tag{3.86}$$

So, we seek $\hat{a}_j$ satisfying the equality

$$g(\hat{a}_j) = E_j, \text{ with } g(a) = \int_0^a \lambda(u,\hat{\boldsymbol{\theta}})Y(u)du. \tag{3.87}$$

It is easy to prove that the limiting distribution of the test statistic does not change in this case.

Denote by $X_{(1)} \leqslant \cdots \leqslant X_{(n)}$ the ordered values from $X_1,\ldots,X_n$. Note that the function

$$g(a) = \sum\nolimits_{i=1}^n \Lambda(X_i \wedge a,\hat{\boldsymbol{\theta}})$$
$$= \sum\nolimits_{i=1}^n \left[(n-i+1)\Lambda(a,\hat{\boldsymbol{\theta}}) + \sum_{l=1}^{i-1} \Lambda(X_{(l)},\hat{\boldsymbol{\theta}})\right] \mathbf{1}_{[X_{(i-1)},X_{(i)}]}(a)$$

is continuous and increasing in $[0,\tau]$; here, $X_{(0)} \equiv 0$, and we use the convention that $\sum_{l=1}^{0} c_l \equiv 0$. Set

$$b_i = (n-i)\Lambda(X_{(i)},\hat{\boldsymbol{\theta}}) + \sum_{l=1}^{i} \Lambda(X_{(l)},\hat{\boldsymbol{\theta}}).$$

If i is the smallest natural number satisfying $E_j \in [b_{i-1},b_i]$, then the equality in (3.87) can be expressed as

$$(n-i+1)\Lambda(a,\hat{\boldsymbol{\theta}}) + \sum_{l=1}^{i-1} \Lambda(X_{(l)},\hat{\boldsymbol{\theta}}) = E_j,$$

and so if $E_j \in [b_{i-1},b_i], j = 1,\ldots,k-1$, then we have

$$\hat{a}_j = \Lambda^{-1}\left((E_j - \sum_{l=1}^{i-1} \Lambda(X_{(l)},\hat{\boldsymbol{\theta}}))/(n-i+1),\hat{\boldsymbol{\theta}}\right), \quad \hat{a}_k = X_{(n)}, \qquad (3.88)$$

where Λ^{-1} is the inverse of the function Λ. We have $0 < \hat{a}_1 < \hat{a}_2 < \cdots < \hat{a}_k = \tau$.

Under this choice of the intervals, we have $E_j = E_k/k$ for any j. Taking into account that

$$\frac{E_j}{n} \xrightarrow{P} \frac{j}{k} \int_0^{\tau} \lambda(u,\boldsymbol{\theta}_0)y(u)du,$$

we have $\hat{a}_j \xrightarrow{P} a_j$, where a_j is defined by

$$\int_0^{a_j} \lambda(u,\boldsymbol{\theta}_0)y(u)du = \frac{j}{k} \int_0^{\tau} \lambda(t,\boldsymbol{\theta}_0)y(u)du.$$

Thus, replacing a_j by $\hat{a}_j$ in the expression of the statistic Y^2, the limiting distribution of the statistic Y^2 is still chi-squared with r degrees of freedom, as in the case of fixed a_j.

3.8.4 Examples

Example 3.81 (Exponential distribution). Let us consider the null hypothesis

$$H_0 : F(t) = 1 - e^{-\lambda t}, \quad t \geqslant 0, \, \lambda > 0,$$

i.e. the distribution of the failure times is exponential, where $\lambda > 0$ is an unknown scale parameter.

Under the hypothesis H_0, we have

$$S(t;\theta,\nu) = \exp\{-\lambda t\}, \quad \Lambda(t;\theta,\nu) = \lambda t,$$
$$\lambda(t;\lambda) = \lambda, \quad \ln\lambda(t;\theta,\nu) = \ln\lambda.$$

Setting $S_n = \sum_{i=1}^{n} X_i$, we have

$$\hat{\lambda} = \delta/S_n, \quad U_j = \sum_{i:X_i\in I_j} \delta_i, \quad \widehat{C}_j = \frac{U_j}{n\hat{\lambda}},$$

$$\hat{i} = \frac{\delta}{n\hat{\lambda}^2}, \quad \widehat{G} = \hat{g}_{11} = \frac{\delta}{n\hat{\lambda}^2} - \sum_{j=1}^{k} \frac{U_j^2}{n^2\hat{\lambda}^2}\frac{n}{U_j} = 0.$$

Remarks 3.6 and 3.7 also imply that $G = g_{11}$ and $\hat{G} = \hat{g}_{11}$ are degenerate, i.e. equal to zero. We cannot therefore consider general inverse of $G = 0$ and $\hat{G} = 0$, and so we find the general inverse $\widehat{\mathbf{V}}^-$ of the matrix $\widehat{\mathbf{V}}$ directly.

Under the exponential distribution, the elements of the matrix $\widehat{\mathbf{V}}$ are

$$\hat{v}_{jj} = \widehat{A}_j - \widehat{C}_j^2\hat{i}^{-1} = \frac{U_j}{n} - \frac{U_j^2}{n\delta},$$

and

$$v_{jj'} = -\widehat{C}_j\hat{i}^{-1}\widehat{C}_{j'} = -\frac{U_jU_{j'}}{n\delta}\text{for} j \neq j'.$$

Now, set

$$\hat{\pi}_j = \frac{U_j}{\delta}, \quad \sum_{j=1}^{k}\hat{\pi}_j = 1, \quad \hat{\boldsymbol{\pi}} = (\hat{\pi}_1,\ldots,\hat{\pi}_k)^T.$$

Denote by $\mathbf{D}$ the diagonal matrix with its diagonal elements as $\hat{\pi}$. The matrix $\widehat{\mathbf{V}}$ and its generalized inverse $\widehat{\mathbf{V}}^-$ are then of the form

$$\widehat{\mathbf{V}} = \frac{\delta}{n}(\hat{\mathbf{D}} - \hat{\boldsymbol{\pi}}\hat{\boldsymbol{\pi}}^T), \quad \widehat{\mathbf{V}}^- = \frac{n}{\delta}(\widehat{\mathbf{D}}^{-1} + \mathbf{1}\mathbf{1}^T),$$

and due to the equalities

$$\mathbf{1}^T\widehat{\mathbf{D}} = \hat{\boldsymbol{\pi}}^T, \quad \mathbf{1}^T\hat{\boldsymbol{\pi}} = \hat{\boldsymbol{\pi}}^T\mathbf{1} = 1, \quad \hat{\mathbf{D}}\mathbf{1} = \hat{\boldsymbol{\pi}}, \quad \hat{\boldsymbol{\pi}}^T\widehat{\mathbf{D}}^{-1} = \mathbf{1}^T,$$

we obtain $\widehat{\mathbf{V}}\widehat{\mathbf{V}}^-\widehat{\mathbf{V}} = \widehat{\mathbf{V}}$.

The quadratic form $Y^2 = \mathbf{Z}^T\widehat{\mathbf{V}}^-\mathbf{Z}$ has the form

$$Y^2 = \frac{n}{\delta}\mathbf{Z}^T\widehat{\mathbf{D}}^{-1}\mathbf{Z} + \frac{n}{\delta}(\mathbf{Z}^T\mathbf{1})^2 = \sum_{j=1}^{k}\frac{(U_j - e_j)^2}{U_j} + \frac{1}{\delta}\left[\sum_{j=1}^{k}(U_j - e_j)\right]^2.$$

Under the hypothesis H_0, the limiting distribution of the statistic Y^2 is chi-square with $Tr(\mathbf{V}^-\mathbf{V}) = k-1$ degrees of freedom, because

$$\widehat{A}_j \xrightarrow{P} A_j > 0, \quad \delta/n \xrightarrow{P} A = \sum_{j=1}^{k} A_j \in (0,1), \quad \hat{\pi}_j \xrightarrow{P} A_j/A = \pi_j, \quad \widehat{\mathbf{D}} \xrightarrow{P} \mathbf{D},$$

so

$$\mathbf{V}^-\mathbf{V} = \frac{1}{A}(\mathbf{D}^{-1} + \mathbf{1}\mathbf{1}^T)A(\mathbf{D} - \boldsymbol{\pi}\boldsymbol{\pi}^T) = \mathbf{I} - \mathbf{1}\boldsymbol{\pi}^T,$$

$$Tr(\mathbf{I} - \mathbf{1}\boldsymbol{\pi}^T) = k - \sum_{j=1}^{k} \pi_j = k-1.$$

But, a careful study of the limiting distribution of Y^2 and its power is warranted. Note that

$$\sum\nolimits_{j=1}^{k} e_j = \hat{\lambda} \int_0^{\tau} Y(u)du = \hat{\lambda} \sum_{i=1}^{n} X_i = \hat{\lambda} S_n = \delta = \sum_{j=1}^{k} U_j.$$

So

$$Q = \frac{1}{\delta}\left[\sum_{j=1}^{k} (U_j - e_j)\right]^2 = 0.$$

Remark 3.9 (About the choice of $\hat{a}_j$). *Set*

$$S_0 \equiv 0, \quad S_i = \sum_{l=1}^{i} X_{(l)} + (n-i)X_{(i)}, \; i = 1, \ldots, n.$$

Then, Remark 3.8 implies that the limits of the intervals I_j are chosen in the following way: if i is the smallest natural number such that

$$S_{i-1} \leqslant \frac{j}{k} S_n \leqslant S_i,$$

then

$$\hat{a}_j = \frac{\frac{j}{k} S_n - \sum_{l=1}^{i-1} X_{(l)}}{n-i+1}, \quad j = 1, \ldots, k-1, \textit{ and } \hat{a}_k = X_{(n)}.$$

The number of expected failures in all intervals are equal in this case, and is

$$e_j = \delta/k \quad \textit{for any } j.$$

It is evident that the null hypothesis will be rejected with approximate significance level α if $Y^2 = \sum_{j=1}^{k} \frac{(U_j - e_j)^2}{U_j} > \chi^2_{\alpha}(k-1)$or $Y^2 = \sum_{j=1}^{k} \frac{(U_j - e_j)^2}{U_j} < \chi^2_{1-\alpha}(k-1)$ depending on an alternative, where $\chi^2_{\alpha}(k-1)$ and $\chi^2_{1-\alpha}(k-1)$ are the upper and lower α percentage points of the χ^2_{k-1} distribution, respectively.

Example 3.82 (Shape and scale families). Let us consider the hypothesis

$$H_0 : F(t) = F_0((t/\theta)^{\nu}), \quad \theta > 0, \nu > 0,$$

meaning that the lifetime distribution belongs to a specified shape and scale family, with F_0 being a specified distribution function, and θ and ν being the unknown scale and shape parameters, respectively.

Set

$$S_0 = 1 - F_0, \quad \Lambda_0 = -\ln S_0, \quad \lambda_0 = -S_0'/S_0.$$

Under the hypothesis H_0, we then have

$$S(t;\theta,\nu) = S_0\left\{\left(\frac{t}{\theta}\right)^{\nu}\right\}, \quad \Lambda(t;\theta,\nu) = \Lambda_0\left\{\left(\frac{t}{\theta}\right)^{\nu}\right\},$$
$$\lambda(t;\theta,\nu) = \frac{\nu}{\theta^{\nu}} t^{\nu-1} \lambda_0\left\{\left(\frac{t}{\theta}\right)^{\nu}\right\},$$
$$\ln\lambda(t;\theta,\nu) = (\nu-1)\ln t - \nu\ln\theta + \ln\nu + \ln\lambda_0\left\{\left(\frac{t}{\theta}\right)^{\nu}\right\}.$$

Denote by $\hat{\theta}$ and $\hat{\nu}$ the ML estimators of the parameters θ and ν, respectively. These estimators maximize the log-likelihood function

$$\ell(\theta,\nu) = \sum_{i=1}^{n} \left\{\delta_i \left[(\nu-1)\ln X_i - \nu\ln\theta + \ln\nu + \ln\lambda_0\left(\left(\frac{X_i}{\theta}\right)^{\nu}\right)\right] - \Lambda_0\left(\left(\frac{X_i}{\theta}\right)^{\nu}\right)\right\}.$$

The estimator $\hat{\boldsymbol{i}} = [\hat{i}_{ls}]_{2\times 2}$ has the form

$$\hat{i}_{11} = \frac{\hat{\nu}^2}{n\hat{\theta}^2} \sum_{i=1}^{n} \delta_i [1 + Y_i g_0(Y_i)]^2,$$

$$\hat{i}_{12} = -\frac{1}{n\hat{\theta}} \sum_{i=1}^{n} \delta_i [1 + Y_i g_0(Y_i)][1 + \ln Y_i (1 + Y_i g_0(Y_i))],$$

$$\hat{i}_{22} = \frac{1}{n\hat{\nu}^2} \sum_{i=1}^{n} \delta_i [1 + \ln Y_i (1 + Y_i g_0(Y_i))]^2,$$

where

$$Y_i = \left(\frac{X_i}{\hat{\theta}}\right)^{\hat{\nu}}, \quad g_0 = (\ln \lambda_0)'.$$

The test statistic for testing H_0 is given by

$$Y^2 = \sum_{j=1}^{k} \frac{(U_j - e_j)^2}{U_j} + Q,$$

where

$$Q = \mathbf{W}^T \widehat{\mathbf{G}}^- \mathbf{W}, \quad \mathbf{W} = (W_1, W_2)^T,$$

$$\widehat{\mathbf{G}} = [\hat{g}_{ii'}]_{2\times 2}, \quad \hat{g}_{ii'} = \hat{i}_{ii'} - \sum_{j=1}^{k} \widehat{C}_{ij} \widehat{C}_{i'j} \widehat{A}_j^{-1},$$

$$W_i = \sum\nolimits_{j=1}^{k} \widehat{C}_{ij} \widehat{A}_j^{-1} Z_j, \quad i,i' = 1,2,$$

$$\widehat{A}_j = U_j/n, \quad \widehat{C}_{1j} = -\frac{\hat{\nu}}{n\hat{\theta}} \sum_{i:X_i \in I_j} \delta_i (1 + Y_i g_0(Y_i)),$$

$$\widehat{C}_{2j} = \frac{1}{n\hat{\nu}} \sum_{i:X_i \in I_j} \delta_i [1 + \ln Y_i (1 + Y_i g_0(Y_i))], \quad Z_j = \frac{1}{\sqrt{n}} (U_j - e_j).$$

A careful study of the limiting distribution of Y^2 and its power is warranted. To construct $\hat{a}_j$, we set

$$b_i = (n - i)\Lambda_0(Y_i) + \sum_{l=1}^{i} \Lambda_0(Y_{(l)}).$$

If i is the smallest natural number such that $E_j \in [b_{i-1}, b_i]$, then

$$\hat{a}_j = \hat{\theta} \left\{ \Lambda_0^{-1} \left((E_j - \sum_{l=1}^{i-1} \Lambda_0(Y_{(l)}))/(n - i + 1) \right) \right\}^{1/\hat{\nu}} \quad \text{and } \hat{a}_k = X_{(n)},$$

where Λ_0^{-1} is the inverse of the function Λ_0.

Under such a choice of the intervals, we have $e_j = E_k/k$ for any j, where $E_k = \sum_{i=1}^{n} \Lambda_0(Y_i)$. Accordingly, the chi-squared test for testing the null hypothesis H_0 will reject the null hypothesis with approximate significance level α if $Y^2 > \chi^2_\alpha(r)$ or $Y^2 < \chi^2_{1-\alpha}(r)$ depending on an alternative, where $\chi^2_\alpha(r)$ and $\chi^2_{1-\alpha}(r)$ are the upper and the lower α percentage points of the χ^2_r distribution, respectively.

The degrees of freedom r in this case is given by

$$r = rank(\mathbf{V}^-) = Tr(\mathbf{V}^-\mathbf{V}), \quad \mathbf{V} = \mathbf{A} - \mathbf{C}^T\mathbf{i}^{-1}\mathbf{C},$$
$$\mathbf{V}^- = \mathbf{A}^{-1} + \mathbf{A}^{-1}\mathbf{C}^T\mathbf{G}^-\mathbf{C}\mathbf{A}^{-1}.$$

It is evident that if $\mathbf{G}$ is non-degenerate, then $r = k$.

Example 3.83 (Chi-squared test for the Weibull distribution). Let us consider the hypothesis

$$H_0 : F(t) = 1 - e^{-(t/\theta)^{\nu}}, t \geqslant 0,$$

which means the lifetime distribution is Weibull; here, $\theta > 0$ and $\nu > 0$ are the unknown scale and shape parameters, respectively. Under the null hypothesis, we have

$$S(t;\theta,\nu) = \exp\left\{-\left(\frac{t}{\theta}\right)^{\nu}\right\}, \quad \Lambda(t;\theta,\nu) = \left(\frac{t}{\theta}\right)^{\nu},$$
$$\lambda(t;\theta,\nu) = \frac{\nu}{\theta^{\nu}}t^{\nu-1}, \quad \ln\lambda(t;\theta,\nu) = (\nu-1)\ln t - \nu\ln\theta + \ln\nu.$$

Denote by $\hat{\theta}$ and $\hat{\nu}$ the ML estimators of the parameters θ and ν, respectively. These estimators then maximize the log-likelihood function

$$\ell(\theta,\nu) = \sum_{i=1}^{n}\{\delta_i[(\nu-1)\ln X_i - \nu\ln\theta + \ln\nu] - (X_i/\theta)^{\nu}\}.$$

For the existence and uniqueness of the ML estimators $\hat{\theta}$ and $\hat{\nu}$, one may refer to Balakrishnan and Kateri (2008). In this case, to construct $\hat{a}_j$, we set

$$Y_i = \left(\frac{X_i}{\hat{\theta}}\right)^{\hat{\nu}}, \quad b_0 = 0, \quad b_i = \sum_{l=1}^{i} Y_{(l)} + (n-i)Y_{(i)}, \; i = 1,\ldots,n.$$

Then, Remark 3.8 implies that the limits of the intervals I_j are chosen in the following way: if i is the smallest natural number such that

$$b_{i-1} \leqslant \frac{j}{k}b_n \leqslant b_i,$$

then

$$\hat{a}_j = \hat{\theta}\left(\frac{\frac{j}{k}b_n - \sum_{l=1}^{i-1} Y_{(l)}}{n-i+1}\right)^{1/\hat{\nu}}, \quad j = 1,\ldots,k-1, \text{ and } \hat{a}_k = X_{(n)}.$$

For such a choice of the intervals, we have

$$e_j = \delta/k \quad \text{for all } j.$$

Remark 3.8 implies that the matrices $\mathbf{G}$ and $\widehat{\mathbf{G}}$ are degenerate and have rank 1, since

$$g_{11} = g_{12} = \hat{g}_{11} = \hat{g}_{12} = 0.$$

So, we need only $\hat{g}_{22}$ to find $\hat{G}^-$. We have

$$\hat{i}_{22} = \frac{1}{n\hat{\nu}^2}\sum_{i=1}^{n}\delta_i[1+\ln Y_i]^2, \quad \widehat{C}_{2j} = \frac{1}{n\hat{\nu}}\sum_{i:X_i\in I_j}\delta_i[1+\ln Y_i],$$

$$\hat{g}_{22} = \hat{i}_{22} - \sum_{j=1}^{k}\widehat{C}_{2j}^2\widehat{A}_j^{-1}, \quad \widehat{A}_j = U_j/n.$$

Then, the matrix $\hat{\mathbf{G}}^-$ has the form

$$\widehat{\mathbf{G}}^- = \begin{pmatrix} 0 & 0 \\ 0 & \hat{g}_{22}^{-1} \end{pmatrix},$$

and consequently

$$Q = \frac{W_2^2}{\hat{g}_{22}}, \quad W_2 = \sum_{j=1}^{k}\widehat{C}_{2j}\widehat{A}_j^{-1}Z_j, \quad Z_j = \frac{1}{\sqrt{n}}(U_j - e_j).$$

According to the chi-squared test based on the statistic Y^2 for testing H_0, the null hypothesis will be rejected with approximate significance level α if $X^2 = \sum_{j=1}^{k}\frac{(U_j-e_j)^2}{U_j} + Q > \chi^2_\alpha(k-1)$ or $X^2 = \sum_{j=1}^{k}\frac{(U_j-e_j)^2}{U_j} + Q < \chi^2_{1-\alpha}(k-1)$ depending on an alternative, where $\chi^2_\alpha(k-1)$ and $\chi^2_{1-\alpha}(k-1)$ are the upper and lower α percentage points of the χ^2_{k-1} distribution, respectively. In fact, a careful study of the limiting distribution of X^2 and its power is warranted.

Example 3.84 (Chi-squared test for the Power Generalized Weibull distribution). Following Bagdonavičius et al. (2010), we suppose that under H_0, the distribution of failure times is Power Generalized Weibull (see Bagdonavičius and Nikulin, 2002), with its survival function

$$S(t,\theta,\nu,\gamma) = \exp\left\{1 - \left(1 + \left(\frac{t}{\theta}\right)^{\nu}\right)^{1/\gamma}\right\}, \quad t \geqslant 0, \quad \theta,\nu,\gamma > 0,$$

the cumulative hazard function

$$\Lambda(t,\theta,\nu,\gamma) = -\left\{1 - \left(1 + \left(\frac{t}{\theta}\right)^{\nu}\right)^{1/\gamma}\right\},$$

and the hazard function

$$\lambda(t,\theta,\nu,\gamma) = \frac{\nu}{\gamma\theta^{\nu}}t^{\nu-1}\left(1 + \left(\frac{t}{\theta}\right)^{\nu}\right)^{(1/\gamma)-1}.$$

The hazard function has following properties:

If $\nu > 1, \nu > \gamma$, then the hazard rate increases from 0 to ∞;
If $\nu = 1, \gamma < 1$, then the hazard rate increases from $(\gamma\theta)^{-1}$ to ∞;
If $0 < \nu < 1, \nu < \gamma$, then the hazard rate decreases from ∞ to 0;
If $0 < \nu < 1, \nu = \gamma$, then the hazard rate decreases from ∞ to $1/\theta$;
If $\gamma > \nu > 1$, then the hazard rate increases from 0 to its maximum value and then decreases to 0, that is, it is $\bigcap$-shaped;
If $0 < \gamma < \nu < 1$, then the hazard rate decreases from ∞ to its minimum value and then increases to ∞, that is, it is $\bigcup$-shaped.

It is known that all the moments of this distribution are finite.

The likelihood function can be expressed as

$$L = \prod_{i=1}^{n} [\lambda(t_i,\theta,\nu,\gamma)]^{\delta_i} [S(t_i,\theta,\nu,\gamma)],$$

with δ_i being the censoring indicator defined by

$$\delta_i = \begin{cases} 1 \text{ if } X_i < C_i \\ 0 \text{ otherwise.} \end{cases}$$

The log-likelihood function in this case is

$$\ell = \ln L = \sum\nolimits_{i=1}^{n} \delta_i \left\{ \ln \nu - \ln \gamma - \nu \ln \theta + (\nu - 1) \ln t_i \right.$$
$$\left. + \left(\frac{1}{\gamma} - 1\right) \ln (1 + Y_i) \right\} + n - \sum_{i=1}^{n} (1 + Y_i)^{1/\gamma},$$

where $Y_i = (t_i/\theta)^{\nu}$. The chi-squared test statistic in this case is

$$Y_n^2 = \sum_{j=1}^{k} \frac{(U_j - e_j)^2}{U_j} + Q,$$

where

$$U_j = \sum\nolimits_{i:X_i \in I_j} \delta_i,$$
$$e_j = \sum\nolimits_{i:X_i > a_{J-1}} \left[\Lambda\left(\left(\frac{a_j \wedge X_i}{\hat{\theta}} \right)^{\hat{\nu}} \right) - \Lambda\left(\frac{a_{j-1}}{\hat{\theta}} \right)^{\hat{\nu}} \right],$$
$$Q = \sum\nolimits_{l=1}^{m} \sum_{l'=1}^{m} W_l g^{ll'} W_{l'},$$

with m being the number of parameters in the lifetime distribution, and

$$W_l = \sum_{j=1}^{k} \widehat{C}_{lj} \widehat{A}_j^{-1} Z_j, \quad \hat{g}_{ll'} = \hat{i}_{ll'} - \sum_{j=1}^{k} \widehat{C}_{lj} \widehat{C}_{l'j} \widehat{A}_j^{-1},$$
$$[\hat{g}^{ll'}]_{m \times m} = [\hat{g}_{ll'}]_{m \times m}^{-1}, \quad \widehat{A}_j = U_j/n,$$

$$\widehat{C}_j = \frac{1}{n} \sum_{i:X_i \in I_j} \delta_i \frac{\partial}{\partial \theta} \ln \lambda(X_i; \hat{\theta}), \qquad \widehat{C}_j = (\widehat{C}_{1j}, \widehat{C}_{2j}, \widehat{C}_{3j})^T,$$

$$\widehat{C}_{1j} = -\frac{\hat{\nu}}{n\hat{\theta}} \sum_{i:X_i \in I_j} \delta_i \left\{ 1 + (\frac{1}{\hat{\gamma}} - 1) \frac{Y_i}{1+Y_i} \right\},$$

$$\widehat{C}_{2j} = \frac{1}{n\hat{\nu}} \sum_{i:X_i \in I_j} \delta_i \left\{ 1 + \ln Y_i + (\frac{1}{\hat{\gamma}} - 1) \frac{Y_i \ln Y_i}{1+Y_i} \right\},$$

$$\widehat{C}_{3j} = -\frac{1}{n\hat{\gamma}} \sum_{i:X_i \in I_j} \delta_i \left\{ 1 + \frac{1}{\hat{\gamma}} \ln(1+Y_i) \right\}.$$

Since

$$\hat{\mathbf{i}}_{ll'} = \frac{1}{n} \sum_{i=1}^{n} \delta_i \frac{\partial \ln \lambda(X_i; \hat{\boldsymbol{\theta}})}{\partial \theta_l} \frac{\partial \ln \lambda(X_i; \hat{\boldsymbol{\theta}})}{\partial \theta_{l'}},$$

we have the elements of the symmetric matrix $\hat{\mathbf{i}}$ as

$$\hat{i}_{11} = \frac{\hat{\nu}}{n\hat{\theta}^2} \sum_{i=1}^{n} \delta_i \left\{ 1 + \left(\frac{1}{\hat{\gamma}} - 1 \right) \frac{Y_i}{1+Y_i} \right\}^2,$$

$$\hat{i}_{22} = \frac{1}{n\hat{\nu}^2} \sum_{i=1}^{n} \delta_i \left\{ 1 + \ln Y_i + \left(\frac{1}{\hat{\gamma}} - 1 \right) \frac{Y_i \ln Y_i}{1+Y_i} \right\}^2,$$

$$\hat{i}_{33} = \frac{1}{n\hat{\gamma}^2} \sum_{i=1}^{n} \delta_i \left\{ 1 + \frac{1}{\hat{\gamma}} \ln(1+Y_i) \right\}^2,$$

$$\hat{i}_{12} = -\frac{1}{n\hat{\theta}} \sum_{i=1}^{n} \delta_i \left\{ 1 + \left(\frac{1}{\hat{\gamma}} - 1 \right) \frac{Y_i}{1+Y_i} \right\} \left\{ 1 + \ln Y_i + \left(\frac{1}{\hat{\gamma}} - 1 \right) \frac{Y_i \ln Y_i}{1+Y_i} \right\},$$

$$\hat{i}_{13} = \frac{\hat{\nu}}{n\hat{\theta}\hat{\gamma}} \sum_{i=1}^{n} \delta_i \left\{ 1 + \left(\frac{1}{\hat{\gamma}} - 1 \right) \frac{Y_i}{1+Y_i} \right\} \left\{ 1 + \frac{1}{\hat{\gamma}} \ln(1+Y_i) \right\},$$

$$\hat{i}_{23} = -\frac{1}{n\hat{\gamma}\hat{\nu}} \sum_{i=1}^{n} \delta_i \left\{ 1 + \ln Y_i + \left(\frac{1}{\hat{\gamma}} - 1 \right) \frac{Y_i \ln Y_i}{1+Y_i} \right\} \left\{ 1 + \frac{1}{\hat{\gamma}} \ln(1+Y_i) \right\}.$$

In fact, a careful study of the limiting distribution of Y_n^2 and its power is warranted.

Evidently, by setting $\gamma = 1$, we can deduce the elements of the estimator $\hat{\mathbf{i}} = [\hat{i}_{ll'}]_{2\times 2}$ and $\widehat{\mathbf{C}}_j$ for the case of the Weibull distribution with survival function

$$S(t; \theta, \nu) = \exp\left\{ -\left(\frac{t}{\theta} \right)^{\nu} \right\},$$

and in this case

$$\hat{i}_{11} = \frac{\hat{\nu}}{n\hat{\theta}^2}\left\{-\delta + (\hat{\nu}+1)\sum_{i=1}^{n} Y_i\right\},$$

$$\hat{i}_{12} = \frac{1}{n\hat{\theta}}\left\{\delta - \sum_{i=1}^{n} Y_i\left(1+\ln Y_i\right)\right\},$$

$$\hat{i}_{22} = \frac{1}{n\hat{\nu}^2}\left\{\delta + \sum_{i=1}^{n} Y_i \ln^2 Y_i\right\},$$

where

$$\sum\nolimits_{i=1}^{n}\delta_i = \delta,\quad \widehat{\mathbf{C}}_j = (\widehat{C}_{1j},\widehat{C}_{2j})^T,$$

$$\widehat{C}_{1j} = -\frac{\hat{\nu}}{\hat{\theta}}\frac{U_j}{n},\quad \widehat{C}_{2j} = \frac{1}{n\hat{\nu}}\sum_{i:X_i\in I_j}\delta_i\left\{1+\ln Y_i\right\}.$$

Example 3.85 (Chi-squared test for Birnbaum-Saunders distribution). Birnbaum and Saunders (1969a,b) proposed a two-parameter distribution as a fatigue life distribution with unimodal hazard rate function; see Kundu et al. (2008) for a formal proof of this property and some inferential issues associated with it. The cumulative distribution function of this distribution is

$$F(t;\alpha,\beta) = \Phi\left[\frac{1}{\alpha}\left\{\left(\frac{t}{\beta}\right)^{\frac{1}{2}} - \left(\frac{\beta}{t}\right)^{\frac{1}{2}}\right\}\right],\quad 0<t<\infty,\quad \alpha,\beta>0,$$

where α is the shape parameter, β is the scale parameter, and $\Phi(x)$ is the standard normal distribution function. The corresponding probability density function is

$$f(t;\alpha,\beta) = \frac{1}{2\sqrt{2\pi}\alpha\beta}\left\{\left(\frac{\beta}{t}\right)^{\frac{1}{2}} + \left(\frac{\beta}{t}\right)^{\frac{3}{2}}\right\}\exp\left[-\frac{1}{2\alpha^2}\left(\frac{t}{\beta}+\frac{\beta}{t}-2\right)\right],$$
$$0<t<\infty,\quad \alpha,\beta>0.$$

The existence and uniqueness of the ML estimators $\hat{\beta}$ and $\hat{\alpha}$, of the parameters β and α, based on complete and censored samples have been established by Birnbaum and Saunders (1969b) and Balakrishnan and Zhu (2012), respectively. In this case, from the log-likelihood function, we can estimate the Fisher information matrix by using the equality

$$\hat{\boldsymbol{i}}_{ll'} = \frac{1}{n}\sum_{i=1}^{n}\delta_i\frac{\partial\ln\lambda(X_i;\hat{\boldsymbol{\theta}})}{\partial\boldsymbol{\theta}_l}\frac{\partial\ln\lambda(X_i;\hat{\boldsymbol{\theta}})}{\partial\boldsymbol{\theta}_{l'}}.$$

The elements of the above Fisher information matrix are as follows:

$$\hat{\boldsymbol{i}}_{11} = \frac{1}{n}\sum_{i=1}^{n}\delta_i\left[-\frac{1}{\alpha}+\frac{1}{\alpha^3}\left(\frac{t_i}{\beta}+\frac{\beta}{t_i}-2\right)\right.$$

$$+\left.\frac{\Phi'_\alpha\left(\frac{1}{\alpha}\left\{\left(\frac{t_i}{\beta}\right)^{\frac{1}{2}}-\left(\frac{\beta}{t_i}\right)^{\frac{1}{2}}\right\}\right)}{1-\Phi\left(\frac{1}{\alpha}\left\{\left(\frac{t_i}{\beta}\right)^{\frac{1}{2}}-\left(\frac{\beta}{t_i}\right)^{\frac{1}{2}}\right\}\right)}\right]^2,$$

$$\hat{\boldsymbol{i}}_{22}=\frac{1}{n}\sum_{i=1}^{n}\delta_i\left[-\frac{1}{\beta}+\frac{\frac{1}{2t}\left\{\left(\frac{t_i}{\beta}\right)^{\frac{1}{2}}+3\left(\frac{\beta}{t_i}\right)^{\frac{1}{2}}\right\}}{\left(\frac{\beta}{t_i}\right)^{\frac{1}{2}}+\left(\frac{\beta}{t_i}\right)^{\frac{3}{2}}}-\frac{1}{2\alpha^2}\left(-\frac{t_i}{\beta^2}+\frac{1}{t_i}\right)\right.$$

$$+\left.\frac{\Phi'_\beta\left(\frac{1}{\alpha}\left\{\left(\frac{t_i}{\beta}\right)^{\frac{1}{2}}-\left(\frac{\beta}{t_i}\right)^{\frac{1}{2}}\right\}\right)}{1-\Phi\left(\frac{1}{\alpha}\left\{\left(\frac{t_i}{\beta}\right)^{\frac{1}{2}}-\left(\frac{\beta}{t_i}\right)^{\frac{1}{2}}\right\}\right)}\right]^2,$$

$$\hat{\boldsymbol{i}}_{12}=\frac{1}{n}\sum_{i=1}^{n}\delta_i\left[-\frac{1}{\alpha}+\frac{1}{\alpha^3}\left(\frac{t_i}{\beta}+\frac{\beta}{t_i}-2\right)\right.$$

$$+\left.\frac{\Phi'_\alpha\left(\frac{1}{\alpha}\left\{\left(\frac{t_i}{\beta}\right)^{\frac{1}{2}}-\left(\frac{\beta}{t_i}\right)^{\frac{1}{2}}\right\}\right)}{1-\Phi\left(\frac{1}{\alpha}\left\{\left(\frac{t_i}{\beta}\right)^{\frac{1}{2}}-\left(\frac{\beta}{t_i}\right)^{\frac{1}{2}}\right\}\right)}\right]$$

$$\times\left[-\frac{1}{\beta}+\frac{\frac{1}{2t_i}\left\{\left(\frac{t_i}{\beta}\right)^{\frac{1}{2}}+3\left(\frac{\beta}{t_i}\right)^{\frac{1}{2}}\right\}}{\left(\frac{\beta}{t_i}\right)^{\frac{1}{2}}+\left(\frac{\beta}{t_i}\right)^{\frac{3}{2}}}-\frac{1}{2\alpha^2}\left(-\frac{t_i}{\beta^2}+\frac{1}{t_i}\right)\right.$$

$$+\left.\frac{\Phi'_\beta\left(\frac{1}{\alpha}\left\{\left(\frac{t_i}{\beta}\right)^{\frac{1}{2}}-\left(\frac{\beta}{t_i}\right)^{\frac{1}{2}}\right\}\right)}{1-\Phi\left(\frac{1}{\alpha}\left\{\left(\frac{t_i}{\beta}\right)^{\frac{1}{2}}-\left(\frac{\beta}{t_i}\right)^{\frac{1}{2}}\right\}\right)}\right],$$

where $\Phi'(x)=\varphi(x)=\frac{1}{\sqrt{2\pi}}e^{-x^2/2}$, so that we have

$$\Phi'_\alpha\left(\frac{1}{\alpha}\left\{\left(\frac{t_i}{\beta}\right)^{\frac{1}{2}}-\left(\frac{\beta}{t_i}\right)^{\frac{1}{2}}\right\}\right)$$

$$=-\frac{1}{\sqrt{2\pi}\alpha^2}\left\{\left(\frac{t_i}{\beta}\right)^{\frac{1}{2}}-\left(\frac{\beta}{t_i}\right)^{\frac{1}{2}}\right\}\exp\left[-\frac{1}{2\alpha^2}\left(\frac{t_i}{\beta}+\frac{\beta}{t_i}-2\right)\right],$$

$$\Phi'_\beta\left(\frac{1}{\alpha}\left\{\left(\frac{t_i}{\beta}\right)^{\frac{1}{2}}-\left(\frac{\beta}{t_i}\right)^{\frac{1}{2}}\right\}\right)$$

$$= -\frac{1}{2\sqrt{2\pi}\alpha t_i}\left\{\left(\frac{t_i}{\beta}\right)^{\frac{3}{2}} + \left(\frac{t_i}{\beta}\right)^{\frac{1}{2}}\right\}\exp\left[-\frac{1}{2\alpha^2}\left(\frac{t_i}{\beta}+\frac{\beta}{t_i}-2\right)\right].$$

Moreover, we have

$$\widehat{\mathbf{C}}_{lj} = \frac{1}{n}\sum_{i:X_i\in I_j}\delta_i\frac{\partial}{\partial\boldsymbol{\theta}}\ln\lambda(X_i,\hat{\boldsymbol{\theta}}),$$

where

$$\widehat{\mathbf{C}}_{1j} = \frac{1}{n}\sum_{i:X_i\in I_j}\delta_i\left[-\frac{1}{\alpha}+\frac{1}{\alpha^3}\left(\frac{t_i}{\beta}+\frac{\beta}{t_i}-2\right)+\frac{\Phi'_\alpha\left(\frac{1}{\alpha}\left\{\left(\frac{t_i}{\beta}\right)^{\frac{1}{2}}-\left(\frac{\beta}{t_i}\right)^{\frac{1}{2}}\right\}\right)}{1-\Phi\left(\frac{1}{\alpha}\left\{\left(\frac{t_i}{\beta}\right)^{\frac{1}{2}}-\left(\frac{\beta}{t_i}\right)^{\frac{1}{2}}\right\}\right)}\right],$$

$$\widehat{\mathbf{C}}_{2j} = \frac{1}{n}\sum_{i:X_i\in I_j}\delta_i\left[-\frac{1}{\beta}+\frac{\frac{1}{2t_i}\left\{\left(\frac{t_i}{\beta}\right)^{\frac{1}{2}}+3\left(\frac{\beta}{t_i}\right)^{\frac{1}{2}}\right\}}{\left(\frac{\beta}{t_i}\right)^{\frac{1}{2}}+\left(\frac{\beta}{t_i}\right)^{\frac{3}{2}}}-\frac{1}{2\alpha^2}\left(-\frac{t_i}{\beta^2}+\frac{1}{t_i}\right)\right.$$

$$\left.+\frac{\Phi'_\beta\left(\frac{1}{\alpha}\left\{\left(\frac{t_i}{\beta}\right)^{\frac{1}{2}}-\left(\frac{\beta}{t_i}\right)^{\frac{1}{2}}\right\}\right)}{1-\Phi\left(\frac{1}{\alpha}\left\{\left(\frac{t_i}{\beta}\right)^{\frac{1}{2}}-\left(\frac{\beta}{t_i}\right)^{\frac{1}{2}}\right\}\right)}\right].$$

3.9 TESTING NORMALITY FOR SOME CLASSICAL DATA ON PHYSICAL CONSTANTS

The classical measurements from Cavendish's determination of the mean density of the earth (relative to that of water), from Millikan's determinations of the charge of electron, and from Michelson and Newcomb's measurements of the velocity of light have all been analyzed by a number of authors including Stigler (1977) and Moore (1984). The main interest of Stigler was to compare 11 different estimators of location parameters by using real data. He concluded that the simple sample mean "compares favorably" with all other estimators considered by him. If an experiment is well organized and all systematic effects are excluded, the measurements, being a sum of numerous independent random effects, should approximately follow a normal distribution. Though Stigler (1977) did not use statistical tests to check for normality of the data, he did comment "that the data set considered tend to have slightly heavier tails than the normal, but that a view of the world through Cauchy-colored glasses may be overly-pessimistic." Moore (1984) carried out a test of normality on the data compiled by Stigler (1977) by using the test of Moore (1971) and Dahiya and

TABLE 3.6 Values of Cavendish's determinations of the density of earth.

5.50	5.55	5.57	5.34	5.42	5.30
5.61	5.36	5.53	5.79	5.47	5.75
4.88	5.29	5.62	5.10	5.63	5.68
5.07	5.58	5.29	5.27	5.34	5.85
5.26	5.65	5.44	5.39	5.46	

Gurland (1972a), viz. the MDG test statistic χ^2_{MDG}, that is, defined in (2.19) with specially constructed grouping intervals (see Section 2.6) and a measure X^2/n for the degree of lack of fit. The main interest of Moore (1984) was to check for the normality of the data using not only the level of significance but also using the measure of degree of lack of fit, comparing its behavior for real and simulated data. To simplify his analysis, Moore used the Pearson-Fisher test with random cells (Moore, 1971) instead of the more powerful NRR statistic. Here, we revisit this analysis by using not only the NRR test, but also the more powerful test $S_n^2(\hat{\boldsymbol{\theta}}_n)$, proposed first by McCulloch (1985), which is easily implemented through Microsoft Excel and VBA codes. In the subsequent analysis, the data compiled by Stigler (1977) and the data from Linnik (1958) are used for this purpose.

3.9.1 Cavendish's measurements

The 29 measurements that Cavendish made on the density of earth (Stigler, 1977, p. 1076) are reproduced in Table 3.6. The boldfaced value of 4.88 in this table deserves some additional discussion. Stigler (1977) noted that "Cavendish erred in taking the mean (5.48) of all 29 (measurements) by treating the value 4.88 as if it were in fact 5.88." He also added that this error of Cavendish was first mentioned and corrected by Bailey in 1843 who gave the mean density of 5.448. Of course, we may consider 4.88 as an outlier, but it seems that 4.88 is a real measurement of Cavendish. Keeping this in mind, Stigler examined the robustness of estimators of the mean for two versions of data from Table 3.6, with and without replacing 4.88 by 5.88.

The main question in any statistical analysis of data is whether we could consider the data as realizations of i.i.d. random variables. To answer this, one has to investigate at least the correlation between measurements and randomness of the data. The sample autocorrelation function of the data in Table 3.6 is shown in Figure 3.12. The non parametric runs test for these data gives a P-value of 0.453, from which we may conclude that Cavendish's data do not provide evidence against the hypothesis that they are realizations of i.i.d. random variables. At this point, it is worth mentioning that the sample autocorrelation function of the data in Table 3.6, with 4.88 replaced by 5.88, is only slightly

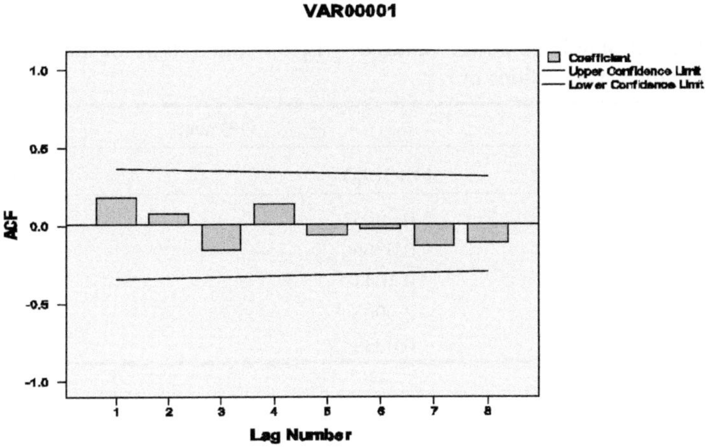

FIGURE 3.12 The sample ACF for the data in Table 3.6.

TABLE 3.7 Values of $Y1_n^2(\hat{\theta}_n)$ and $S_n^2(\hat{\theta}_n)$, and the corresponding P-values, as functions of r. The observation 4.88 was retained in the data.

				P-values	
r	$\chi^2_{r-1}(0.05)$	$Y1_n^2(\hat{\theta}_n)$	$S_n^2(\hat{\theta}_n)$	$Y1_n^2(\hat{\theta}_n)$	$S_n^2(\hat{\theta}_n)$
4	7.815	**0.585**	**0.521**	0.900	0.771
5	9.488	**2.807**	**1.760**	0.591	0.415
6	11.070	**4.796**	**2.460**	0.441	0.292
7	12.592	**2.239**	**1.804**	0.896	0.406
8	14.067	**1.875**	**0.321**	0.966	0.852

different from that of Figure 3.12. The corresponding P-value of the runs test in this case is 0.132 which also does not contradict the null hypothesis that those data are realizations of i.i.d. random variables.

In Table 3.7, the values of the NRR test $Y1_n^2(\hat{\boldsymbol{\theta}}_n)$ and the test $S_n^2(\hat{\boldsymbol{\theta}}_n)$, and the P-values of these tests, as functions of the number of equiprobable intervals used, are given. Here, $\hat{\boldsymbol{\theta}}_n$ stands for the MLE of $\boldsymbol{\theta} = (\theta_1,\theta_2)^T$, where $\theta_1 = \mu$ and $\theta_2 = \sigma^2$ of the normal distribution. The values of statistics presented in bold mean that the null hypothesis is not rejected at level 0.05.

Taking into account that for the test $S_n^2(\hat{\boldsymbol{\theta}}_n)$, the number of degrees of freedom is 2, and $\chi_2^2(0.05) = 5.991$, we observe that the test does not reject the

TABLE 3.8 Simulated power values of $Y1_n^2(\hat{\theta}_n)$ and $S_n^2(\hat{\theta}_n)$ for the logistic alternative, as functions of *r*.

	Power	
r	$Y1_n^2(\hat{\theta}_n)$	$S_n^2(\hat{\theta}_n)$
4	0.0903	0.0884
5	0.0966	0.0930
6	0.1041	0.1056
7	0.0966	0.1053
8	0.0935	0.1107

hypothesis of the normal distribution for Cavendish's determinations of the density of earth. Incidentally, using the MDG test with $r = 7$ intervals, Moore (1984) obtained the same result. It is now of interest to compare the power of $Y1_n^2(\hat{\boldsymbol{\theta}}_n), S_n^2(\hat{\boldsymbol{\theta}}_n)$, and χ^2_{MDG} tests for the logistic alternative which is quite close to the normal null hypothesis. With $n = 29, \hat{\mu} = 5.448, \hat{\sigma} = 0.2171$ (see also Stigler, 1977), and the simulated critical value of χ^2_{MDG} test for $\alpha = 0.05$ which is 10.038 as found in Table 2 of Dahiya and Gurland (1972a), the power is determined to be $\mathbf{P}(\chi^2_{MDG} > 10.038) = 0.059$. This is only slightly higher than the nominal level of the test. Based on $N = 10{,}000$ simulations, the powers of NRR and $S_n^2(\hat{\boldsymbol{\theta}}_n)$ tests were determined under the same conditions and these values are presented in Table 3.8.

From Table 3.8, we see that the power of the NRR and $S_n^2(\hat{\boldsymbol{\theta}}_n)$ tests, when the number of equiprobable cells $r = 7$, are higher than that of the χ^2_{MDG} test. Moreover, the power of the $S_n^2(\hat{\boldsymbol{\theta}}_n)$ test is higher than that of the NRR test for $r = 6,7,8$. It is worth mentioning here that the DN test possesses no power at all in this case. So, the most powerful $S_n^2(\hat{\boldsymbol{\theta}})$ test for the logistic alternative confirms Moore's (1984) conclusion that the normal distribution fits Cavendish's data very well.

From the viewpoint of robustness of modified chi-squared tests, it is of interest to calculate $Y1_n^2(\hat{\boldsymbol{\theta}}_n)$ and $S_n^2(\hat{\boldsymbol{\theta}}_n)$ statistics when 4.88 is replaced by 5.88. These results are presented in Table 3.9.

From Table 3.9, we see that if we replace the value 4.88 by 5.88, the result is obtained to be the same, and these data do not contradict normality either. From the results in Tables 3.7 and 3.9, it also follows that the NRR and $S_n^2(\hat{\boldsymbol{\theta}}_n)$ tests are robust to the presence of an "outlier."

3.9.2 Millikan's measurements

Millikan obtained 58 values for the charge of the electron, and these are presented in Table 3.10. They are reproduced from the book of Linnik (1958).

TABLE 3.9 Values of $Y1_n^2(\hat{\theta}_n)$ and $S_n^2(\hat{\theta}_n)$, and the corresponding *P*-values, as functions of *r*. The observation 5.88 is used in place of 4.88.

				P-values	
r	$\chi^2_{r-1}(0.05)$	$Y1_n^2(\hat{\theta}_n)$	$S_n^2(\hat{\theta}_n)$	$Y1_n^2(\hat{\theta}_n)$	$S_n^2(\hat{\theta}_n)$
4	7.815	**1.831**	**1.827**	0.608	0.401
5	9.488	**1.867**	**1.190**	0.760	0.551
6	11.070	**2.399**	**0.172**	0.792	0.918
7	12.592	**3.229**	**0.172**	0.780	0.918
8	14.067	**6.540**	**1.138**	0.478	0.566

TABLE 3.10 Millikan's values of the charge of the electron.

4.781	4.764	4.777	4.809	4.761	4.769
4.795	4.776	4.765	4.790	4.792	4.806
4.769	4.771	4.785	4.779	4.758	4.779
4.792	4.789	4.805	4.788	4.764	4.785
4.779	4.772	4.768	4.772	4.810	4.790
4.775	4.789	4.801	4.791	4.799	4.777
4.772	4.764	4.785	4.788	4.799	4.749
4.791	4.774	4.783	4.783	4.797	4.781
4.782	4.778	4.808	4.740	4.790	
4.767	4.791	4.771	4.775	4.747	

The sample autocorrelation function for the data in Table 3.10 is displayed in Figure 3.13. The non parametric runs test for these data gives a *P*-value of 0.784, from which we may conclude that Millikan's data do not contradict the hypothesis that they are realizations of i.i.d. random variables. In Table 3.11, the values of the NRR test $Y1_n^2(\hat{\boldsymbol{\theta}}_n)$ and the test $S_n^2(\hat{\boldsymbol{\theta}}_n)$, and the *P*-values of these tests, as functions of the number of equiprobable random intervals used, are presented. Here, $\hat{\boldsymbol{\theta}}_n$ denotes the MLE of $\boldsymbol{\theta} = (\theta_1,\theta_2)^T$, where $\theta_1 = \mu$ and $\theta_2 = \sigma^2$ of the normal distribution. For Millikan's data in Table 3.10, we find $\hat{\theta}_1 = 4.7808$ and $\hat{\theta}_2 = \hat{\sigma}^2 = 0.00023$.

Taking into account that, for the test $S_n^2(\hat{\boldsymbol{\theta}}_n)$, the number of degrees of freedom is 2, and $\chi_2^2(0.05) = 5.991$, we observe that the test does not reject the hypothesis of the normal distribution for Millikan's determinations of the charge of the electron.

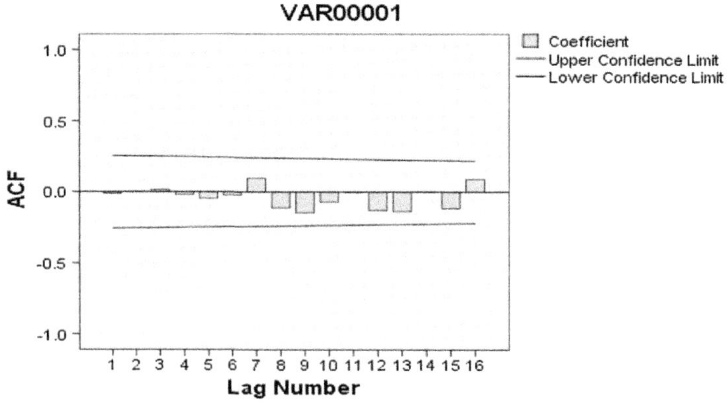

FIGURE 3.13 The sample ACF for the data in Table 3.10.

TABLE 3.11 Values of $Y1_n^2(\hat{\theta}_n)$ and $S_n^2(\hat{\theta}_n)$, and the corresponding *P*-values, as functions of *r*.

				P-values	
r	$\chi_{r-1}^2(0.05)$	$Y1_n^2(\hat{\theta}_n)$	$S_n^2(\hat{\theta}_n)$	$Y1_n^2(\hat{\theta}_n)$	$S_n^2(\hat{\theta}_n)$
4	7.815	**2.100**	**1.757**	0.552	0.415
8	14.067	**6.199**	**3.202**	0.517	0.202
10	16.919	**6.448**	**1.791**	0.694	0.408
15	23.685	**13.959**	**3.775**	0.453	0.151

TABLE 3.12 Simulated power of $Y1_n^2(\hat{\theta}_n)$ and $S_n^2(\hat{\theta}_n)$ tests for the logistic alternative, as functions of *r*.

	Power	
r	$Y1_n^2(\hat{\theta}_n)$	$S_n^2(\hat{\theta}_n)$
4	0.122	0.125
8	0.133	0.169
10	0.139	0.178
15	0.128	0.207

TABLE 3.13 The values below +299000 are Michelson's determinations of the velocity of light in km/s.

850	960	880	890	890	740	940	880	810	840
900	960	880	810	780	1070	940	860	820	810
930	880	720	800	760	850	800	720	770	810
950	850	620	760	790	980	880	860	740	810
980	900	970	750	820	880	840	950	760	850
1000	830	880	910	870	980	790	910	920	870
930	810	850	890	810	650	880	870	860	740
760	880	840	880	810	810	830	840	720	940
1000	800	850	840	950	1000	790	840	850	800
960	760	840	850	810	960	800	840	780	870

Based on $N = 10{,}000$ simulations, the powers of NRR and $S_n^2(\hat{\boldsymbol{\theta}}_n)$ tests for the logistic alternative were determined and these values are presented in Table 3.12.

From Table 3.12, we see that the maximum gain in power for the $S_n^2(\hat{\boldsymbol{\theta}}_n)$ test compared to that of the NRR test for the logistic alternative is $0.207/0.128 = 1.617$. This ratio is higher than the ratio $0.1107/0.0935 = 1.184$ found from Table 3.8. Evidently, the gain in power increases with an increase in the number of measurements (see Figure 3.1).

3.9.3 Michelson's measurements

Michelson's data of 100 determinations of the velocity of light (see Stigler, 1977, Table 6) are reproduced here in Table 3.13.

The sample autocorrelation function for the data in Table 3.13 is displayed in Figure 3.14. This sample ACF shows a statistically significant positive

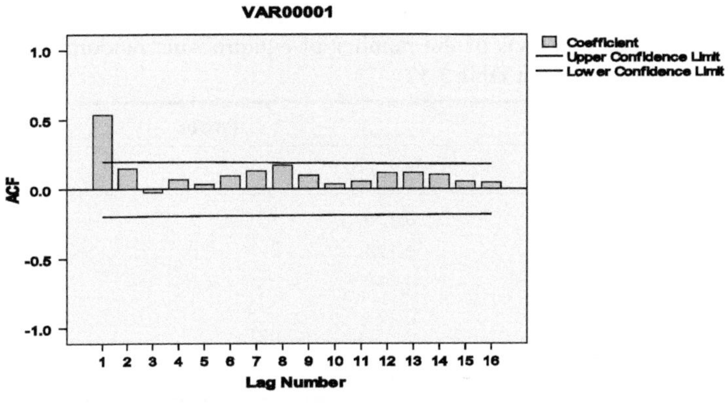

FIGURE 3.14 The sample ACF for the data in Table 3.13.

TABLE 3.14 Values of $Y1_n^2(\hat{\theta}_n)$ and $S_n^2(\hat{\theta}_n)$, and the corresponding P-values, as functions of r, for Michelson's data in Table 3.13.

				P-values	
r	$\chi^2_{r-1}(0.05)$	$Y1_n^2(\hat{\theta}_n)$	$S_n^2(\hat{\theta}_n)$	$Y1_n^2(\hat{\theta}_n)$	$S_n^2(\hat{\theta}_n)$
7	12.591	**10.325**	**3.054**	0.112	0.217
8	14.067	**12.990**	**0.503**	0.072	0.778
10	16.919	20.171	**0.180**	0.017	0.914
15	23.685	**17.852**	**0.901**	0.214	0.637
20	30.143	46.133	**2.626**	0.001	0.269
25	36.415	49.005	**1.147**	0.002	0.563

correlation of lag 1 and not a statistically significant but systematic positive correlation for other lags. The fact of "positive dependence among the observations" was mentioned earlier by Moore (1984). From this and the fact that the P-value of the runs test is close to zero, we may conclude that Michelson's measurements cannot be considered as realizations of i.i.d. random variables. If, however, we proceed to apply the NRR test $Y1_n^2(\hat{\theta}_n)$ and the test $S_n^2(\hat{\theta}_n)$ for these data, we obtain the results as presented in Table 3.14. From Table 3.14, we observe that the NRR test rejects the null hypothesis for $r = 10,20,25$. Also, the more powerful $S_n^2(\hat{\theta}_n)$ test does not reject it for all values of equiprobable random intervals r considered.

Based on $N = 10{,}000$ simulations, the power of NRR and $S_n^2(\hat{\theta}_n)$ tests for the logistic alternative for Michelson's data (with $\hat{\mu} = 852.4$ and $\hat{\sigma} = 78.6145$) were determined and these results are presented in Table 3.15.

TABLE 3.15 Simulated power of $Y1_n^2(\hat{\theta}_n)$ and $S_n^2(\hat{\theta}_n)$ for the logistic alternative, as functions of the number of equiprobable random cells r, for Michelson's data in Table 3.13.

	Power	
r	$Y1_n^2(\hat{\theta}_n)$	$S_n^2(\hat{\theta}_n)$
7	0.178	0.231
8	0.174	0.233
10	0.177	0.253
15	0.173	0.269
20	0.176	0.288
25	0.173	0.374

From Table 3.15, we observe that for the logistic alternative the power of $S_n^2(\hat{\boldsymbol{\theta}}_n)$ is 1.64 times more than that of the NRR test if $r = 20$, and is 2.16 times more if $r = 25$. Using the MDG test χ^2_{MDG} with $r = 7$ intervals, we obtain $\chi^2_{MDG} = 7.52$ (see also Moore, 1984) which does not reject the hypothesis of normality at level $\alpha = 0.05$. Based on $N = 10{,}000$ simulations and the critical value of 10.03 of the χ^2_{MDG} test for $\alpha = 0.05$ level, we obtain the power $\mathbf{P}(\chi^2_{MDG} > 10.03) = 0.108$ which 2.14 times less than the power of 0.231 of the $S_n^2(\hat{\boldsymbol{\theta}}_n)$ test; see Table 3.15.

Moore (1982) commented that "for testing of fit to a special normal law, it is shown that when observations come from a quite general class of Gaussian stationary processes, positive correlation among the observations is confounded with lack of normality." Gleser and Moore (1983) further mentioned that "confounding of positive dependence with lack of fit is a general phenomenon in use of omnibus tests of fit." This is especially so with the modified chi-squared tests of the form $V^T(\hat{\boldsymbol{\theta}}_n)W_n(\hat{\boldsymbol{\theta}}_n)V(\hat{\boldsymbol{\theta}}_n)$ where $W_n(\hat{\boldsymbol{\theta}}_n)$ converges to $W_n(\boldsymbol{\theta}_0)$ in probability. Evidently, the $S_n^2(\hat{\boldsymbol{\theta}}_n)$ test belongs to this class, but it does not confound the positive dependence with lack of fit with the null hypothesis formulated under an assumption that observations are realizations of i.i.d. random variables. This contradiction can possibly be explained by the fact that Gleser and Moore (1983) did not take into account the power of the tests used.

3.9.4 Newcomb's measurements

Newcomb made a series of measurements on the passage of time for light to pass over a distance of 3721 meters and back, from Fort Myer on the west bank of the Potomac to a fixed mirror at the base of the Washington monument. These data are presented in Table 3.16. The sample autocorrelation function for the

TABLE 3.16 Newcomb's measurements on the passage time for light, taken from Stigler (1977, Table 5).

28	29	24	37	36	26	29
26	22	20	25	23	32	27
33	24	36	28	27	32	28
24	21	32	26	27	24	29
34	25	36	30	28	39	16
−44	30	28	32	27	28	23
27	23	25	36	31	24	
16	29	21	26	27	25	
40	31	28	30	26	32	
−2	19	29	22	33	25	

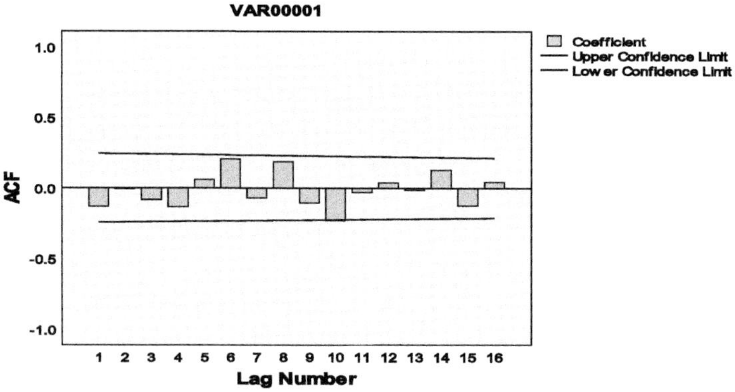

FIGURE 3.15 The sample ACF for the data in Table 3.16.

data in Table 3.16 is displayed in Figure 3.15, which does not show significant correlation among Newcomb's measurements. Since the P-value of the runs test is 0.078 in this case, we may conclude that, at level 0.05, these data do not contradict the hypothesis that they are realizations of i.i.d. random variables. The MDG test, with $r = 11$ intervals and all 66 observations in Table 3.16, yields $\chi^2_{MDG} = 43.333$ that rejects the hypothesis of normality; refer to Table 1 of Moore (1984). The same result is obtained by the usage of the NRR and $S_n^2(\hat{\theta}_n)$ tests as well. Moore (1984) noted that there is "one egregious outlier (-44), which Newcomb himself eliminated." After the exclusion of that outlier, and with $r = 7$ and $n = 65$, we obtain the value of $\chi^2_{MDG} = 7.908$ using which we do not reject the hypothesis of normality. Due to the low power of the χ^2_{MDG}, we may also apply the NRR and $S_n^2(\hat{\theta}_n)$ tests, with $r = 7,8,10,12,15$ equiprobable random cells and $n = 65$. The results so obtained are presented in Table 3.17.

TABLE 3.17 Values of $Y1_n^2(\hat{\theta}_n)$ and $S_n^2(\hat{\theta}_n)$, as functions of r, for 65 measurements from Newcomb's data in Table 3.16.

r	$\chi^2_{r-1}(0.05)$	$Y1_n^2(\hat{\theta}_n)$	$S_n^2(\hat{\theta}_n)$
7	12.591	16.079	11.694
8	14.067	**10.855**	7.063
10	16.919	**14.323**	8.345
12	19.675	27.962	11.119
15	23.685	**22.942**	19.826

TABLE 3.18 Values of $Y1_n^2(\hat{\theta}_n)$ and $S_n^2(\hat{\theta}_n)$, as functions of r, for 64 measurements from Newcomb's data in Table 3.16.

r	$\chi^2_{r-1}(0.05)$	$Y1_n^2(\hat{\theta}_n)$	$S_n^2(\hat{\theta}_n)$
7	12.591	**3.228**	**1.587**
8	14.067	**6.056**	**1.144**
10	16.919	**12.989**	**0.119**
12	19.675	**6.414**	**2.294**
15	23.685	**16.099**	**1.032**

From Table 3.17, we see that the NRR test does not reject the null hypothesis for $r = 8,10,15$ while it rejects for $r = 7,12$. Also, the $S_n^2(\hat{\theta}_n)$ tests rejects H_0 for all grouping intervals considered. For this reason, we may consider excluding one more "outlier," from the data, viz. -2. The results obtained from the remaining 64 observations are presented in Table 3.18.

From Table 3.18, we observe that the normal distribution fits the data of these 64 measurements obtained by removing the two smallest outliers from Table 3.16.

3.10 TESTS BASED ON DATA ON STOCK RETURNS OF TWO KAZAKHSTANI COMPANIES

Suppose P_t is the price of a security at time t. Then, $R_t = \ln(P_t/P_{t-1})$ is called the return for period t. It is usually assumed that $R_t, t = 1,2,\cdots$, are independent normal random variables. Though this model has been criticized, "much financial theory has continued to employ the simpler normal model"; see Moore and Stubblebine (1981) and the references therein. A popular opinion is that weekly and monthly returns follow normality very well, but this may not be true for daily returns.

We shall consider here daily and weekly returns for two Kazakhstani companies: Kazkommerts Bank and Kazakhmys (data have been taken from the site: http://finance.yahoo.com, secured on 18.02.2010). Kazkommerts bank is one of the largest private banks in CIS. Kazakhmys is an international natural resources company. It processes, refines, and produces metals such as copper, zinc, gold, and silver. Kazakhmys is also the largest power provider in Kazakhstan. From Figure 3.16, we observe two evidently different patterns—one before the financial crisis started in August 2008 (observation 439 on the graph) and another during the crisis from August 2008 till February 2010. The behavior of daily returns of Kazakhmys is observed to have the same pattern. We observe more or less stable volatility of returns before the crisis. Then, at the

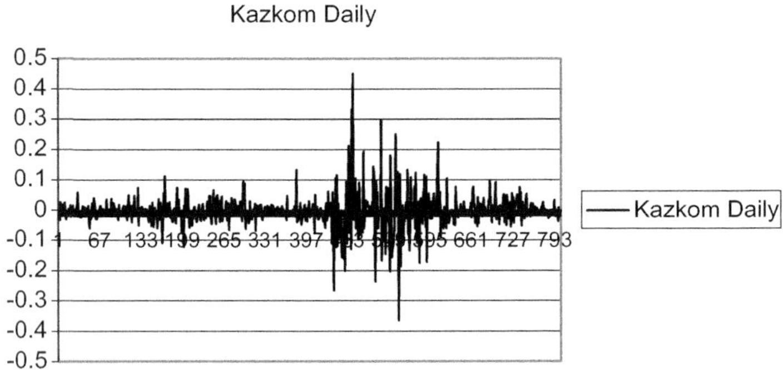

FIGURE 3.16 Daily returns of Kazkommerts bank for the period December 2006–February 2010.

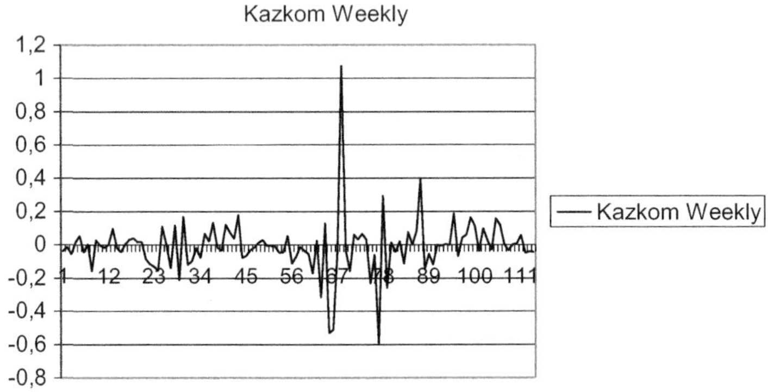

FIGURE 3.17 Weekly returns of Kazkommerts bank for the period December 2006–February 2010.

first stage of the crisis, it has increased much and then has gradually decreased. The same picture is seen for weekly returns in Figure 3.17. For this reason, we shall analyze these data separately for these two periods.

3.10.1 Analysis of daily returns

To be able to use the test statistics discussed in the preceding sections, we have to ensure that the data can be considered as realizations of at least uncorrelated random variables. The nonparametric runs test yielded the following P-values: 0.064 for daily returns of Kazkommerts bank before the crisis, and 0.206 during the crisis. Therefore, at level $\alpha = 0.05$, the hypothesis of randomness is not rejected. The sample ACFs for those two time periods are shown in Figures 3.18 and 3.19, respectively.

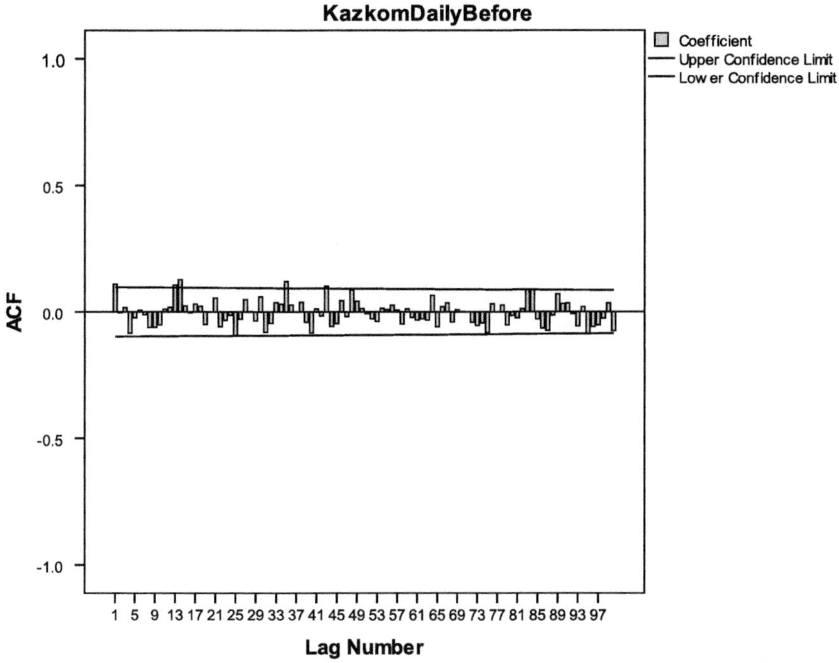

FIGURE 3.18 The sample ACF of Kazkommerts bank daily returns for the period December 2006–August 2008.

Figures 3.18 and 3.19 do not show a significant correlation for almost all lags. For the data from Kazakhmys company, the P-values are found to be 0.434 and 0.530, and consequently, the hypothesis of randomness is not rejected in this case as well. A significant correlation of daily returns is not observed either.

An application of the NRR $Y1_n^2(\hat{\boldsymbol{\theta}}_n)$ and McCh $S_n^2(\hat{\boldsymbol{\theta}}_n)$ tests to 420 observations (from 15.12.2006 till 14.08.2008) of Kazkommerts bank daily returns before the crisis yields nearly zero P-values, thus rejecting normality. But, normality can also be rejected if the process is not stationary. To check for this, we divided the data of 420 observations before crisis into four segments: first 100 values, second 100, third 100, and last 120 values. The standard deviations of daily returns for those segments are found to be 0.018, 0.030, 0.035, and 0.023, respectively. From these, we indeed see that, for Kazkommerts bank's daily returns before the crisis, the process is not stationary. A similar partitioning was done for the remaining 382 observations (from 15.08.2008 till 18.02.2010) of Kazkommerts bank's daily returns during the crisis, with the last segment containing 82 values. In this case, the standard deviations of daily returns for these segments are 0.103, 0.090, 0.032, and 0.024, respectively. Here again, the process cannot be considered as being stationary.

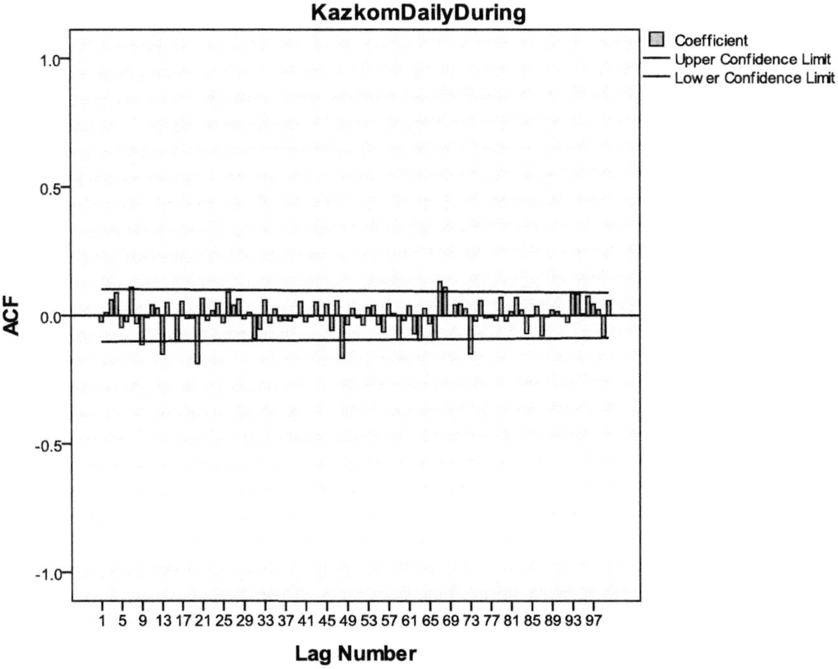

FIGURE 3.19 The sample ACF of Kazkommerts bank daily returns for the period August 2008–February 2010.

An application of the NRR $Y1_n^2(\hat{\theta}_n)$ and McCh $S_n^2(\hat{\theta}_n)$ tests for partitioned data gives P-values as presented in Table 3.19. To make the results comparable, we used in all cases $r = 10$ equiprobable random grouping intervals. From Table 3.19, we observe that for two segments (first and third) the normality of Kazkommerts bank's daily returns before the crisis is not rejected at level $\alpha = 0.05$. The situation changes for the crisis period (see the last two columns of Table 3.19). The NRR test rejects normality for all four segments, while the McCh test rejects at level $\alpha = 0.05$ for the first and fourth segments. In fact,

TABLE 3.19 ***P*-values of $Y1_n^2(\hat{\theta}_n)$ and $S_n^2(\hat{\theta}_n)$ for Kazkom Daily data.**

	Before crisis		During crisis	
	$Y1_n^2(\hat{\theta}_n)$	$S_n^2(\hat{\theta}_n)$	$Y1_n^2(\hat{\theta}_n)$	$S_n^2(\hat{\theta}_n)$
First 100	**0.058**	**0.309**	0.000	0.000
Second 100	0.001	0.002	0.000	**0.071**
Third 100	**0.504**	**0.228**	0.000	0.019
Last 120/82	0.000	0.000	0.000	**0.089**

TABLE 3.20 ***P*-values of $Y1_n^2(\hat{\theta}_n)$ and $S_n^2(\hat{\theta}_n)$ for Kazakhmys Daily data.**

	Before crisis		During crisis	
	$Y1_n^2(\hat{\theta}_n)$	$S_n^2(\hat{\theta}_n)$	$Y1_n^2(\hat{\theta}_n)$	$S_n^2(\hat{\theta}_n)$
First 100	**0.128**	**0.852**	**0.121**	**0.070**
Second 100	**0.502**	**0.192**	**0.774**	**0.584**
Third 100	0.009	**0.403**	**0.570**	**0.701**
Last 120/82	**0.168**	**0.634**	**0.148**	0.007

the McCh test rejects normality for all four segments at level $\alpha = 0.1$. Thus, we observe a clear violation of normality of Kazkommerts bank's daily returns during the crisis period.

An analogous analysis for Kazakhmys' daily returns are given in Table 3.20. From Table 3.20, we observe that normality of daily returns before and during crisis for the Kazakhmys company is not rejected at level $\alpha = 0.05$ in 14 cases

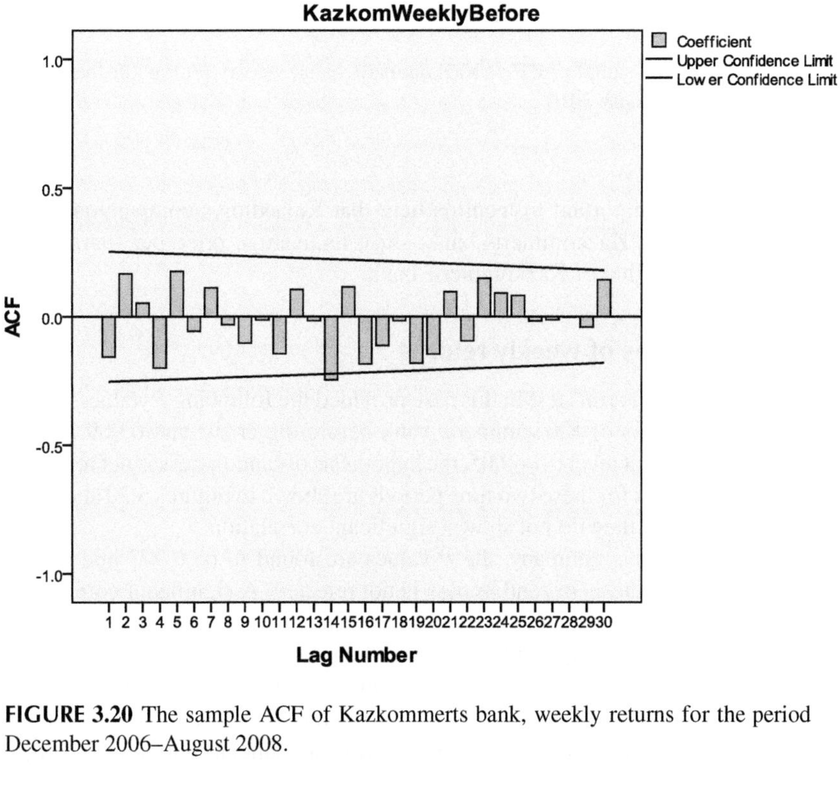

FIGURE 3.20 The sample ACF of Kazkommerts bank, weekly returns for the period December 2006–August 2008.

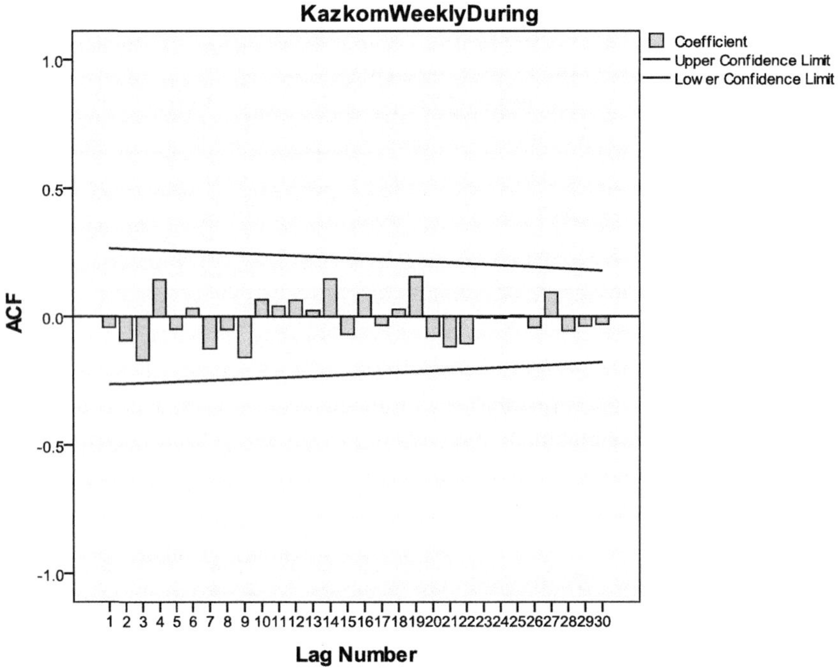

FIGURE 3.21 The sample ACF of Kazkommerts bank, weekly returns for the period August 2008–February 2010.

out of 16. It is important to mention here that Kazakhmys company is much more stable than Kazkommerts bank, since its average price per share is 91 times more than that of Kazkommerts bank.

3.10.2 Analysis of weekly returns

The nonparametric runs test in this case provided the following P-values: 0.120 for weekly returns of Kazkommerts bank before the crisis, and 0.909 for the crisis period. So, at level $\alpha = 0.05$, the hypothesis of randomness is not rejected. The sample ACFs for these two time periods are shown in Figures 3.20 and 3.21, respectively, and they do not show a significant correlation.

For Kazakhmys company, the P-values are found to be 0.972 and 0.252, and so, the hypothesis of randomness is not rejected. A significant correlation of weekly returns is also not observed and we can, therefore, proceed to use the goodness of fit tests discussed in the preceding sections.

The results so obtained are presented in Table 3.21. From Table 3.21, we observe that under normal business conditions, the weekly stock returns follow normality very well. However, during the period of financial crisis, both $Y1_n^2(\hat{\boldsymbol{\theta}}_n)$

TABLE 3.21 ***P*-values of $Y1_n^2(\hat{\theta}_n)$ and $S_n^2(\hat{\theta}_n)$ for Kazkom and Kazakhmys weekly data.**

	Kazkom		Kazakhmys	
	$Y1_n^2(\hat{\theta}_n)$	$S_n^2(\hat{\theta}_n)$	$Y1_n^2(\hat{\theta}_n)$	$S_n^2(\hat{\theta}_n)$
Before crisis	**0.846**	**0.901**	**0.197**	**0.507**
During crisis	0.000	0.000	**0.091**	**0.052**

and $S_n^2(\hat{\boldsymbol{\theta}}_n)$ reject normality for Kazkommerts bank's weekly returns at level $\alpha = 0.05$, but do not reject normality for Kazakhmys' weekly returns. This again confirms the higher stability of Kazakhmys company.

Remark 3.10. *From the above analysis, it may also be noted that if normality is actually present in the data, then the NRR test $Y1_n^2(\hat{\boldsymbol{\theta}}_n)$ rejects it more often than the more powerful and more stable McCh test $S_n^2(\hat{\boldsymbol{\theta}}_n)$ (see Remark 3.3).*

Chapter 4

Wald's Method and Hsuan-Robson-Mirvaliev Test

4.1 WALD'S METHOD AND MOMENT-TYPE ESTIMATORS

Let $X_1, \cdots, X_n$ be i.i.d. random variables following a continuous family of distributions with cdf

$$F(x;\boldsymbol{\theta}), x \in R^1, \boldsymbol{\theta} = (\theta_1, \cdots, \theta_s)^T \in \boldsymbol{\Theta} \subset R^s,$$

and $f(x;\boldsymbol{\theta})$ be the corresponding pdf.

Let a function $\boldsymbol{g}(x) = (g_1(x), \cdots, g_s(x))^T$ be such that the system of s equations

$$\mathbf{m}(\boldsymbol{\theta}) = \overline{\mathbf{g}},$$

where $\overline{\mathbf{g}} = (\overline{g}_1, \cdots, \overline{g}_s)^T, \mathbf{m}(\boldsymbol{\theta}) = (m_1(\boldsymbol{\theta}), \cdots, m_s(\boldsymbol{\theta}))^T, \overline{g}_j = \frac{1}{n}\sum_{i=1}^{n} g_j(X_i)$ for $j = 1, \cdots, s$, and $m_j(\boldsymbol{\theta}) = \int g_j(x) f(x,\boldsymbol{\theta})dx$ is uniquely solved to obtain $\overline{\boldsymbol{\theta}}_n = \mathbf{m}^{-1}(\overline{\mathbf{g}})$. Then, $\overline{\boldsymbol{\theta}}_n$ is the method of moments estimator (MME) for the parameter $\boldsymbol{\theta}$. It is customary to use the functions $g_i(x) = x^i, i = 1, \cdots, s$, even though it may be convenient to use some other functions in some situations; see Balakrishnan and Cohen (1991), for example.

Let $\mathbf{K}$ be a $s \times s$ matrix with its elements as

$$K_{ij}(\boldsymbol{\theta}) = \int x^i \frac{\partial f(x,\boldsymbol{\theta})}{\partial \theta_j} dx, \quad i,j = 1, \cdots, s, \tag{4.1}$$

and $\mathbf{V}$ be a $s \times s$ matrix with its elements as

$$V_{ij}(\boldsymbol{\theta}) = m_{ij}(\boldsymbol{\theta}) - m_i(\boldsymbol{\theta})m_j(\boldsymbol{\theta}), \tag{4.2}$$

Chi-Squared Goodness of Fit Tests with Applications. http://dx.doi.org/10.1016/B978-0-12-397194-4.00004-1

where $m_i(\boldsymbol{\theta}) = \mathbf{E}_{\boldsymbol{\theta}}(X^i)$ and $m_{ij}(\boldsymbol{\theta}) = \mathbf{E}_{\boldsymbol{\theta}}(X^{i+j}), i,j = 1,\cdots,s$, are population moments. Consider the partition of x-axis into r disjoint fixed or random intervals $\Delta_j, j = 1,\ldots,r$. Let $p_j(\boldsymbol{\theta})$ be the probability for the jth interval and consider the $r \times s$ matrix $\mathbf{C}$ with its elements as

$$C_{jk}(\boldsymbol{\theta}) = \frac{1}{\sqrt{p_j(\boldsymbol{\theta})}} \left(\int_{\Delta_j} x^k f(x,\boldsymbol{\theta})dx - p_j(\boldsymbol{\theta}) m_k(\boldsymbol{\theta}) \right), \quad j = 1,\ldots,r,\ k = 1,\cdots,s. \tag{4.3}$$

Hsuan and Robson (1976) showed that, for both fixed and random grouping intervals, the limiting covariance matrix of the vector $\mathbf{V}^{(n)}(\overline{\boldsymbol{\theta}}_n)$ of standardized frequencies with components $v_j(\overline{\boldsymbol{\theta}}_n) = (N_j^{(n)} - np_j(\overline{\boldsymbol{\theta}}_n))/\sqrt{np_j(\overline{\boldsymbol{\theta}}_n)}$, $j = 1,\cdots,r$, is

$$\boldsymbol{\Sigma} = \mathbf{I} - \mathbf{q}\mathbf{q}^T + \mathbf{B}\mathbf{K}^{-1}\mathbf{V}(\mathbf{K}^{-1})^T\mathbf{B}^T - \mathbf{C}(\mathbf{K}^{-1})^T\mathbf{B}^T - \mathbf{B}\mathbf{K}^{-1}\mathbf{C}^T, \tag{4.4}$$

where $\mathbf{q} = \left(\sqrt{p_1(\boldsymbol{\theta})},\cdots,\sqrt{p_r(\boldsymbol{\theta})}\right)^T$ and the elements of the matrix $\mathbf{B}$ are as defined in (3.9) or (3.10) depending on whether fixed or random cells are used. They showed that Wald's statistic $\mathbf{Q}(\overline{\boldsymbol{\theta}}_n) = \mathbf{V}^{(n)T}(\overline{\boldsymbol{\theta}}_n)\boldsymbol{\Sigma}_n^{-}\mathbf{V}^{(n)}(\overline{\boldsymbol{\theta}}_n)$, where $\boldsymbol{\Sigma}_n^{-}$ is the generalized matrix inverse of the estimate of $\boldsymbol{\Sigma}$, will follow the chi-squared distribution with $r-1$ degrees of freedom in limit. Hsuan and Robson (1976) have given the test statistic $\mathbf{Q}(\overline{\boldsymbol{\theta}}_n)$ explicitly for the exponential family of distributions in the case of equiprobable grouping cells. Later, in Section 4.3, we shall show that this result coincides with the NRR test. They did not present the statistic $\mathbf{Q}(\overline{\boldsymbol{\theta}}_n)$ explicitly in the general case.

Mirvaliev (2001) presented $\boldsymbol{\Sigma}$ in the following equivalent form

$$\boldsymbol{\Sigma} = \mathbf{I} - \mathbf{q}\mathbf{q}^T - \mathbf{C}\mathbf{V}^{-1}\mathbf{C}^T + (\mathbf{C} - \mathbf{B}\mathbf{K}^{-1}\mathbf{V})\mathbf{V}^{-1}(\mathbf{C} - \mathbf{B}\mathbf{K}^{-1}\mathbf{V})^T, \tag{4.5}$$

and then obtained the Moore-Penrose matrix inverse $\boldsymbol{\Sigma}^+$ as follows:

$$\boldsymbol{\Sigma}^+ = \mathbf{A} - \mathbf{A}(\mathbf{C} - \mathbf{B}\mathbf{K}^{-1}\mathbf{V})\mathbf{L}^{-1}(\mathbf{C} - \mathbf{B}\mathbf{K}^{-1}\mathbf{V})^T\mathbf{A}, \tag{4.6}$$

where

$$\mathbf{A} = \mathbf{I} - \mathbf{q}\mathbf{q}^T + \mathbf{C}(\mathbf{V} - \mathbf{C}^T\mathbf{C})^{-1}\mathbf{C}^T, \tag{4.7}$$

and

$$\mathbf{L} = \mathbf{V} + (\mathbf{C} - \mathbf{B}\mathbf{K}^{-1}\mathbf{V})^T\mathbf{A}(\mathbf{C} - \mathbf{B}\mathbf{K}^{-1}\mathbf{V}). \tag{4.8}$$

Direct matrix manipulations show that the matrix $\boldsymbol{\Sigma}^+$ satisfies the conditions $\boldsymbol{\Sigma}\boldsymbol{\Sigma}^+\boldsymbol{\Sigma} = \boldsymbol{\Sigma}, \boldsymbol{\Sigma}^+\boldsymbol{\Sigma}\boldsymbol{\Sigma}^+ = \boldsymbol{\Sigma}^+, (\boldsymbol{\Sigma}\boldsymbol{\Sigma}^+)^T = \boldsymbol{\Sigma}\boldsymbol{\Sigma}^+$, and $(\boldsymbol{\Sigma}^+\boldsymbol{\Sigma})^T = \boldsymbol{\Sigma}^+\boldsymbol{\Sigma}$, and is therefore the Moore-Penrose matrix inverse of matrix $\boldsymbol{\Sigma}$ in (4.5). By using the invariance principle of Rao and Mitra (1971), Wald's test statistic in this case can be written as

$$Y2_n^2(\overline{\boldsymbol{\theta}}_n) = X_n^2(\overline{\boldsymbol{\theta}}_n) + R_n^2(\overline{\boldsymbol{\theta}}_n) - Q_n^2(\overline{\boldsymbol{\theta}}_n), \tag{4.9}$$

where

$$R_n^2(\overline{\boldsymbol{\theta}}_n) = \mathbf{V}^{(n)T}(\overline{\theta}_n)\mathbf{C}(\mathbf{V} - \mathbf{C}^T\mathbf{C})^{-1}\mathbf{C}^T\mathbf{V}^{(n)}(\overline{\boldsymbol{\theta}}_n), \qquad (4.10)$$

and

$$Q_n^2(\overline{\boldsymbol{\theta}}_n) = \mathbf{V}^{(n)T}(\overline{\theta}_n)\mathbf{A}(\mathbf{C} - \mathbf{B}\mathbf{K}^{-1}\mathbf{V})\mathbf{L}^{-1}(\mathbf{C} - \mathbf{B}\mathbf{K}^{-1}\mathbf{V})^T\mathbf{A}\mathbf{V}^{(n)}(\overline{\theta}_n); \qquad (4.11)$$

here, in all matrices involved, we replace $\boldsymbol{\theta}$ by its MME $\overline{\boldsymbol{\theta}}_n$. The quadratic form in (4.9) is henceforth referred to as Hsuan-Robson-Mirvaliev (HRM) statistic that follows the chi-squared distribution with $r-1$ degrees of freedom in the limit.

Using the formulas in (2.25)–(2.27), the HRM test in (4.9) can be presented as

$$Y2_n^2(\overline{\boldsymbol{\theta}}_n) = U_n^2(\overline{\boldsymbol{\theta}}_n) + S1_n^2(\overline{\boldsymbol{\theta}}_n), \qquad (4.12)$$

where $U_n^2(\overline{\boldsymbol{\theta}}_n)$ is the DN statistic and

$$S1_n^2(\overline{\boldsymbol{\theta}}_n) = W_n^2(\overline{\boldsymbol{\theta}}_n) + R_n^2(\overline{\boldsymbol{\theta}}_n) - Q_n^2(\overline{\boldsymbol{\theta}}_n) \qquad (4.13)$$

is asymptotically independent of the $U_n^2(\overline{\boldsymbol{\theta}}_n)$ part of $Y2_n^2(\overline{\boldsymbol{\theta}}_n)$ in (4.12). Just as in the case of the NRR test, the statistic $S1_n^2(\overline{\boldsymbol{\theta}}_n)$, for fixed or random equiprobable cells, as a rule possesses higher power than the HRM statistic (see also Remark 3.3).

Singh (1987) suggested a modification of the NRR test valid for any $\sqrt{n}$-consistent estimator including the MME (see the comments in the preceding Chapter in this regard).

4.2 DECOMPOSITION OF HSUAN-ROBSON-MIRVALIEV TEST

Consider the limiting covariance matrix in (4.5) of the standardized frequencies. If

$$\boldsymbol{\Sigma}_k = \mathbf{I}_k - \mathbf{q}_k\mathbf{q}_k^T - \mathbf{C}_k\mathbf{V}^{-1}\mathbf{C}_k^T + (\mathbf{C}_k - \mathbf{B}_k\mathbf{K}^{-1}\mathbf{V})\mathbf{V}^{-1}(\mathbf{C}_k - \mathbf{B}_k\mathbf{K}^{-1}\mathbf{V})^T,$$

then

$$|\boldsymbol{\Sigma}_k| = \left(1 - \sum_{i=1}^{k} p_i\right)|\mathbf{V} - \mathbf{V}_k||\mathbf{L}_k|/|\mathbf{V}|^2, \qquad (4.14)$$

and

$$\boldsymbol{\Sigma}_k^{-1} = \mathbf{A}_k - \mathbf{A}_k(\mathbf{C}_k - \mathbf{B}_k\mathbf{K}^{-1}\mathbf{V})\mathbf{L}_k^{-1}(\mathbf{C}_k - \mathbf{B}_k\mathbf{K}^{-1}\mathbf{V})^T\mathbf{A}_k, \qquad (4.15)$$

where

$$\mathbf{A}_k = \mathbf{M}_k + \mathbf{M}_k\mathbf{C}_k(\mathbf{V} - \mathbf{V}_k)^{-1}\mathbf{C}_k^T\mathbf{M}_k,$$

$$\mathbf{M}_k = \mathbf{I}_k + \left(1 - \sum_{i=1}^{k} p_i\right)^{-1}\mathbf{q}_k\mathbf{q}_k^T,$$

$$\mathbf{L}_k = \mathbf{V} + (\mathbf{C}_k - \mathbf{B}_k\mathbf{K}^{-1}\mathbf{V})^T\mathbf{A}_k(\mathbf{C}_k - \mathbf{B}_k\mathbf{K}^{-1}\mathbf{V}),$$

and

$$\mathbf{V}_k = \mathbf{C}_k^T\mathbf{C}_k + \left(1-\sum_{i=1}^{k} p_i\right)^{-1}(\mathbf{C}_k^T\mathbf{q}_k)(\mathbf{C}_k^T\mathbf{q}_k)^T, \quad k=1,\cdots,r-1.$$

Consider the matrix $\mathbf{R}_{r-1}$ with its elements as

$$r_{ii} = \sqrt{|\boldsymbol{\Sigma}_{i-1}|/|\boldsymbol{\Sigma}_i|}, \quad i=1,\cdots,r-1,$$
$$r_{ij} = -r_{ii}(\mathbf{d}_{i(i-1)}^*)^T(\boldsymbol{\Sigma}_{i-1}^{-1})_j, \quad j=1,\cdots,i-1,\ i\neq j, i\geqslant 2,$$

where

$$(\mathbf{d}_{i(i-1)}^*)^T = -(\sqrt{p_i}\mathbf{q}_{i-1}^T + \mathbf{c}_i\mathbf{V}^{-1}\mathbf{C}_{i-1}^T - (\mathbf{c}_i-\mathbf{b}_i\mathbf{K}^{-1}\mathbf{V})\mathbf{V}^{-1}(\mathbf{C}_i-\mathbf{B}_i\mathbf{K}^{-1}\mathbf{V})^T),$$

$\mathbf{c}_i$ and $\mathbf{b}_i$ the ith row of matrices $\mathbf{C}$ and $\mathbf{B}$, and $(\boldsymbol{\Sigma}_{i-1}^{-1})_j$ being the jth column of $\boldsymbol{\Sigma}_{i-1}^{-1}$.

From Lemma 2.1 and the transformation $\delta_{r-1}(\overline{\boldsymbol{\theta}}) = \mathbf{RV}(\overline{\boldsymbol{\theta}})$, where $\mathbf{R} = (\mathbf{R}_{r-1}\vdots\mathbf{0})$, with the use of MME in place of $\boldsymbol{\theta}$ in all the matrices, we obtain the following result.

Theorem 4.1. *Under suitable regularity conditions (Hsuan and Robson, 1976), the expansion*

$$Y2_n^2(\overline{\boldsymbol{\theta}}_n) = \delta_1^2(\overline{\boldsymbol{\theta}}_n) + \cdots + \delta_{r-1}^2(\overline{\boldsymbol{\theta}}_n)$$

of the HRM statistic holds and, in the limit, under H_0, *the statistics* $\delta_i^2(\overline{\boldsymbol{\theta}}_n)$, $i=1,\cdots,r-1$, *are distributed independently as* χ_1^2 *and the statistic* $Y2_n^2(\overline{\boldsymbol{\theta}}_n)$ *is distributed as* χ_{r-1}^2.

4.3 EQUIVALENCE OF NIKULIN-RAO-ROBSON AND HSUAN-ROBSON-MIRVALIEV TESTS FOR EXPONENTIAL FAMILY

Consider the exponential family of distributions with density

$$f(x;\boldsymbol{\theta}) = h(x)\exp\left\{\sum_{j=1}^{s}\theta_j x^j + V(\boldsymbol{\theta})\right\}, \quad x\in\mathcal{X}\subseteq R^1, \tag{4.16}$$

$\mathcal{X}$ is open in R^1, $\mathcal{X} = \{x : f(x,\boldsymbol{\theta}) > 0\}$, and $\boldsymbol{\theta}\in\boldsymbol{\Theta}\subset R^s$. The family in (4.16) is quite rich, and contains many important distributions such as the family of Poisson distributions and the family of normal distributions. In this case, it is known that, under H_0, the statistic

$$\mathbf{U}_n = \left(\sum_{i=1}^{n} X_i, \sum_{k=1}^{n} X_i^2, \cdots, \sum_{i=1}^{n} X_i^s\right)^T$$

is minimal sufficient for the parameter $\boldsymbol{\theta}$; see Voinov and Nikulin (1993), for example.

Let us now assume that

(1) the support $\mathcal{X}$ does not depend on $\boldsymbol{\theta}$;
(2) the $s \times s$ matrix with elements

$$H_{ij} = -\frac{\partial^2}{\partial \theta_i \partial \theta_j} V(\boldsymbol{\theta}), \quad i,j = 1, \cdots, s,$$

is positive definite on $\boldsymbol{\Theta}$;
(3) The population moments $m_j(\boldsymbol{\theta}) = \mathbf{E}_{\boldsymbol{\theta}}(X_1^j), j = 1, \cdots, s$, all exist.

Differentiating the evident equality $\int_{x \in \mathcal{X}} f(x, \boldsymbol{\theta}) dx = 1$ with respect to θ_j, we obtain

$$\int\limits_{x \in \mathcal{X}} \frac{\partial f(x,\boldsymbol{\theta})}{\partial \theta_j} dx = \int\limits_{x \in \mathcal{X}} h(x) \exp\left\{ \sum_{j=1}^{s} \theta_j x^j + V(\boldsymbol{\theta}) \right\} \left(x^j + \frac{\partial V(\boldsymbol{\theta})}{\partial \theta_j} \right) dx$$
$$= m_j(\boldsymbol{\theta}) + \frac{\partial V(\boldsymbol{\theta})}{\partial \theta_j} = 0.$$

From this, it follows that the population moments of the distribution in (4.16) are

$$m_j(\boldsymbol{\theta}) = -\frac{\partial V(\boldsymbol{\theta})}{\partial \theta_j}, \quad j = 1, \cdots, s. \tag{4.17}$$

The statistic $\mathbf{T}_n = \mathbf{U}_n / n$ is the best unbiased estimator for $\mathbf{m}(\boldsymbol{\theta}) = (m_1(\boldsymbol{\theta}), \cdots, m_s(\boldsymbol{\theta}))^T$, i.e. $\mathbf{E}_{\boldsymbol{\theta}}(\mathbf{T}_n) = \mathbf{m}(\boldsymbol{\theta})$. It is possible to find the method of moments estimator $\bar{\boldsymbol{\theta}}_n$ for $\boldsymbol{\theta}$ in a unique way from the equation $\mathbf{T}_n = \mathbf{m}(\boldsymbol{\theta})$. This estimator is expressed in terms of the sufficient statistic $\mathbf{U}_n$ as $\bar{\boldsymbol{\theta}}_n = \bar{\boldsymbol{\theta}}_n(\mathbf{U}_n)$.

On the other hand (see Theorem 5.1.2 of Zacks, 1971), Conditions (1)–(3) provide the existence of the maximum likelihood estimator $\hat{\boldsymbol{\theta}}_n = \hat{\boldsymbol{\theta}}_n(\mathbf{U}_n)$ which is the root of the same equation $\mathbf{T}_n = \mathbf{m}(\boldsymbol{\theta})$, and so for the exponential family in (4.16), the method of maximum likelihood and the method of moments provide the same estimator for $\boldsymbol{\theta}$, viz., $\bar{\boldsymbol{\theta}}_n \equiv \hat{\theta}_n$.

Theorem 4.2 (Voinov and Pya, 2004). *Assume that Conditions (1)–(3) stated above hold. Then, for the exponential family of distributions in (4.16), the NRR $Y1_n^2(\hat{\boldsymbol{\theta}}_n)$ and the HRM $Y2_n^2(\bar{\boldsymbol{\theta}}_n)$ statistics are identical.*

Proof. For the model in (4.16), the elements of the $r \times s$ matrix **B** are

$$B_{jk} = \frac{1}{\sqrt{p_j(\boldsymbol{\theta})}} \int\limits_{\Delta_j} \frac{\partial f(x,\boldsymbol{\theta})}{\partial \theta_k} dx = \frac{1}{\sqrt{p_j(\boldsymbol{\theta})}} \int\limits_{\Delta_j} f(x,\boldsymbol{\theta}) \left(x^k + \frac{\partial V(\boldsymbol{\theta})}{\partial \theta_k} \right) dx$$

$$= \frac{1}{\sqrt{p_j(\boldsymbol{\theta})}} \left(\int\limits_{\Delta_j} x^k f(x,\boldsymbol{\theta}) dx - p_j(\boldsymbol{\theta}) m_k(\boldsymbol{\theta}) \right),$$

$$j = 1, \cdots, r, \ k = 1, \cdots, s.$$

Comparing this with the elements of the matrix **C** in (4.3), we see that the matrices **B** and **C** coincide for the exponential family in (4.16).

Consider the elements of matrices **K** in (4.1) and **V** in (4.2). For the model in (4.16), we have

$$K_{ij}(\boldsymbol{\theta}) = \int\limits_{x \in \mathcal{X}} x^i \frac{\partial f(x,\boldsymbol{\theta})}{\partial \theta_j} dx = \int\limits_{x \in \mathcal{X}} x^{i+j} f(x,\boldsymbol{\theta}) dx + \int\limits_{x \in \mathcal{X}} x^i f(x,\boldsymbol{\theta}) \frac{\partial V(\boldsymbol{\theta})}{\partial \theta_j} dx$$
$$= m_{ij}(\boldsymbol{\theta}) - m_i(\boldsymbol{\theta}) m_j(\boldsymbol{\theta}) = \mathbf{V}_{ij}(\boldsymbol{\theta}), \quad i,j = 1, \cdots, s.$$

Since **B** equals **C** and **K** equals **V**, we readily have

$$\mathbf{C} - \mathbf{B}\mathbf{K}^{-1}\mathbf{V} = \mathbf{C} - \mathbf{B} = \mathbf{0}.$$

From this identity and Eq. (4.11), it follows that $Q^2(\boldsymbol{\theta}) = 0$.

Since

$$\frac{\partial \ln f(x,\boldsymbol{\theta})}{\partial \theta_i} = \frac{\partial}{\partial \theta_i} \left[\ln (h(x)) + \sum_{i=1}^{s} \theta_i x^i + V(\boldsymbol{\theta}) \right] = x^i + \frac{\partial V(\boldsymbol{\theta})}{\partial \theta_i} = x^i - m_i(\boldsymbol{\theta}),$$

it is easy to observe that the Fisher information matrix

$$\mathbf{J} = \mathbf{E}_{\boldsymbol{\theta}} \left[\left(\frac{\partial \ln f(x,\boldsymbol{\theta})}{\partial \theta_1}, \cdots, \frac{\partial \ln f(x,\boldsymbol{\theta})}{\partial \theta_s} \right) \right.$$
$$\left. \times \left(\frac{\partial \ln f(x,\boldsymbol{\theta})}{\partial \theta_1}, \cdots, \frac{\partial \ln f(x,\boldsymbol{\theta})}{\partial \theta_s} \right)^T \right] = \mathbf{V}.$$

From the above identity and (4.10), it readily follows that

$$\mathbf{C}(\mathbf{V} - \mathbf{C}^T\mathbf{C})^{-1}\mathbf{C}^T = \mathbf{B}(\mathbf{J} - \mathbf{B}^T\mathbf{B})^{-1}\mathbf{B}^T,$$
$$R_n^2(\boldsymbol{\theta}) = P_n^2(\boldsymbol{\theta}),$$

where $P_n^2(\boldsymbol{\theta})$ is as defined in (3.8). Finally, since $\hat{\boldsymbol{\theta}}_n$ equals $\overline{\boldsymbol{\theta}}_n$ for the exponential family in (4.16), we have $Y1_n^2(\hat{\boldsymbol{\theta}}_n) \equiv Y2_n^2(\overline{\boldsymbol{\theta}}_n)$.

Hsuan and Robson (1976) derived explicitly their test statistic for the family in (4.16), but they did not mention that in this case MMEs coincide with MLEs and, consequently, their test coincides with the already known NRR test.

4.4 COMPARISONS OF SOME MODIFIED CHI-SQUARED TESTS

In this section, we present some simulated results of powers of the modified chi-squared tests (see Voinov et al., 2009) introduced in the preceding sections. It is well known that if the random cells converge to a certain limit in probability, then they will not change the limiting distributions of chi-squared type statistics (Chibisov, 1971 and Moore and Spruill, 1975). In the empirical study here, we observed no difference between the simulated results for both random and fixed cells within statistical errors of simulation. Under H_0, the simulated levels of tests considered, defined with the use of theoretical critical value of level $\alpha = 0.05$ of a corresponding chi-squared distribution, always remained within the 95% confidence interval [0.046, 0.054] for $N = 10{,}000$ runs. For this reason, we determined the simulated power of tests by using the theoretical critical values. For the nonparametric Anderson and Darling (1954) test, we simulated the power by using simulated critical values at level $\alpha = 0.05$, since analytical expressions of limiting distributions of this test are unavailable.

4.4.1 Maximum likelihood estimates

Many results are known for testing composite null hypothesis about normality using chi-squared type tests; see, for example, Dahiya and Gurland (1973), McCulloch (1985), and Lemeshko and Chimitova (2003), as discussed earlier in Chapter 3. Figure 4.1 provides a graphical plot of the power of these tests. The dramatic increase in the power of $S_n^2(\hat{\boldsymbol{\theta}}_n)$ compared to the NRR $Y1_n^2(\hat{\boldsymbol{\theta}}_n)$ test was first mentioned by McCulloch (1985). This means that the statistic $S_n^2(\hat{\boldsymbol{\theta}}_n)$ in (3.24) recovers and uses the largest part of the Fisher sample information lost due to grouping by equiprobable random or fixed intervals. Note that the DN $U_n^2(\hat{\boldsymbol{\theta}}_n)$ test and the Pearson-Fisher test in (2.19), which use a very small part of the sample information, possess almost no power for any number of equiprobable cells. The power of the XR^2 test of Dahiya and Gurland (1972a), Dahiya and Gurland (1973) is maximal for the smallest number of cells r, but is still less than that of the NRR $Y1_n^2(\hat{\boldsymbol{\theta}}_n)$ test, which in turn is less than that of $S_n^2(\hat{\boldsymbol{\theta}}_n)$ for any r. If $r > 40$, the expected cell frequencies become small and limiting distributions of chi-squared type tests may differ from those of χ^2, and for this reason we restricted r to the range of 4–40.

Analogous behavior of powers for all the tests is also observed for some other symmetrical alternatives such as the triangular, uniform, and double-exponential (Laplace); see, for example, Figure 4.2 for the triangular alternative.

It seems that for both heavy-tailed and short-tailed symmetrical alternatives, in the case of equiprobable cells, the statistic $S_n^2(\hat{\boldsymbol{\theta}}_n)$ is the superior one for the normal null.

It has to be noted that the relation between powers of different chi-squared type statistics depend not only on the alternative, but also on the null hypothesis. Consider, for example, the null hypothesis as the two-parameter exponential

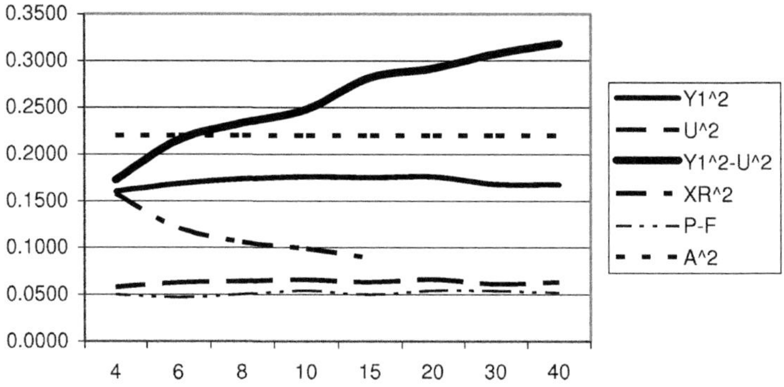

FIGURE 4.1 Estimated powers as functions of the number of equiprobable cells r when testing H_0: Normal against the logistic alternative for NRR $(Y1\hat{}2)$, DN$(U\hat{}2)$, $S_n^2(\hat{\boldsymbol{\theta}}_n)(Y1\hat{}2-U\hat{}2)$, Dahiya and Gurland $(XR\hat{}2)$, Pearson-Fisher (PF), and Anderson–Darling $(A\hat{}2)$ tests, based on the number of runs $N = 10{,}000$, sample size $n = 100$, and level $\alpha = 0.05$.

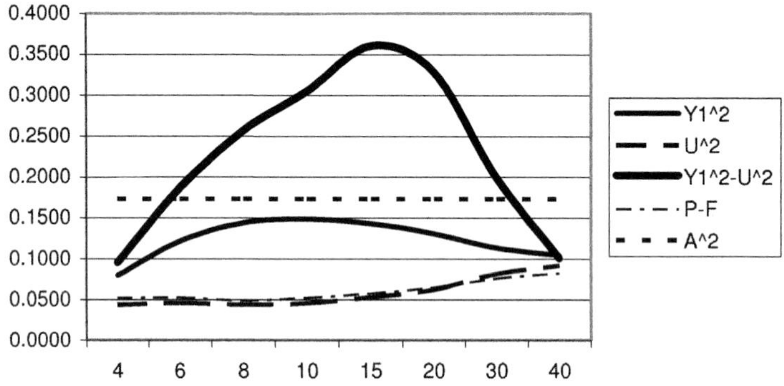

FIGURE 4.2 Estimated powers as functions of the number of equiprobable cells r when testing H_0: Normal against the triangular alternative for NRR $(Y1\hat{}2)$, DN$(U\hat{}2)$, $S_n^2(\hat{\boldsymbol{\theta}}_n)(Y1\hat{}2-U\hat{}2)$, Dahiya and Gurland $(XR\hat{}2)$, Pearson-Fisher (PF), and Anderson–Darling $(A\hat{}2)$ tests, based on the number of runs $N = 10{,}000$, sample size $n = 100$, and level $\alpha = 0.05$.

distribution with pdf $f(x,\mu,\theta) = \frac{1}{\theta}e^{-(x-\mu)/\theta}, x \geqslant \mu$. The MLEs $\hat{\theta}_n$ and $\hat{\mu}_n$, of the parameters θ and μ, in this case are X and $X_{(1)}$, $\frac{1}{n}\sum_{i=2}^{n}(X_{(i)} - X_{(1)})$, respectively. The simulated power of the NRR test for the semi-normal alternative with density

$$f(x,\mu,\theta) = \frac{\sqrt{2}}{\sqrt{\pi}\theta}\exp\left\{-\frac{(x-\mu)^2}{2\theta^2}\right\}, \quad x \geqslant \mu,\ \theta > 0,\ \mu \in R^1, \qquad (4.18)$$

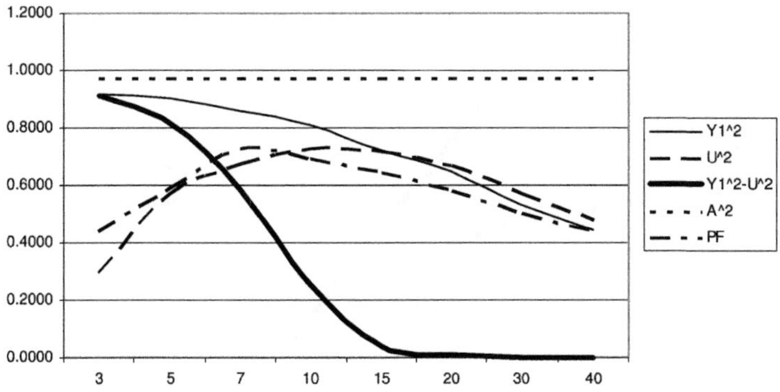

FIGURE 4.3 Estimated powers as functions of the number of equiprobable cells r when testing H_0: Exp against semi-normal alternative for NRR ($Y1\hat{}2$),DN($U\hat{}2$), $S_n^2(\hat{\boldsymbol{\theta}}_n)(Y1\hat{}2 - U\hat{}2)$, Pearson-Fisher (PF), and Anderson–Darling ($A\hat{}2$) tests, based on the number of runs $N = 10{,}000$, sample size $n = 200$, and level $\alpha = 0.05$.

is presented in Figure 4.3. We see from Figure 4.3 that in this case both the NRR test and $S_n^2(\hat{\boldsymbol{\theta}}_n)$ test possess the highest power for small number of equiprobable random cells. Analogous behavior of powers for these tests has also been observed for the triangular alternative with pdf $f(x,\mu,\theta) = 2(\theta - x)/(\theta - \mu)^2, \mu \leqslant x \leqslant \theta, \theta > \mu, \mu \in R^1$, and uniform alternative with pdf $f(x,\mu,\theta) = 1/(\sqrt{12}\theta), \mu \leqslant x \leqslant \mu + \sqrt{12}\theta, \theta > 0, \mu \in R^1$. From Figure 4.3, we observe that the DN and PF tests use much larger part of the Fisher sample information compared to the normal null hypothesis, and that their power are comparable with that of the NRR test when $r \geqslant 15$. We also see that the $S_n^2(\hat{\boldsymbol{\theta}}_n)$ test possesses less power than the $Y1_n^2(\hat{\boldsymbol{\theta}}_n)$ test for any $r > 3$.

4.4.2 Moment-type estimators

In some cases like the logistic family of distributions, the computation of the MLEs is not simple, and in these cases the implementation of the NRR test becomes difficult; see Aguirre and Nikulin (1994a,b). In such cases, it may then be convenient to use MMEs instead though they are not as efficient as the MLEs. To illustrate the applicability of the HRM test $Y2_n^2(\overline{\boldsymbol{\theta}}_n)$ (see Eqs. (4.9)–(4.11)) based on the MMEs, consider the logistic distribution as the null hypothesis. Since the regularity conditions of Hsuan and Robson (1976) are satisfied for the logistic distribution, the statistic in (4.9) can be used for testing the validity of this null hypothesis. The behavior of the power of different chi-squared type statistics in this case is similar to that for the normal null hypothesis (see Figure 4.4). The most powerful test is the statistic $S1_n^2(\overline{\boldsymbol{\theta}})$ in (4.13) that recovers and uses a large part of the Fisher information lost while grouping the data by equiprobable intervals.

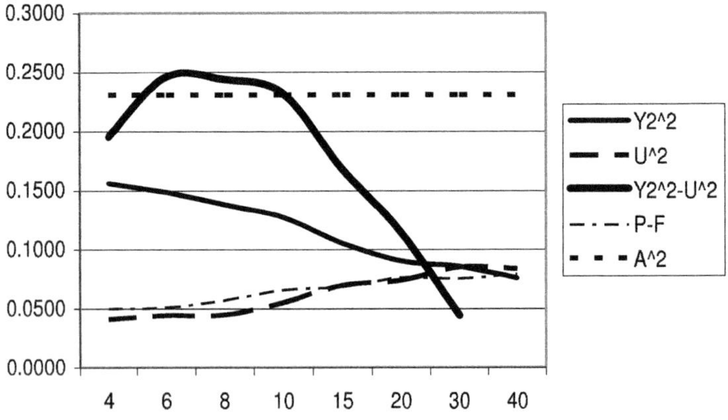

FIGURE 4.4 Estimated powers as functions of the number of equiprobable cells r when testing H_0: Logistic against normal alternative for HRM ($Y2\hat{}2$), DN ($U\hat{}2$), $S1_n^2(\overline{\boldsymbol{\theta}}_n)(Y2\hat{}2 - U\hat{}2)$, Dahiya and Gurland ($XR\hat{}2$), Pearson-Fisher (P-F), and Anderson–Darling ($A\hat{}2$) tests, based on the number of runs $N = 10{,}000$, sample size $n = 100$, and level $\alpha = 0.05$.

Fisher (1952b) (see also Dzhaparidze, 1983 and Paardekooper et al., 1989) proposed the following iterative procedure of obtaining an asymptotically efficient estimator based on any $\sqrt{n}$-consistent estimator $\tilde{\boldsymbol{\theta}}_n$. This approach describes another way of implementing the NRR test: find $\tilde{\boldsymbol{\theta}}_n$ first, improve it by using the idea of Fisher, and then use it in the NRR statistic. Fisher's iterative formula is

$$\tilde{\boldsymbol{\theta}}_n^{i+1} = \tilde{\boldsymbol{\theta}}_n^{i} + \frac{1}{n}\left(\mathbf{J}^{-1}(\boldsymbol{\theta})\right)_{\tilde{\boldsymbol{\theta}}_n^{i}} \left(\frac{\partial L_n}{\partial \boldsymbol{\theta}}\right)_{\tilde{\boldsymbol{\theta}}_n^{i}}, \quad i = 0,1,\cdots, \tag{4.19}$$

where $L_n = \sum_{i=1}^{n} \log f(X_i,\boldsymbol{\theta})$ and $\left(\partial L_n/\partial \boldsymbol{\theta}\right) = \left(\partial L_n/\partial \theta_1, \cdots ,\partial L_n/\partial \theta_s\right)^T$. Then, Fisher showed that, for any starting value of $\tilde{\boldsymbol{\theta}}_n^{0}$, the result of the very first iteration $\tilde{\boldsymbol{\theta}}_n^{1}$ from (4.19) is an estimator as efficient as the MLE $\hat{\boldsymbol{\theta}}_n$ asymptotically.

Consider the logistic null hypothesis and the normal distribution as an alternative. For this case, Figure 4.5 presents the simulated powers of the HRM test $Y2_n^2(\overline{\boldsymbol{\theta}}_n)$, $S1_n^2(\overline{\boldsymbol{\theta}}_n) = Y2_n^2(\overline{\boldsymbol{\theta}}_n) - U_n^2(\overline{\boldsymbol{\theta}}_n)$ and $Y1_n^2(\hat{\boldsymbol{\theta}}^1)$, where $\hat{\boldsymbol{\theta}}_n^1$ is obtained from (4.19) as the first iterate (see Voinov and Pya, 2004). From Figure 4.5, we observe that for the same number of intervals $r = 6 - 14$, the implementation of these improved estimates has resulted in an increase in power as compared to the HRM test. We also note that the improvement is not as large as the one produced by the use of $S1_n^2(\overline{\boldsymbol{\theta}}_n)$ test which recovers much more information lost due to data grouping. Incidentally, this serves as a good example to demonstrate

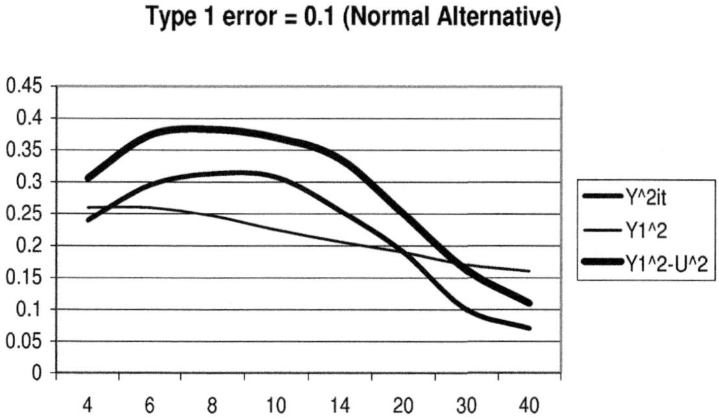

FIGURE 4.5 Simulated powers of $Y1_n^2(\hat{\theta}_n^1) = Y2\hat{}2, Y2_n^2(\overline{\theta}_n) = Y1\hat{}\,2$ and $S1_n^2(\overline{\theta}_n) = Y1\hat{}2 - U\hat{}2$ tests based on $n = 200$ and $N = 10{,}000$.

that sometimes tests based on non-efficient MMEs may possess higher power than tests based on efficient MLEs.

As in Section 3.2, in the case of equiprobable random or fixed intervals, we see here the uselessness of DN and PF tests and the superiority of the $S1_n^2(\overline{\theta}_n)$ (see also Remark 3.3). It is also of interest to note that the power of the nonparametric Anderson-Darling test A^2 can be lower or higher than that of $S_n^2(\hat{\theta}_n)$ or $S1_n^2(\overline{\theta}_n)$.

4.5 NEYMAN-PEARSON CLASSES

Since Mann and Wald (1942) recommended the use of equiprobable partitioning of a sample space, it has been used by many researchers. However, some authors disagree on this; see Ivchenko and Medvedev (1980) and Boero et al. (2004a,b). The latter have remarked that "The choice of non-equiprobable classes can result in substantial gains in power." Using simulations, they have investigated power of the classical Pearson's test for non-equiprobable cells, but the partitions used were rather artificial without proper motivation. Their comment that the power of the test depends on the closeness of the intersection points of two density functions to the class boundaries is, as we shall see below, of great importance. To illustrate this, let us consider Neyman-Pearson classes (Greenwood and Nikulin, 1996). These classes correspond to the minimal partitioning of a sample space that increases or even maximizes Pearson's measure of the distance between the null and alternative hypotheses (see also Remark 3.3). It is the case when the class boundaries coincide with points of intersection. Let $f(x)$ and $g(x)$ be the densities of the null and the alternative hypotheses, respectively. Then, Greenwood and Nikulin (1996) defined Neyman-Pearson classes as follows:

$$I_1 = \{x : f(x) < g(x)\} \text{ and } I_2 = \{x : f(x) \geqslant g(x)\}. \tag{4.20}$$

For several points of intersection, it is possible to define more than two classes with the intersecting points being the class boundaries. We shall refer to such cells as Neyman-Pearson type classes.

4.5.1 Maximum likelihood estimators

Suppose we are testing the standard normal null hypothesis $H_0 : X \sim n(x,0,1)$ with pdf

$$n(x,0,1) = \frac{1}{\sqrt{2\pi}} \exp\left(-\frac{x^2}{2}\right), \quad |x| < \infty, \tag{4.21}$$

against the standard logistic alternative $H_a : X \sim l(x,0,1)$ with pdf

$$l(x,0,1) = \frac{\pi}{\sqrt{3}} \frac{\exp\left(-\frac{\pi x}{\sqrt{3}}\right)}{\left\{1+\exp\left(-\frac{\pi x}{\sqrt{3}}\right)\right\}^2}, \quad x \in R^1. \tag{4.22}$$

To construct Neyman-Pearson classes, we then have to solve the equation $n(x,0,1) = l(x,0,1)$. Microsoft Excel Solver gives the following four points of intersection: ± 0.682762, $\pm$ 2.374686 (see Figure 4.6).

The following four Neyman-Pearson type classes can then be considered:

$$\begin{aligned}
\Delta_1 &= (-\infty, -2.374686] \cup [2.374686, +\infty),\\
\Delta_2 &= [-2.374686, -0.682762],\\
\Delta_3 &= [-0.682762, 0.682762],\\
\Delta_4 &= [0.682762, 2.374686].
\end{aligned} \tag{4.23}$$

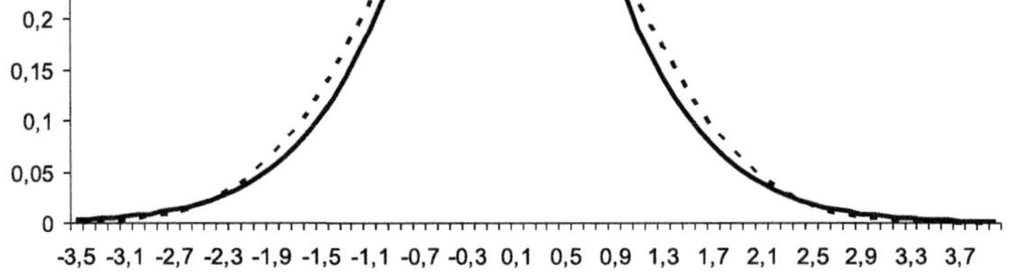

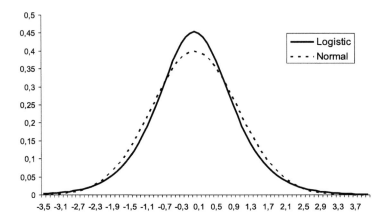

FIGURE 4.6 Probability density functions of $n(x,0,1)$ and $l(x,0,1)$.

TABLE 4.1 Simulated power of $Y1_n^2(\hat{\theta}_n)$ test for two Neyman-Pearson type classes, and powers of NRR $Y1_n^2(\hat{\theta}_n)$, DN$U_n^2(\hat{\theta}_n)$, $S_n^2(\hat{\theta}_n)$, and Pearson-Fisher (PF) tests for four Neyman-Pearson type classes, and of Anderson-Darling A^2 test, based on $N = 10{,}000$ simulations, sample size $n = 200$, and level $\alpha = 0.05$.

	$Y1_n^2(\hat{\theta}_n)(r=2)$	$Y1_n^2(\hat{\theta}_n)$	$U_n^2(\hat{\theta}_n)$	$S_n^2(\hat{\theta}_n)$	PF	A^2
Logist.	0.393	0.290	0.374	0.088	0.374	0.394
Triang.	0.451	0.414	0.623	0.049	0.657	0.173
Laplace	0.977	0.948	0.969	0.242	0.972	0.982

One may also consider the following two Neyman-Pearson classes:

$$\Delta_1 = (-\infty, -2.374686] \cup [-0.682762, 0.682762] \cup [2.374686, +\infty),$$
$$\Delta_2 = [-2.374686, -0.682762] \cup [0.682762, 2.374686]. \tag{4.24}$$

Let us now consider the triangular distribution with pdf

$$t(x,0,1) = \frac{1}{\sqrt{6}} - \frac{|x|}{6}, \quad -\sqrt{6} \leqslant x \leqslant \sqrt{6},$$

and the double-exponential (Laplace) distribution with pdf

$$d(x,0,1) = \frac{1}{\sqrt{2}} \exp(-\sqrt{2}|x|), \quad -\infty < x < \infty.$$

The four Neyman-Pearson classes for these two distributions can also be readily determined.

Considering the parameters of the considered distributions to be unknown, powers of different tests for the logistic, triangular, and Laplace alternatives were all estimated by simulation, and these results are presented in Table 4.1.

4.5.2 Moment-type estimators

Suppose we are testing the standard logistic null hypothesis $H_0 : X \sim l(x,0,1)$ against standard normal, triangular $t(x,0,1)$, and Laplace $d(x,0,1)$ alternatives with the parameter $\boldsymbol{\theta}$ being estimated by the MME $\overline{\boldsymbol{\theta}}_n$. Considering the model parameters to be unknown, powers of different tests for the normal, triangular and Laplace alternatives were all estimated by simulation, and these results are presented in Table 4.2.

Comparing the results displayed in Figures 4.1, 4.2, and 4.4 with those in Tables 4.1 and 4.2, we see a dramatic difference between the two sets of results. For equiprobable cells, the DN and PF statistics possess no or very low power

TABLE 4.2 Simulated power of $Y2_n^2(\overline{\theta}_n)$ test for two Neyman-Pearson type classes, and powers of HRM $Y2_n^2(\overline{\theta}_n)$, DN$U_n^2(\overline{\theta}_n)$, $S1_n^2(\overline{\theta})$ and Pearson-Fisher (PF) tests for four Neyman-Pearson type classes, and of Anderson-Darling A^2 test, based on $N = 10{,}000$ simulations, sample size $n = 200$, and level $\alpha = 0.05$.

	$Y2_n^2(\overline{\theta})(r=2)$	$Y2_n^2(\overline{\theta}_n)$	$U_n^2(\overline{\theta}_n)$	$S_n^2(\hat{\theta}_n)$	PF	A^2
Normal	0.334	0.204	0.357	0.034	0.320	0.231
Triang.	0.908	0.912	0.982	0.018	0.979	0.775
Laplace	0.725	0.582	0.656	0.159	0.649	0.659

and the tests $S_n^2(\hat{\theta}_n)$ and $S1_n^2(\overline{\theta}_n)$ are observed to be the most powerful ones. For Neyman-Pearson type classes, we observe the reverse with DN and PF tests becoming powerful and $S_n^2(\hat{\theta}_n)$ and $S1_n^2(\overline{\theta}_n)$ losing their power. The power of DN and PF tests for four Neyman-Pearson type classes are comparable and in addition are comparable with those of $Y1_n^2(\hat{\boldsymbol{\theta}}_n)$ and $Y2_n^2(\overline{\boldsymbol{\theta}})$ for two classes.

It is of importance to note that the power of $Y1_n^2(\hat{\boldsymbol{\theta}}_n)$ and $Y2_n^2(\overline{\boldsymbol{\theta}})$ for two NP classes (see Tables 4.1 and 4.2) are essentially higher than the maximal power for equiprobable cells (Figures 4.1 and 4.2). This means that equiprobable partitioning of a sample space can be recommended if and only if alternative hypothesis cannot be specified.

Remark 4.1. *Comments made earlier in Remark 3.3 hold true for the tests $Y1_n^2(\hat{\boldsymbol{\theta}}_n)$ and $Y2_n^2(\overline{\boldsymbol{\theta}})$ based on two Neyman-Pearson classes, since their variances are the smallest possible. Naturally, the stability of these test statistics is the highest.*

4.6 MODIFIED CHI-SQUARED TEST FOR THREE-PARAMETER WEIBULL DISTRIBUTION

The Weibull distribution plays an important role in the analysis of lifetime or response data in reliability and survival studies. To extend applications of the distribution to a wider class of failure rate models, several modifications of the classical Weibull model have been proposed in the literature. Mudholkar et al. (1995), Mudholkar et al. (1996) introduced the Exponentiated and Generalized Weibull families, which include distributions with unimodal and bathtub failure rates, and in addition possess a broad class of monotone hazard rates. Furthermore, Bagdonavičius and Nikulin (2002) proposed another generalization—the Power Generalized Weibull family as described earlier in Section 3.7.

In this section, we first consider parameter estimation and modifications of chi-squared type tests that may be used when testing for the three-parameter Weibull distribution (Voinov et al., 2008a). Then, we use Monte Carlo simulation to study the power of these tests for equiprobable random cells against two modifications of the Weibull family of distributions. An application of the NP classes is then considered and finally some discussion of the results and concluding comments are made.

4.6.1 Parameter estimation and modified chi-squared tests

Consider the three-parameter Weibull distribution with pdf

$$f(x;\theta,\mu,p)=\frac{p}{\theta}\left(\frac{x-\mu}{\theta}\right)^{p-1}\exp\left\{-\left(\frac{x-\mu}{\theta}\right)^{p}\right\},$$
$$x>\mu,\theta,\ p>0,\ \mu\in R^1. \tag{4.25}$$

It is well known that there are numerical problems in determining the maximum likelihood estimates (MLEs) of the distribution in (4.25) if all three parameters are unknown. Sometimes, there is no local maximum for the likelihood function, and in some situations the likelihood can be infinite (Lockhart and Stephens, 1994). If we wish to apply the NRR test, then we need the Fisher information matrix $\mathbf{J}$ with its elements as follows:

$$J_{11}=\frac{1}{p^2}\left[(C-1)^2+\frac{\pi^2}{6}\right],\quad J_{12}=J_{21}=\frac{C-1}{\theta},$$
$$J_{13}=J_{31}=-\frac{1}{\theta}\Gamma\left(2-\frac{1}{p}\right)\left[\psi\left(1-\frac{1}{p}\right)+1\right],\quad J_{22}=\frac{p^2}{\theta^2},$$
$$J_{23}=J_{32}=\frac{p^2}{\theta^2}\Gamma\left(2-\frac{1}{p}\right),\quad J_{33}=\frac{(p-1)^2}{\theta^2}\Gamma\left(1-\frac{2}{p}\right),$$

where $C=0.577215665$ is the Euler's constant, and $\psi(x)$ is the psi or digamma function. From the above expression, we see that $\mathbf{J}$ does not exist for infinitely many values of the unknown shape parameter p (for $p=1/(2+k)$ and $p=2/(2+k),k=0,1,2,\cdots$). Because this problem associated with the MLEs, the NRR test based on MLEs is not easy to apply, and one may instead use the HRM test based on moment-type estimates (MMEs) $\bar{\boldsymbol{\theta}}_n=(\bar{\theta}_{1n},\bar{\theta}_{2n},\bar{\theta}_{3n})^T$ of $\bar{\boldsymbol{\theta}}=(\theta_1,\theta_2,\theta_3)^T$, where $\theta_1=\mu,\theta_2=\theta,\theta_3=p$, which in this case can be found by solving the following system of equations (see McEwen and Parresol, 1991):

$$\sum_{l=0}^{i}\binom{i}{l}\theta^{i-l}\mu^l\Gamma\left(1+\frac{i-l}{p}\right)=\frac{1}{n}\sum_{k=1}^{n}X_k^i,\quad i=1,2,3,$$

by using, say, the Microsoft Excel Solver. The above estimators exist for any $p>0$ (see Figure 4.7). One may also refer to Balakrishnan and Cohen (1991) for some other forms of moment-type estimators.

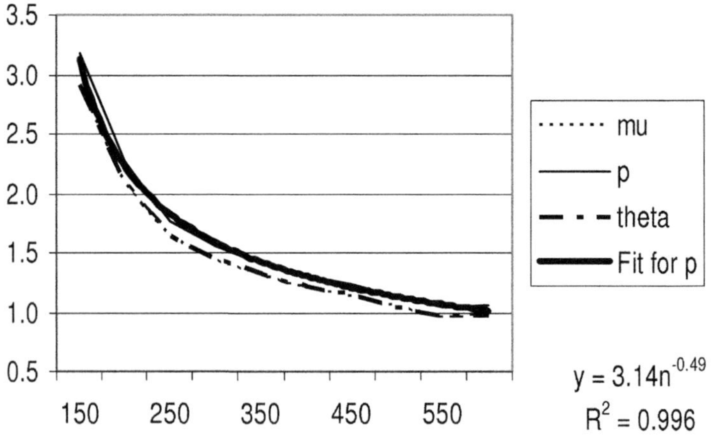

FIGURE 4.7 Simulated average absolute errors of the MMEs $\bar{\mu},\bar{\theta},\bar{p}$ against the true parameter values, as a function of the sample size n.

The simulated average absolute errors of these MMEs are plotted in Figure 4.7 against the true parameter values, as a function of the sample size n. From Figure 4.7, it can be seen that the MMEs are approximately $\sqrt{n}$-consistent. The regularity conditions of Hsuan and Robson (1976) needed for implementing the HRM test based on these MMEs are as follows:

(1) The MMEs are $\sqrt{n}$-consistent;
(2) Matrix **K** (see Section 9.2) is non-singular;
(3) $\int_{x>\mu} g_i(x) f(x;\boldsymbol{\theta})dx$, $\int_{x>\mu} g_i(x)\frac{\partial f(x;\boldsymbol{\theta})}{\partial \theta_j}dx$, $\int_{x>\mu} g_i(x)\frac{\partial^2 f(x;\boldsymbol{\theta})}{\partial \theta_j \partial \theta_k}dx$,

where $g_i(x) = x^i$, all exist and are finite and continuous in $\boldsymbol{\theta}$ for $i,j,k = 1,2,3$, in a neighborhood of the true value of the parameter $\boldsymbol{\theta}$. It can be verified that the conditions in (1)–(3) are satisfied for the three-parameter Weibull family in (4.25) if $p > 2$ (Voinov et al., 2008a). This allows us to use the HRM test in (4.9), the DN test in (2.27), and the $S1_n^2(\overline{\boldsymbol{\theta}}_n)$ test in (4.13). Explicit expressions for all the elements of matrices **K**,**B**,**C**, and **V** needed in the computation of these statistics are presented in Section 9.2.

4.6.2 Power evaluation

To investigate the power of DN $U_n^2(\overline{\boldsymbol{\theta}}_n)$, HRM $Y2_n^2(\overline{\boldsymbol{\theta}}_n)$, and $S1_n^2(\overline{\boldsymbol{\theta}}_n)$ tests for the three-parameter Weibull null hypothesis against two useful reliability models as alternatives, we conducted Monte Carlo simulations with different number of equiprobable random cells. Figures 4.8 and 4.9 present these simulated power values.

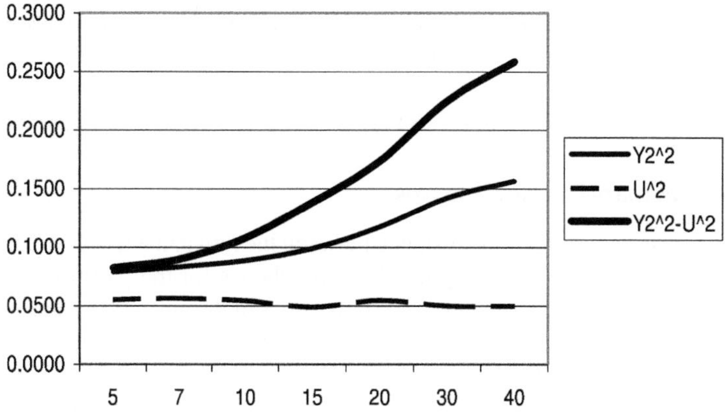

FIGURE 4.8 Simulated powers, as functions of the number of equiprobable cells r, of the tests $Y2_n^2(\overline{\boldsymbol{\theta}}_n)(Y2\hat{}2), U_n^2(\overline{\boldsymbol{\theta}}_n)(U\hat{}2)$, and $S1_n^2(\overline{\boldsymbol{\theta}}_n)(Y2\hat{}2 - U\hat{}2)$ tests for the Exponentiated Weibull alternative with cdf $F(x) = [1-\exp(1-(x/\alpha)^{\beta})]^{\gamma}, x, \alpha, \beta, \gamma > 0$, based on $N = 10{,}000$ simulations, sample size $n = 200$, and level $\alpha = 0.05$.

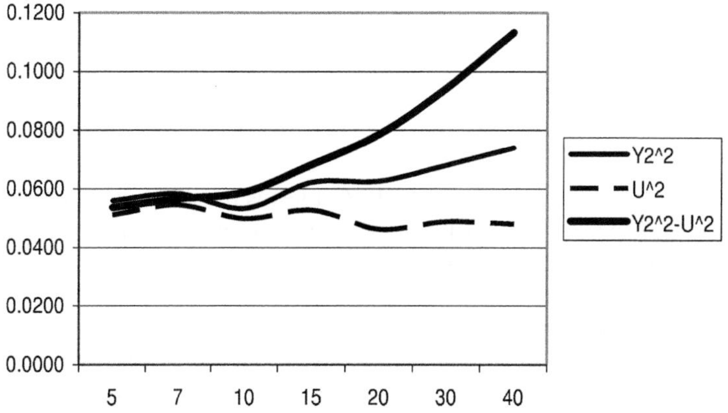

FIGURE 4.9 Simulated powers, as functions of the number of equiprobable cells r, of the tests $Y2_n^2(\overline{\boldsymbol{\theta}}_n)(Y2\hat{}2), U_n^2(\overline{\boldsymbol{\theta}}_n)(U\hat{}2)$, and $S1_n^2(\overline{\boldsymbol{\theta}}_n)(Y2\hat{}2 - U\hat{}2)$ tests for the Power Generalized Weibull alternative with cdf $F(x) = 1 - \exp\{1 - [1 + (x/\sigma)^{\nu}]^{1/\gamma}\}, x, \sigma, \nu, \gamma > 0$, under $N = 10{,}000$ simulations, sample size $n = 200$, and level $\alpha = 0.05$.

From these two figures, we observe that the DN $U_n^2(\overline{\boldsymbol{\theta}}_n)$ test for equiprobable random cells possesses no power for both Exponentiated and Power Generalized Weibull alternatives. On the other hand, the $S1_n^2(\overline{\boldsymbol{\theta}}_n)$ test is the most powerful test for both alternatives considered, especially with larger number of cells. The same behavior between the powers of $U_n^2(\overline{\boldsymbol{\theta}}_n)$ and $S1_n^2(\overline{\boldsymbol{\theta}}_n)$ tests was observed when testing the logistic null hypothesis against the normal distribution alternative (see Section 4.4). The case $r > 40$ was not considered since the expected cell probabilities become small in this case and the limiting distributions of the considered tests may not follow the chi-square distribution.

4.6.3 Neyman-Pearson classes

To maximize a measure of the distance between the null and alternative density functions, one may use the Neyman-Pearson classes for the partitioning of a sample space. Let $f(x;\boldsymbol{\theta})$ and $g(x;\boldsymbol{\varphi})$ be densities of the null and the alternative hypotheses, respectively. Given the parameters $\boldsymbol{\theta}$ and $\boldsymbol{\varphi}$, define two Neyman-Pearson classes as follows: $I_1 = \{x : f(x;\boldsymbol{\theta}) < g(x;\boldsymbol{\varphi})\}$ and $I_2 = \{x : f(x;\boldsymbol{\theta}) \geqslant g(x;\boldsymbol{\varphi})\}$. Suppose the density functions intersect at three points, say, x_1,x_2 and x_3, then $I_1 = (0,x_1] \cup [x_2,x_3]$ and $I_2 = (x_1,x_2) \cup (x_3, +\infty)$.

Then, the power of the HRM statistic $Y2_n^2(\overline{\boldsymbol{\theta}}_n)$ was simulated for the three-parameter Weibull family in (4.25) as the null hypothesis and for the two alternatives mentioned for the case when the test was applied with equiprobable cells for the same sample size n and level $\alpha = 0.05$. The results so obtained are presented in Table 4.3. The statistical errors shown in Table 4.3 are of one simulated standard deviation.

4.6.4 Discussion

From Figures 4.8 and 4.9, we may conclude that if an alternative hypothesis is not specified and we use equiprobable cells, then the DN $U_n^2(\overline{\boldsymbol{\theta}}_n)$ test is not satisfactory and the more powerful $S1_n^2(\overline{\boldsymbol{\theta}}_n)$ test is recommended in this case. On the other hand, if the alternative hypothesis is specified, then from Table 4.3, we see that the HRM test $Y2_n^2(\overline{\boldsymbol{\theta}}_n)$ for two Neyman-Pearson classes possesses higher power and is therefore the one we would recommend.

Consider the numerical data of Sample 1 from Smith and Naylor, 1987 that describe the strength of glass fibers of length 1.5 cm. The global maximum of the likelihood function for the three-parameter continuous Weibull probability distribution is $+\infty$ and is achieved at $\mu = X_{(1)}$. This creates problems in obtaining the MLEs. To overcome this problem, Smith and Naylor (1987) proposed to consider observations as being integers, and by adopting this approach, they estimated the parameters of the hypothesized three-parameter Weibull distribution as $\hat{\mu} = -1.6, \hat{\boldsymbol{\theta}} = 3.216$, $\hat{p} = 11.9$, by the local maximum

TABLE 4.3 Power of the HRM test for two Neyman-Pearson classes.

	W-PGW	W-ExpW
$\alpha = 0.05$	0.141 ± 0.025	0.294 ± 0.015

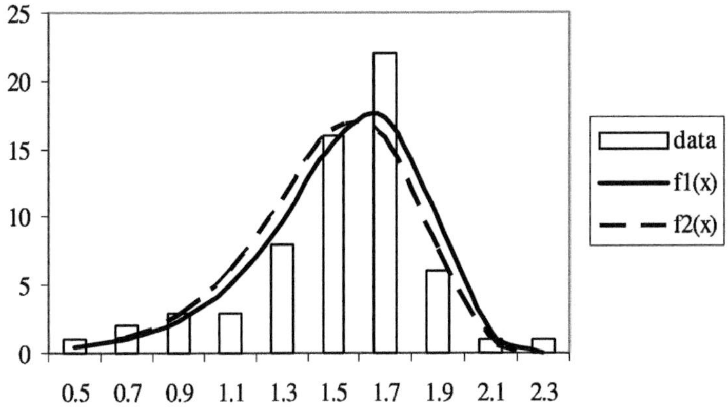

FIGURE 4.10 The histogram of the data of McEwen and Parresol (1991) and the density in (4.25) calculated with the MMEs ($f1(x)$) and the MLEs ($f2(x)$).

likelihood estimators. To test the null hypothesis, they then compared the empirical distribution function (EDF) of the original data with the approximate 95% confidence limits formed from the 19 EDFs based on pseudorandom samples. The MMEs in this case are quite different and are $\bar{\mu} = -4.4, \bar{\boldsymbol{\theta}} = 6.063, \bar{p} = 22.92$, but the density function in (4.25) calculated for these two sets of estimates do not differ much (see Figure 4.10).

The fit based on the MMEs ($\chi_9^2 = 49.5$) is slightly better than that based on the MLEs ($\chi_9^2 = 58.0$). Having calculated the statistics $Y2_n^2(\overline{\boldsymbol{\theta}}_n)$ and $S1_n^2(\overline{\boldsymbol{\theta}}_n)$ with the number of equiprobable cells as $r = 12$, we obtained the P-values to be 0.101 and 0.043, respectively. From this, it follows that the null hypothesis of the three-parameter Weibull distribution is not rejected at level $\alpha = 0.05$ by $Y2_n^2(\overline{\boldsymbol{\theta}}_n)$, but is rejected by the more powerful $S1_n^2(\overline{\boldsymbol{\theta}}_n)$ test.

4.6.5 Concluding remarks

From the results obtained, it follows that the shapes of the Three-parameter Weibull, Power Generalized Weibull, and Exponentiated Weibull distributions are quite close to each other even though their hazard rate functions could be different. This suggests that it would be useful to develop some goodness of fit tests which will directly compare the observed and hypothetical failure rate functions (see Hjort, 1990).

Chapter 5

Modifications Based on UMVUEs

5.1 TESTS FOR POISSON, BINOMIAL, AND NEGATIVE BINOMIAL DISTRIBUTIONS

Bol'shev and Mirvaliev (1978) were possibly the first to construct the modified chi-squared tests based on Uniformly Minimum Variance Unbiased Estimators (UMVUEs) of the unknown parameters for Poisson, binomial, and negative binomial distributions. The probability mass functions of Poisson, binomial, negative binomial, hypergeometric, and negative hypergeometric distributions are as follows:

$$\mathbf{P}(X = x,\theta) = \frac{\theta^x}{x!}\exp(-\theta), \quad \theta > 0,$$

$$\mathbf{P}_1(X = x,n,\theta) = \binom{n}{x}\left(\frac{\theta}{n}\right)^x\left(1-\frac{\theta}{n}\right)^{n-x},$$
$$0 \le \theta \le n,\ n = 1,2,\ldots,$$

$$\mathbf{P}_2(X = x,m,\theta) = \binom{m+x-1}{m-1}\left(\frac{\theta}{m+\theta}\right)^x$$
$$\times\left(1-\frac{\theta}{m+\theta}\right)^m, \theta > 0,\ m > 0,$$

Chi-Squared Goodness of Fit Tests with Applications. http://dx.doi.org/10.1016/B978-0-12-397194-4.00005-3

$$\mathbf{P}_3(X = x, N, M, n) = \frac{\binom{n}{x}\binom{N-n}{M-x}}{\binom{N}{M}},$$
$$M = 0, \dots, N,\ n = 1, \dots, N,\ N = 1,2, \dots,$$
$$\mathbf{P}_4(X = x, N, M, m) = \frac{\binom{m+x-1}{m-1}\binom{N-m-x}{M-m}}{\binom{N}{M}},$$
$$M = 0, \dots, N,\ 0 < m \leqslant M,\ N = 1,2, \dots$$

Let X be a random variable with its probability mass function as

$$\begin{aligned}\mathbf{P}(X = x) &= w_x = w_x(a,n,m,\theta)\\ &= \frac{1}{2}a(a+1)\mathbf{P}_1(X = x,n,\theta)\\ &\quad +(1-a^2)\mathbf{P}(X = x,\theta) + \frac{1}{2}a(a-1)\mathbf{P}_2(X = x,m,\theta),\end{aligned} \tag{5.1}$$

where $a \in \{-1,0,1\}, n \in \{1,2,\dots\}$, and $m > 0$. The values of a,n,m are assumed to be known, while θ is a nuisance parameter. If $a = 0$ or -1, then $\theta > 0$; otherwise, if $a = 1$, then $0 \leqslant \theta \leqslant n$. Note that if $a = 0$, then (5.1) follows the Poisson distribution; if $a = 1$, it follows the binomial distribution; if $a = -1$, it follows the negative binomial distribution; and if $a = -1$ and $m = 1$, then it follows the geometric distribution.

Let $X_1, \dots, X_t, t \geqslant 2$, be i.i.d. random variables from the distribution in (5.1). In this case, the sum $S = X_1 + \cdots + X_t$ is a complete sufficient statistic for the nuisance parameter θ, and, for $s \in \{0,1,\dots\}, i \in \{1,\dots,t\}$, and $x \in \{0,1,\dots,s\}$, we have

$$\begin{aligned}\mathbf{P}(S = s) &= w_s(a,tn,tm,t\theta),\\ \mathbf{P}(X_i = x|S = s) &= W_x(a,n,m,s,t)\\ &= \frac{1}{2}a(a+1)\mathbf{P}_3(X = x,tn,s,n)\\ &\quad +(1-a^2)\mathbf{P}_1(X = x,s,s/t)\\ &\quad +\frac{1}{2}a(a-1)\mathbf{P}_4(X = x,mt+s-1,mt-1,m).\end{aligned} \tag{5.2}$$

Since $\mathbf{EP}(X_i = x|S = s) = \mathbf{P}(X_i = x)$, the distribution in (5.2) is the UMVUE for (5.1) (see Voinov and Nikulin, 1993). In other words, $W_x = W_x(a,n,m,S,t), x = 0,1,\dots,S$, is the UMVUE of w_x (see Eq. (5.1)).

Let $N_{ij} = 1$ if $X_i = i$, and $N_{ij} = 0$ if $X_i \neq i$ for $i = 0,1,\ldots,$ $j = 1,2,\ldots,t$, and let

$$N_i = N_{i1} + \cdots + N_{it}, \quad i = 0,\ldots,S. \tag{5.3}$$

Then, the frequencies in (5.3) satisfy the following linear conditions:

$$N_0 + N_1 + \cdots + N_S = t \quad \text{and} \quad N_1 + 2N_2 + \cdots + SN_S = S.$$

Let $I_1, I_2, \ldots, I_r$ be non-empty non-intersecting subsets of the set $I = \{0,1,\ldots,S\}$ such that $\bigcup I_i = I$. Let us assume that $S \in I_r$ and $r \geqslant 2$.

Consider the vectors

$$\mathbf{N}^{(0)} = (N_1^{(0)},\ldots,N_r^{(0)})^T \quad \text{and} \quad \mathbf{W}^{(c)} = (W_1^{(c)},\ldots,W_r^{(c)})^T$$

with their components as

$$N_j^{(0)} = \sum_{i \in I_j} N_i, \quad W_j^{(c)} = \sum_{i \in I_j} i^c W_i, \qquad j = 1,\ldots,r,\ c = 0,1.$$

Under this setup, Bol'shev and Mirvaliev (1978) established the following result.

Theorem 5.1. *If the sample size t increases unboundedly such that $S \asymp t$, then conditional on S, the distribution of the vector $\mathbf{Z} = t^{-1/2}(\mathbf{N}^{(0)} - t\mathbf{W}^{(0)})$ is asymptotically normal with zero mean vector, $E(\mathbf{Z}|S) = \mathbf{0}$, and covariance matrix as $\mathbf{E}(\mathbf{Z}\mathbf{Z}^T|S) = \mathbf{B} + o(1)$ $(t \to \infty)$, where*

$$\mathbf{B} = \mathbf{D}^{(0)} - \mathbf{W}^{(0)}\mathbf{W}^{(0)T} - \frac{1}{\beta}\left[\frac{S}{t}\mathbf{W}^{(0)} - \mathbf{W}^{(1)}\right]\left[\frac{S}{t}\mathbf{W}^{(0)} - \mathbf{W}^{(1)}\right]^T,$$

$\mathbf{D}^{(0)}$ is a diagonal matrix with elements $W_1^{(0)},\ldots,W_r^{(0)}$ on the main diagonal,

$$\beta = \frac{S}{t} + \left(\frac{S}{t}\right)^2\left[-\frac{a(a+1)}{2n} + \frac{a(a-1)}{2m}\right],$$

and the rank of $\mathbf{B}$ equals $r - 1$.

This theorem generalizes the earlier results of Sevast'yanov and Chistyakov (1964) and Kolchin (1968) for the Poisson ($a = 0$) distribution, and the result of Park (1973) for the geometric distribution ($a = -1, m = 1$).

Then, by using Wald's method (see Chapter 3), Bol'shev and Mirvaliev (1978) showed that the quadratic form

$$Y^2 = \sum_{j=1}^{r} \frac{(N_j^{(0)} - tW_j^{(0)})^2}{tW_j^{(0)}} + \frac{1}{t\gamma}\left(\sum_{j=1}^{r} \frac{W_j^{(1)}N_j^{(0)}}{W_j^{(0)}} - S\right)^2, \tag{5.4}$$

where

$$\gamma = \beta + \left(\frac{S}{t}\right)^2 - \sum_{j=1}^{r} \frac{(W_j^{(1)})^2}{W_j^{(0)}},$$

will asymptotically ($t \to \infty, S \asymp t$) follow the chi-square distribution with $r-1$ degrees of freedom.

Remark 5.1. *The limiting distributions of the vector* $\mathbf{Z}$ *and the statistic in (5.4) will not change if* W_i *will be replaced by* $\tilde{w}_i = w_i(a,n,m,S/t)$.

Remark 5.2. *The test statistic in (5.4) is useful when the number of grouping cells* $r \geqslant 2$ *and an alternative hypothesis is not specified. Suppose we have to test one of the following three hypotheses:* $a = -1,0,+1$, *against an alternative of the same kind (e.g.* $a = 0$ *against* $a = -1$*). In this case, the power of the test can be increased by using the Neyman-Pearson type classes (see Section 2.3).*

Using Stirling's formula, we can show that for the Poisson, binomial, and negative binomial distributions, we have

$$\begin{aligned} 2n \ln \frac{\mathbf{P}(X = x,\theta)}{\mathbf{P}_1(X = x,n,\theta)} &\sim 2m \ln \frac{\mathbf{P}_2(X = x,m,\theta)}{P(X = x,\theta)} \\ &\sim \frac{2nm}{n+m} \ln \frac{\mathbf{P}_2(X = x,m,\theta)}{\mathbf{P}_1(X = x,n,\theta)} \\ &\sim (x - x_1)(x - x_2), \quad n,m \to \infty, \end{aligned}$$

where

$$x_1 = \theta + 0.5 - \sqrt{\theta + 0.25}, \quad x_2 = \theta + 0.5 + \sqrt{\theta + 0.25}. \tag{5.5}$$

If $x < x_1$ or $x > x_2$, then

$$\mathbf{P}_1(X = x,n,\theta) < P(X = x,\theta) < \mathbf{P}_2(X = x,m,\theta).$$

If $x_1 < x < x_2$, then

$$\mathbf{P}_2(X = x,m,\theta) < P(X = x,\theta) < \mathbf{P}_1(X = x,n,\theta).$$

Replacing θ in (5.5) by S/t, we can define two NP type classes

$$I_1 = \{k_1 + 1, \ldots ,k_2\}, \quad I_2 = \{0,1, \ldots ,k_1,k_2 + 1,k_2 + 2, \ldots\},$$

where $k_1 = [x_1]$ and $k_2 = [x_2]$. Then, Bol'shev and Mirvaliev (1978) showed in this case that the statistic

$$Y = \frac{N_1^{(0)} - tW_1^{(0)}}{\sqrt{t\left[W_1^{(0)}(1 - W_1^{(0)}) - \frac{1}{\beta}\left(\frac{S}{t}W_1^{(0)} - W_1^{(1)}\right)^2\right]}}, \tag{5.6}$$

under H_0 when $t \to \infty$ and $S \asymp T$, will asymptotically follow the standard normal distribution $\Phi(\,\cdot\,)$. Let us denote the null hypotheses $a = -1,0,1$ by symbols $\bar{b}$ (negative binomial), p (Poisson), and b (binomial), respectively, and let $\langle H_0|H_1\rangle$ mean that we are testing H_0 against H_1. Using (5.6), in the cases of $\langle p|\bar{b}\rangle, \langle b|\bar{b}\rangle, \langle b|p\rangle$, and $\langle b|\bar{b} \cup p\rangle$, we have to reject H_0 if $Y < -\Phi^{-1}(1-\alpha)$, where α is the nominal level of the test. In the cases of $\langle p|b\rangle, \langle \bar{b}|b\rangle, \langle \bar{b}|p\rangle$, and $\langle \bar{b}|p \cup b\rangle$, the rejection region becomes $Y > \Phi^{-1}(1-\alpha)$.

5.2 CHI-SQUARED TESTS FOR ONE-PARAMETER EXPONENTIAL FAMILY

Consider a continuous random variable X belonging to the one-parameter exponential distribution with density

$$f(x,\theta) = g(x)\exp\{V_1(\theta)T(x) + V_2(\theta)\}, \quad \theta \in \Theta \subset R^1, \; x \in R^1, \tag{5.7}$$

where $g(x)$ and $T(x)$ are known functions, and $V_1(\theta)$ and $V_2(\theta)$ are continuously differentiable functions of a parameter θ. If $X_1,\ldots,X_n$ are i.i.d. from the family in (5.7), then the sum $S = \sum_{i=1}^n T(X_i)$ is a complete sufficient statistic for the parameter θ.

Let us assume the following conditions:

(i) A random variable $T(X)$ possesses finite population moments up to the fourth order;
(ii) The random variable $Z = (S - na)/(b\sqrt{n})$, where $a = \mathbf{E}[T(X)]$ and $b = \sqrt{Var[T(X)]}$, possesses a bounded probability density function;
(iii) The support of the density in (5.7) does not depend on the parameter θ.

Under the conditions in (i)–(iii), the UMVUEs $\hat{f}(x|S)$ and $\hat{f}(x_1,x_2|S)$ of the densities $f(x,\theta)$ and $f(x_1,x_2,\theta)$ are given by (see Voinov and Nikulin, 1993)

$$\hat{f}(x|S) = \frac{f(x,\theta)f_{n-1}(S - T(x),\theta)}{f_n(S,\theta)} \tag{5.8}$$

and

$$\hat{f}(x_1,x_2|S) = \frac{f(x_1,x_2,\theta)f_{n-2}(S - T(x_1) - T(x_2),\theta)}{f_n(S,\theta)}, \tag{5.9}$$

respectively, where $f_n(S,\theta)$ is the density of S. Let $I_1,I_2,\ldots,I_r$ be non-intersecting intervals such that $\bigcup I_j = R^1$. Let $N_j^{(n)}$ denote the number of elements of the sample $X_1,\ldots,X_n$ that fall into the interval $I_j, j = 1,\ldots,r$, and

$$A_j = \frac{a_j - a\mathbf{P}_j}{b}, \tag{5.10}$$

where $\mathbf{P}_j = P(X \in I_j), a_j = \int_{I_j} T(x)f(x,\theta)dx, j = 1,\ldots,r$. It is well known (see Voinov and Nikulin, 1993) that the UMVUEs $\hat{p}_j$ and $\hat{p}_{ij}$ of $\mathbf{P}_j$ and $\mathbf{P}_{ij}$ are given by

$$\hat{p}_j = \int\limits_{x \in I_j} \hat{f}(x|S)dx \tag{5.11}$$

and

$$\hat{p}_{ij} = \int\limits_{x_1 \in I_i} \int\limits_{x_2 \in I_j} \hat{f}(x_1,x_2|S)dx_1,dx_2, \quad i,j = 1,\ldots,r, \tag{5.12}$$

respectively. Under this setting, Chichagov (2006) proved the following result.

Theorem 5.2. *Under the regularity conditions in (i)–(iii), the random vector*

$$\mathbf{Y}(n) = \left(\frac{N_1^{(n)} - n\hat{p}_1}{\sqrt{n}}, \ldots, \frac{N_{r-1}^{(n)} - n\hat{p}_{r-1}}{\sqrt{n}} \right)^T$$

asymptotically will follow the multivariate normal distribution with zero mean vector and a non-degenerate covariance matrix $\boldsymbol{\Sigma} = ||\sigma_{ij}||$, *with its elements as* $\sigma_{ij} = p_i\delta_{ij} - p_i p_j - A_i A_j, i,j = 1,\ldots,r-1$. *Moreover, the UMVUE* $\hat{\sigma}_{ij}$ *of* σ_{ij} *is*

$$\hat{\sigma}_{ij} = \hat{p}_i(\delta_{ij} - \hat{p}_j) - (n-1)(\hat{p}_i\hat{p}_j - \hat{p}_{ij}). \tag{5.13}$$

Corollary 5.1. *Let* $\hat{\boldsymbol{\Sigma}}_n$ *be the matrix with its elements as in (5.13). Then, the quadratic form*

$$Y_{Ch}^2 = \mathbf{Y}^T(n)\hat{\boldsymbol{\Sigma}}_n^{-1}\mathbf{Y}(n) \tag{5.14}$$

will asymptotically follow the chi-square distribution with $r-1$ *degrees of freedom. The statistic in (5.14) can be written explicitly as (Chichagov, 2006)*

$$Y_{Ch}^2 = \sum_{j=1}^{r} \frac{(N_j^{(n)} - n\hat{p}_j)^2}{n\hat{p}_j} + \left(1 - \sum_{j=1}^{r} \frac{\hat{A}_j^2}{\hat{p}_j}\right)^{-1} \left(\sum_{j=1}^{r} \frac{N_j^{(n)}\hat{A}_j}{\sqrt{n}\hat{p}_j}\right)^2. \tag{5.15}$$

From (5.15), we see that for implementing this test, we need the MVUEs of A_j and p_j, for $j = 1,\ldots,r$. Chichagov (2006) gave several examples of these MVUEs in the case when the density function of $T(X)$ is gamma distributed, with known shape parameter ν and unknown scale parameter σ, of the form

$$f_g(y,\sigma,\nu) = \frac{y^{\nu-1}}{\sigma^\nu \Gamma(\nu)} \exp(-y/\sigma), \quad y > 0. \tag{5.16}$$

This assumption is satisfied for several known one-parameter continuous distributions such as normal, inverse Gaussian, Weibull, and Pareto. Chichagov (2006) proved the following general result.

Theorem 5.3. *Let the positive half-line* $[0, +\infty)$ *be partitioned into intervals* $J_T(j) = [J_{T1}(j), J_{T2}(j)]$ *such that* $\{x \in R^1 : T(x) \in J_T(j)\} = J(j)$, *and* $\cup J_T(j) = [0, +\infty), j = 1, \ldots, r$. *Then,*

$$A_j = \frac{1}{\sqrt{\nu}}[J_{T1}(j) f_g(J_{T1}(j),\sigma,\nu) - J_{T2}(j) f_g(J_{T2}(j),\sigma,\nu)]$$

and its UMVUE is

$$\hat{A}_j = \frac{f_B\left(J_{T1}(j)/S,\nu+1,(n-1)\nu\right) - f_B\left(J_{T2}(j)/S,\nu+1,(n-1)\nu\right)}{n\sqrt{\nu}}, \tag{5.17}$$

where

$$f_B(x,\nu_1,\nu_1) = \frac{\Gamma(\nu_1+\nu_2)}{\Gamma(\nu_1)\Gamma(\nu_2)} x^{\nu_1-1}(1-x)^{\nu_2-1}, \quad 0 < x < 1, \ \nu_1 > 0, \ \nu_2 > 0.$$

Similarly, the UMVUE of p_j *is*

$$\hat{p}_j = \int\limits_{y \in J_T(j)} \hat{f}_g(y,\sigma,\nu)dy, \tag{5.18}$$

where

$$\hat{f}_g(y,\sigma,\nu) = \frac{1}{S} f_B(y/S,\nu,(n-1)\nu).$$

Example 5.2.1. Assume that a random variable X follows the Pareto distribution with density

$$f(x,\theta) = \frac{\theta\lambda^\theta}{x^{\theta+1}} = \frac{\theta}{x}\exp\{\theta(\ln\lambda - \ln x)\}, \quad x \geqslant \lambda, \ \theta > 0.$$

It evidently belongs to the family in (5.7) with $V_1(\theta) = -\theta, V_2(\theta) - \ln\theta$ and $g(x) = 1/x$.

Suppose the shape parameter λ is known, while the nuisance parameter θ is unknown. Assume $T(X) = \ln X - \ln\lambda$ and $\sigma = 1/\theta$, then $a = 1/\theta = \sigma$ and $b^2 = 1/\theta^2 = \sigma^2$. For simplicity, let us define grouping intervals as

$$J(j) = [\lambda\exp(J_{j-1}), \lambda\exp(J_j)),$$
$$J_0 = 0 < J_1 < J_2 < \cdots < J_{r-1} < J_r = \infty, \quad j = 0,1,\ldots,r.$$

In this case (Chichagov, 2006), we have

$$p_j = \int\limits_{J(j)} \frac{\theta\lambda^\theta}{x^{\theta+1}}dx = \exp\left(-\frac{J_{j-1}}{\sigma}\right) - \exp\left(-\frac{J_j}{\sigma}\right),$$

$$A_j = \int\limits_{J(j)} \frac{(\ln x/\lambda - 1/\theta)}{1/\theta}\frac{\theta\lambda^\theta}{x^{\theta+1}}dx = \int\limits_{J_{j-1}}^{J_j} \frac{y-\sigma}{\sigma^2}\exp\left(-y/\sigma\right)dy$$
$$= J_{j-1} f_g(J_{j-1},\sigma,1) - J_j f_g(J_j,\sigma,1).$$

Using Eqs. (5.17) and (5.18), and the sufficient statistic $S_1 = \sum_{i=1}^{n} (\ln X_i - \ln \lambda)$, the UMVUEs of p_j and A_j can be expressed as

$$\hat{p}_j = \left(1 - \frac{J_{j-1}}{S_1}\right)^{n-1} I(J_{j-1} < S_1) - \left(1 - \frac{J_j}{S_1}\right)^{n-1} I(J_j < S_1), \quad (5.19)$$

$$\begin{aligned}\hat{A}_j &= \frac{1}{n}\left[f_B\left(\frac{J_{j-1}}{S_1},2,n-1\right) - f_B\left(\frac{J_j}{S_1},2,n-1\right)\right] \\ &= (n-1)\left[\frac{J_{j-1}}{S_1}\left(1 - \frac{J_{j-1}}{S_1}\right)^{n-2} I(J_{j-1} < S_1)\right. \\ &\quad \left. - \frac{J_j}{S_1}\left(1 - \frac{J_j}{S_1}^{n-2}\right) I(J_j < S_1)\right], \end{aligned} \quad (5.20)$$

where $I(A)$ is the indicator function of the event A. Upon substituting (5.19) and (5.20) into (5.15), we obtain the needed test statistic.

An alternative test statistic for the Pareto null hypothesis was suggested by Gulati and Shapiro (2008).

Remark 5.3. *Under conditions (i)–(iii), the above result is valid for discrete distributions as well.*

Remark 5.4. *Numerous chi-squared tests based on UMVUEs have been proposed for Modified Power Series Distributions, Discrete Multivariate Exponential Distributions, General Continuous Exponential Distributions, and Natural Exponential Families with Power Variance Functions; for pertinent details, one may refer to the book of Greenwood and Nikulin (1996).*

5.3 REVISITING CLARKE'S DATA ON FLYING BOMBS

As an example of the spatial distribution of random points on a plane, consider the data of hits of flying bombs in the southern part of London during World War II. The entire territory was divided into $t = 576$ pieces of land with 1/4 km^2 each. Frequencies of hits are then as presented in Table 5.1.

The total number of flying bombs was $t = \sum N_k = 576$ and $S = \sum kN_k = 537$. For the Poisson null hypothesis $P(k) = \theta^k \exp(-\theta)/k!, k = 0,1,\ldots,$ Clarke (see Feller, 1964) used the standard Pearson's test in (2.3) with the parameter θ replaced by its MLE based on non-grouped data. He then showed that this test does not reject the null hypothesis of the Poisson model. The only error in this analysis is that, when using the MLEs based on non-grouped data, the statistic in (2.3) does not follow in the limit the chi-squared probability distribution (Chernoff and Lehmann, 1954; Bol'shev and Mirvaliev, 1978). Formally, it is possible to use the Pearson-Fisher test in (2.19) with $\hat{\theta}_n$ being an estimator of the parameter θ based on grouped data, but as already pointed out, this test possesses low power. Let us use the powerful test Y in (5.6) or, equivalently Y^2, which follows chi-squared distribution with one degree of freedom.

Consider two null hypotheses for describing these data: the Poisson distribution $P(k) = \theta^k \exp(-\theta)/k!$, $k = 0, 1, \ldots$, and the binomial distribution

$$P_1(k) = \binom{S}{k}\left(\frac{\theta}{S}\right)^x\left(1-\frac{\theta}{S}\right)^{S-k}, \quad k = 0, \ldots, S.$$

Since S is a complete sufficient statistic, the UMVUE of θ will be $\hat{\theta} = S/t = 0.9323$. From Eq. (5.2), the UMVUEs W_k of $P(k)$ and $W1_k$ of $P_1(k)$ will be

$$W_k = \binom{S}{k}\left(\frac{\theta}{S}\right)^x\left(1-\frac{\theta}{S}\right)^{S-k}, \quad k = 0, \ldots, S,$$

and

$$W1_k = \frac{\binom{S}{k}\binom{St-S}{S-k}}{\binom{St}{S}}, \quad k = 0, \ldots, S,$$

respectively, and the corresponding results are given in Table 5.1. Consider two Neyman-Pearson type classes: $\Delta_1 = \{1,2\}$, where $P_1(k)$ is more than $P(k)$, and $\Delta_2 = \{0\} \cup \{3, \ldots\}$, where $P_1(k)$ is less than $P(k)$. Table 5.2 presents results for these cases.

Since the P-values of the Y^2 test are large enough (0.566 and 0.548), we see that both null hypotheses are not rejected and so both of them can be used for describing the data. This is not surprising since the distributions $P_1(k)$ and $P(k)$ (see columns 3 and 4 of Table 5.1) are close to each other. In other words,

TABLE 5.1 The observed frequencies N_k, the MVUEs W_k of the hypothesized Poisson probabilities, and the MVUEs $W1_k$ of the hypothesized binomial probabilities for the data on hits of flying bombs.

k	N_k	$tP(k)$	$tP_1(k)$	W_k	$W1_k$	kW_k	$kW1_k$
0	229	226.74	226.56	0.393	0.393	0	0
1	**211**	**211.39**	**211.59**	**0.367**	**0.368**	**0.367**	**0.368**
2	**93**	**98.54**	**98.62**	**0.171**	**0.171**	**0.342**	**0.343**
3	35	30.62	30.59	0.053	0.053	0.159	0.159
4	7	7.14	7.10	0.012	0.012	0.049	0.049
>4	1	1.57	1.55	0.003	0.003	0.013	0.013

TABLE 5.2 Values of $N_1^{(0)} = N_1 + N_2$, $W_1^{(0)} = W_1 + W_2$, $W_1^{(1)} = W_1 + 2W_2$, $\beta = S/t$ for $P(k)$, and $\beta = S/t - S/t^2$ for $P_1(k)$, for the data on hits of flying bombs.

	Under H_0	
	$P(k)$	$P_1(k)$
$N_1^{(0)}$	304	304
$W_1^{(0)}$	0.5386	0.5390
$tW_1^{(0)}$	310.205	310.480
$W_1^{(1)}$	0.7091	0.7104
β	0.9323	0.9307
Y^2	0.3301	0.3607
P-value	0.566	0.548

the sample size in Clarke's data is high enough that for a small parameter θ/S, the binomial distribution $P_1(k)$ almost attains its limiting Poisson distribution $P(k)$. This is also in conformance with Clarke's finding that the data on flying bombs can be well described by the Poisson model.

Chapter 6

Vector-Valued Tests

6.1 INTRODUCTION

Different ways of improving goodness of fit tests have been discussed in the literature. For obtaining more powerful and suitable tests, Cochran (1954) proposed to use single degree of freedom, or groups of degrees from the total chi-squared sum. This has been materialized for different classical and modified chi-squared statistics, for which independent chi-squared components with one degree of freedom in the limit have been developed; see Anderson (1994), Rayner (2002), Voinov et al. (2007), Voinov (2010), and also Section 2.3. Lemeshko and Postovalov (1998) and Lemeshko (1998) succeeded in increasing the power of the PF test by a special data grouping method. Their idea was to construct grouping cells in such a way that the determinant of the Fisher information matrix would be maximized.

Sometimes, one may combine test statistics to obtain a more powerful test, or a test that is sensitive to a specific alternative, or even a test for checking the consensus of a set of tests. Combined test statistics can be either dependent or independent. Many results are available regarding the combination of independent test statistics based on the probability integral transformation; see, for example, Van Zwet and Oosterhoff (1967), Wilk and Shapiro (1968), Littell and Folks (1971), Koziol and Perlman (1978), Marden (1982), Rice (1990), Mathew et al. (1993), and Sarkar and Chang (1997). Combining independent tests in linear models was considered by Zhou and Mathew (1993). Brown (1975) considered the same problem when the combined tests are not independent, and developed a method for combining non-independent one-sided tests about a location parameter.

Chi-Squared Goodness of Fit Tests with Applications. http://dx.doi.org/10.1016/B978-0-12-397194-4.00006-5

To combine information from several sources, the above-mentioned works, incidentally, did not use vector-valued statistics which may also be useful for obtaining more powerful tests.

Zhakharov et al. (1969) proposed a sequential m-dimensional chi-squared vector-valued test

$$X_n^2 = (X_{n_1}^2, X_{n_2}^2, \ldots, X_{n_m}^2)^T$$

based on m subsamples, embedded into each other, of a sample of size n such that $n_1 < n_2 < \cdots < n_m \leqslant n$, where $X_{n_i}^2, i = 1, \ldots, m$, are standard Pearson sums. The null hypothesis is accepted if one of the events $A_k, k = 1, \ldots, m$, occurs, where

$$A_k = \{X_{n_1}^2 > x_{1,\alpha}, \ldots, X_{n_{k-1}}^2 > x_{k-1,\alpha}, X_{n_k}^2 \leqslant x_{k,\alpha}\}$$

and $x_{i,\alpha}$ $(i = 1, \ldots, m)$ are the critical values. If, for all $k = 1, \ldots, m, X_{n_k}^2 > x_{k,\alpha}$, then the null hypothesis is rejected.

Mason and Schuenemeyer (1983) proposed a vector-valued test statistic of the form

$$(\omega_1 L_{n,1}, \omega_2 L_{n,2}, K_n, \omega_3 U_{n,1}, \omega_4 U_{n,2})^T,$$

where K_n is the Kolmogorov-Smirnov test, $L_{n,1}, L_{n,2}, U_{n,1}, U_{n,2}$ are Rényi-type tests, $\omega_1, \ldots, \omega_4$ are non-negative weights, and $0 < c < \infty$ is a constant that depends on the level of significance α. The null hypothesis is rejected if

$$\max\{\omega_1 L_{n,1}, \omega_2 L_{n,2}, K_n, \omega_3 U_{n,1}, \omega_4 U_{n,2}\} > c.$$

Note here that all components of the proposed statistic are based on the same sample.

Voinov and Grebenyk (1989) used a two-dimensional vector-valued test $\mathbf{V}_n = (K_n, R_n)^T$, where K_n is the Kolmogorov-Smirnov statistic and R_n is the signed rank statistic. In spite of the correlation between K_n and R_n, the test based on a rejection region, which is the intersection of corresponding rejection regions of the two components of $\mathbf{V}_n$, enabled well the recognition of a pattern of an image with signal/noise ratio less than one.

6.2 VECTOR-VALUED TESTS: AN ARTIFICIAL EXAMPLE

Consider the following artificial example of testing a simple null hypothesis about a probability distribution against a simple alternative:

$$H_0 : P(X \leqslant x) = F(x) \text{ vs. } H_a : P(X \leqslant x) = G(x). \quad (6.1)$$

Let Y_{1n}^2 and Y_{2n}^2 be two independent statistics such that

$$\lim_{n\to\infty} P(Y_{1n}^2 \leqslant y|H_0) = \lim_{n\to\infty} P(Y_{2n}^2 \leqslant y|H_0) = P(\chi_4^2 \leqslant y), \quad (6.2)$$

$$\lim_{n\to\infty} P(Y_{1n}^2 \leqslant y|H_a) = \lim_{n\to\infty} P(Y_{2n}^2 \leqslant y|H_a) = P(\chi_4^2(3.5) \leqslant y), \quad (6.3)$$

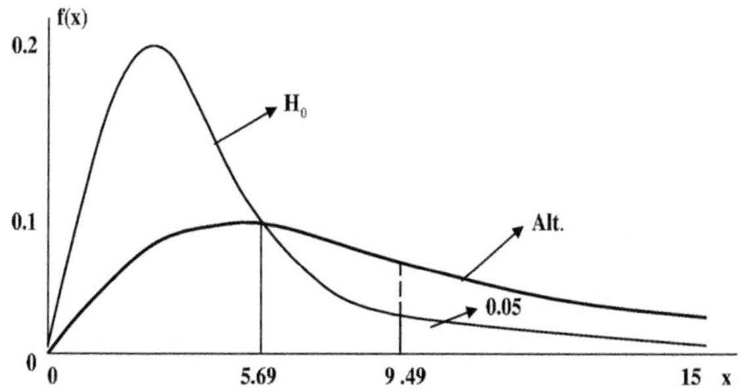

FIGURE 6.1 Probability density functions of $Y_{in}^2, i = 1,2$, under H_0 (thin solid line) and under H_a (bold solid line). The density function under H_a was calculated by using formula (1) of Cohen (1988).

where χ_4^2 is the central chi-square random variable with 4 degrees of freedom, and $\chi_4^2(3.5)$ is the non-central chi-square random variable with 4 degrees of freedom and non-centrality parameter 3.5. For brevity, the sign of the limit will be omitted in the sequel.

Consider a two-dimensional vector-valued test $\mathbf{U}_n = (Y_{1n}^2, Y_{2n}^2)^T$ with rejection region S_1 of the intersection-type, i.e. $S_1 = (Y_{1n}^2 > y_1) \cap (Y_{2n}^2 > y_1)$. Since Y_{1n}^2 and Y_{2n}^2 are independent and identical, the probability of falling into S_1 under H_0 (or the level of significance of the test) will be (see Figure 6.1)

$$P\{(Y_{1n}^2 > y_1|H_0) \cap (Y_{2n}^2 > y_1|H_0)\} = P(Y_{1n}^2 > y_1|H_0)P(Y_{2n}^2 > y_1|H_0)$$
$$= \alpha_1^2 = \alpha,$$

where α_1 is the level of significance of any component Y_{in}^2, $i = 1,2$, and α is the level of significance of the vector statistic $\mathbf{U}_n$. Suppose we wish to set $\alpha = 0.05$, then $\alpha_1 = 0.2236$, in which case the critical value y_1 of χ_4^2 variable will be $y_1 = 5.69$. In this case, the power of the vector-valued test $\mathbf{U}_n$ is determined as

$$P(\mathbf{U}_n \in S_1|H_a) = P(Y_{1n}^2 > y_1|H_a)P(Y_{2n}^2 > y_1|H_a) = (0.586)^2 = 0.343.$$

The values of the probabilities $P(Y_{in}^2 > y_1|H_a), i = 1,2$, were calculated by using formula (1) of Ding (1992). At the same time, power of each component of $\mathbf{U}_n$ for the same level of significance $\alpha = 0.05$ is

$$P(Y_{1n}^2 > y_2|H_a) = P(Y_{2n}^2 > y_2|H_a) = 0.282,$$

where $y_2 = 9.49$. Thus, we observe that the power of the components Y_{in}^2, $i = 1,2$, implemented independently, is 1.216 times less than the power of $\mathbf{U}_n$.

Consider the vector-valued test $\mathbf{U}_n = (Y_{1n}^2, Y_{2n}^2)^T$ with the rejection region S_2 of the union-type, i.e. $S_2 = (Y_{1n}^2 > y_3) \cup (Y_{2n}^2 > y_3)$. Suppose we set

$\alpha = 0.05$ once again. Then,

$$\begin{aligned} P(\mathbf{U}_n \in S_2|H_0) &= P\{(Y_{1n}^2 > y_3|H_0) \cup (Y_{2n}^2 > y_3|H_0)\} \\ &= P(Y_{1n}^2 > y_3|H_0) + P(Y_{2n}^2 > y_3|H_0) \\ &\quad - P(Y_{1n}^2 > y_3|H_0)P(Y_{2n}^2 > y_3|H_0) \\ &= 0.05, \end{aligned}$$

which means $P(Y_{1n}^2 > y_3|H_0) = 0.02532$, and so $y_3 = 11.1132$. Since $P(Y_{in}^2 > 11.1132|H_a) = 0.19525$, the power of $\mathbf{U}_n$ for this rejection region of union-type is

$$\begin{aligned} P(\mathbf{U}_n \in S_2|H_a) &= P\{(Y_{1n}^2 > y_3|H_a) \cup (Y_{2n}^2 > y_3|H_a)\} \\ &= P(Y_{1n}^2 > y_3|H_a) + P(Y_{21n}^2 > y_3|H_a) \\ &\quad - P(Y_{1n}^2 > y_3|H_a)P(Y_{2n}^2 > y_3|H_a) \\ &= 0.352, \end{aligned}$$

which is 1.25 times more powerful than $Y_{in}^2, i = 1,2$, implemented individually.

This simple artificial example shows that the use of two-dimensional vector-valued tests may result in an increase in power as compared to the powers of individual components of the vector-valued statistic.

Consider now the three-dimensional vector-valued test

$$\mathbf{T}_n = (Y_{1n}^2, Y_{2n}^2, Y_{3n}^2)^T,$$

where $Y_{in}^2, i = 1,2,3$, are three identical independent statistics with the same limiting distributions as in (6.2) and (6.3), for testing the hypotheses in (6.1). Let S_3 be an intersection-type rejection region of the form

$$S_3 = (Y_{1n}^2 > y_4) \cap (Y_{2n}^2 > y_4) \cap (Y_{3n}^2 > y_4).$$

The probability of $\mathbf{T}_n$ falling into S_3 under H_0 (or the level of significance of the test) will be

$$\begin{aligned} P(\mathbf{T}_n \in S_3|H_0) &= P(Y_{1n}^2 > y_4|H_0)P(Y_{2n}^2 > y_4|H_0)P(Y_{3n}^2 > y_4|H_0) \\ &= \alpha_1^3 = \alpha, \end{aligned}$$

where α_1 is the level of significance for any component $Y_{in}^2, i = 1,2,3$, and α is the level of significance of the vector statistic $\mathbf{T}_n$. Suppose we set $\alpha = 0.05$, then $\alpha_1 = 0.3684$, in which case the critical value y_4 of χ_4^2 variable will be $y_4 = 4.288$. Since $P(Y_{in}^2 > 4.288|H_a) = 0.723$, the power of the vector-valued test $\mathbf{T}_n$ is determined as

$$\begin{aligned} P(\mathbf{T}_n \in S_3|H_a) &= P(Y_{1n}^2 > y_4|H_a)P(Y_{2n}^2 > y_4|H_a)P(Y_{3n}^2 > y_4|H_a) \\ &= 0.723^3 = 0.378. \end{aligned}$$

Thus, we observe that the power of the components $Y_{in}^2, i = 1,2,3$, implemented individually is 1.34 times less than the power of $\mathbf{T}_n$.

Now, let us consider the vector-valued test $\mathbf{T}_n = (Y_{1n}^2, Y_{2n}^2, Y_{3n}^2)^T$ with the rejection region S_4 of the union-type, i.e. $S_4 = (Y_{1n}^2 > y_5) \cup (Y_{2n}^2 > y_5) \cup (Y_{3n}^2 > y_5)$. If we set $\alpha = 0.05$, then from

$$\begin{aligned}
P(\mathbf{T}_n \in S_4|H_0) = &P(Y_{1n}^2 > y_5|H_0) + P(Y_{2n}^2 > y_5|H_0) + P(Y_{3n}^2 > y_5|H_0) \\
&- P(Y_{1n}^2 > y_5|H_0)P(Y_{2n}^2 > y_5|H_0) \\
&- P(Y_{1n}^2 > y_5|H_0)P(Y_{3n}^2 > y_5|H_0) \\
&- P(Y_{2n}^2 > y_5|H_0)P(Y_{3n}^2 > y_5|H_0) \\
&+ P(Y_{1n}^2 > y_5|H_0)P(Y_{2n}^2 > y_5|H_0)P(Y_{3n}^2 > y_5|H_0) \\
= &0.05,
\end{aligned}$$

it follows that $P(Y_{in}^2 > y_3|H_0) = 0.016952$, and so $y_5 = 12.0543$. Since $P(Y_{in}^2 > 12.0543|H_a) = 0.15601$, the power of $\mathbf{T}_n$ for the rejection region of union-type is given by

$$\begin{aligned}
P(\mathbf{T}_n \in S_4|H_a) = &P(Y_{1n}^2 > y_5|H_a) + P(Y_{2n}^2 > y_5|H_a) + P(Y_{3n}^2 > y_5|H_a) \\
&- P(Y_{1n}^2 > y_5|H_a)P(Y_{2n}^2 > y_5|H_a) \\
&- P(Y_{1n}^2 > y_5|H_a)P(Y_{3n}^2 > y_5|H_a) \\
&- P(Y_{2n}^2 > y_5|H_a)P(Y_{3n}^2 > y_5|H_a) \\
&+ P(Y_{1n}^2 > y_5|H_a)P(Y_{2n}^2 > y_5|H_a)P(Y_{3n}^2 > y_5|H_a) \\
= &0.399,
\end{aligned}$$

which is 1.415 times more powerful than $Y_{in}^2, i = 1,2,3$, implemented individually.

The results of the above two artificial examples are summarized in Table 6.1. From Table 6.1, we observe that the power of the vector-valued test depends not only on the structure of the rejection region, but also on the dimensionality of the vector. We see that the gain in power, measured as the ratio of corresponding powers, is higher for the union-type rejection regions, and is also higher for the

TABLE 6.1 Gain in power for two- and three-dimensional tests for two types of rejection region (with $\alpha = 0.05$).

Type of the rejection region	Gain in power with two components	Gain in power with three components
$\cap$	0.343/0.282 = 1.216	0.378/0.282 = 1.340
$\cup$	0.352/0.282 = 1.248	0.399/0.282 = 1.415

three-dimensional vector compared to the two-dimensional one. Of course, the situation can be quite different for other forms of alternatives. Voinov and Pya (2010) showed that if the alternative hypothesis (under the same assumptions as above) is the central χ_6^2 distribution, then the gain in power is less for the union-type rejection regions, but is still higher for the three-dimensional vector test as compared to the two-dimensional one.

6.3 EXAMPLE OF SECTION 2.3 REVISITED

Using the notation of Section 2.3, consider the vector-valued test $\mathbf{U} = (U_1, U_2)^T$, where $U_1 = X_n^2$ is defined for the four Neyman-Pearson type intervals in (2.18), and $U_2 = \delta_1^2 + \delta_3^2$ is defined for another variant of four Neyman-Pearson type intervals

$$\Delta_{13} = (-\infty; -2.28121) \cup (2.28121; +\infty),$$
$$\Delta_{14} = (-0.827723; 0.827723), \quad \Delta_{23} = (-2.28121; -0.827723),$$
$$\Delta_{24} = (0.827723; 2.28121).$$

Over the intervals Δ_{13} and Δ_{14}, we have $t(x,0,1) < l(x,0,1)$, while over the intervals Δ_{23} and $\Delta_{24}, t(x,0,1) \geqslant l(x,0,1)$. The statistic U_1 is distributed in limit as χ_3^2 with power 0.851, and the statistic U_2 is distributed in limit as χ_2^2 with power 0.729. Note that both U_1 and U_2 are based on the same sample. Since the components of the vector $\mathbf{U} = (U_1, U_2)^T$ are evidently correlated, we have to use simulation for assessing the power of the vector statistic $\mathbf{U}$. Since $\mathbf{U} = (U_1, U_2)^T$ is two-dimensional, it is possible to construct and consider infinitely many different rejection regions. For example, let us consider the rejection region of the test $\mathbf{U} = (U_1, U_2)^T$ to be the intersection of two sets $S = (U_1 > U_{1cr}) \cap (U_2 > U_{2cr})$ displayed in Figure 6.2. Such a rejection region seems to be reasonable because scalar tests U_1 and U_2 take into account

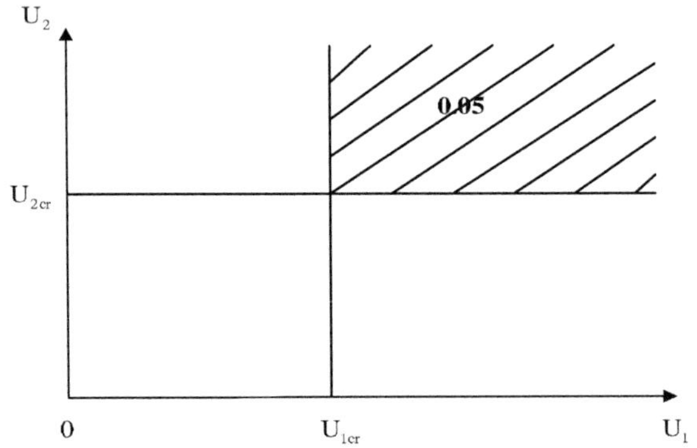

FIGURE 6.2 A sketch of the rejection region S of $\mathbf{U} = (U_1, U_2)^T$ (dashed area).

different distributional characteristics of the hypotheses under test, and so one can expect that, under suitable conditions, the vector-valued test **U** will possess larger power than its components.

Since the level $\alpha = 0.05$ of the vector-valued test **U** can be achieved for different combinations of $U_{1\text{cr}}$ and $U_{2\text{cr}}$, on which the power will depend, we produced by simulation the power values for six pairs of critical values for U_1 and U_2 ($n = 200$ and the number of runs $N = 5000$). For each pair, the power (probability of falling into the rejection region under the triangular alternative) so determined are presented in Table 6.2.

We observe that the power of the vector-valued test **U** depends on the choice of critical values of U_1 and U_2, and is maximized for $U_{1\text{cr}} = 7.4$ and $U_{2\text{cr}} = 2.147$ (being equal to 0.882) which is larger than those of the components U_1 and U_2 (being 0.851 and 0.729, respectively). Note that the simulated correlation coefficient between U_1 and U_2 is $\rho = 0.75$ for H_0, and $\rho = 0.94$ for H_a, both being significant at level 0.01.

Consider another vector-valued test $\mathbf{U}_1 = (U_1, U_{21})^T$, where $U_1 = X_n^2$ (the same as for $\mathbf{U} = (U_1, U_2)^T$) but $U_{21} = X_n^2$ for two Neyman-Pearson classes defined in Section 2.3. The power of U_{21} in this case is 0.855. Through simulation, we produced for five pairs of critical values for U_1 and U_2 ($n = 200$ and the number of runs $N = 5000$), and for each pair the power of $\mathbf{U}_1$ was determined, and these results are presented in Table 6.3.

Here again, we see that the power of the vector-valued test $\mathbf{U}_1$ depends on the choice of critical values $U_{1\text{cr}}$ and $U_{21\text{cr}}$, and gets maximized for $U_{1\text{cr}} = 6.4$ and $U_{21\text{cr}} = 1.95$ (being equal to 0.922) which is larger than that of the components U_1 and U_{21} (being 0.851 and 0.855, respectively). This power is almost two times more than

TABLE 6.2 Critical values of the test $\mathbf{U} = (U_1, U_2)^T$ with $\alpha = 0.05$, and corresponding powers.

$U_{1\text{cr}}$	7.2	7.3	7.4	7.5	7.6	7.7
$U_{2\text{cr}}$	3.232	2.824	2.147	1.938	1.354	0.865
Power of **U**	0.864	0.869	0.882	0.881	0.880	0.876

TABLE 6.3 Critical values of the test $\mathbf{U}_1 = (U_1, U_{21})^T$ with $\alpha = 0.05$, and corresponding powers.

$U_{1\text{cr}}$	6.2	6.4	6.6	6.8	7.0
$U_{2\text{cr}}$	2.30	1.950	1.627	1.242	1.155
Power of $\mathbf{U}_1$	0.915	0.922	0.915	0.913	0.909

TABLE 6.4 A comparison of power values of different test statistics.

Test statistic	Partitioning of the sample space	Power
X_n^2 of Pearson	4 equiprobable cells	0.320
δ_2^2 of Anderson (1994)	4 equiprobable cells	0.470
$\delta_1^2+\delta_3^2$	4 NP type classes	0.729
X_n^2	2 NP classes	0.855
$\mathbf{U}=(U_1,U_2)^T$	4 different NP type classes	0.882
$\mathbf{U}_1=(U_1,U_{21})^T$	2 and 4 NP type classes	0.922

the power of Anderson's δ_2^2 (which is 0.47) for equiprobable cells. Note that the simulated correlation coefficient between U_1 and U_{21} is $\rho = 0.57$ for H_0 and is $\rho = 0.94$ for H_a, both significant at level 0.01. It is of interest to note also that the power of $\mathbf{U}_1$ is larger than that of $\mathbf{U}$. All the above results and also those from Section 2.3 are summarized in Table 6.4.

From this table, we see how much improvement can be achieved in the power of a test by using different approaches. We do not insist that the test $\mathbf{U}_1$ for the example considered is the best one, because many other possibilities exist for the construction of a test. For example, one may use different kinds of rejection regions, different dimensionality for vector-valued tests, different test statistics as individual components, and so on.

6.4 COMBINING NONPARAMETRIC AND PARAMETRIC TESTS

Consider a two-dimensional vector-valued test with correlated components comprising the modified chi-squared and the nonparametric statistics as its components (Voinov and Pya, 2010):

$$\mathbf{V}_n = (S1_n^2(\bar{\boldsymbol{\theta}}_n), A_n^2)^T = (Y2_n^2(\bar{\boldsymbol{\theta}}_n) - U_n^2(\bar{\boldsymbol{\theta}}_n), A_n^2)^T, \tag{6.4}$$

where $Y2_n^2(\bar{\boldsymbol{\theta}}_n)$ is the HRM statistic (see Section 4.1), $U_n^2(\bar{\boldsymbol{\theta}}_n)$ is the DN test in (2.21), $\bar{\boldsymbol{\theta}}_n$ is the MME of the true parameter $\boldsymbol{\theta}$, and A_n^2 is the Anderson-Darling test statistic (Anderson and Darling, 1954).

Under some regularity conditions, the HRM test $Y2_n^2(\bar{\boldsymbol{\theta}}_n)$ possesses in the limit χ_{r-1}^2 distribution under the null hypothesis, with r being the number of grouping intervals. For any $\sqrt{n}$-consistent estimator $\tilde{\boldsymbol{\theta}}_n$ of the parameter $\boldsymbol{\theta}$ including the MME $\bar{\boldsymbol{\theta}}_n$, the DN test $U_n^2(\tilde{\boldsymbol{\theta}}_n)$ follows in the limit χ_{r-s-1}^2 distribution, where s is the number of unknown parameters. The modified chi-squared test $S1_n^2(\bar{\boldsymbol{\theta}}_n)$ is distributed in the limit as χ_s^2. It is known (Voinov et al., 2009) that, for equiprobable intervals of grouping, the latter test is more powerful than $Y2_n^2(\bar{\boldsymbol{\theta}}_n)$.

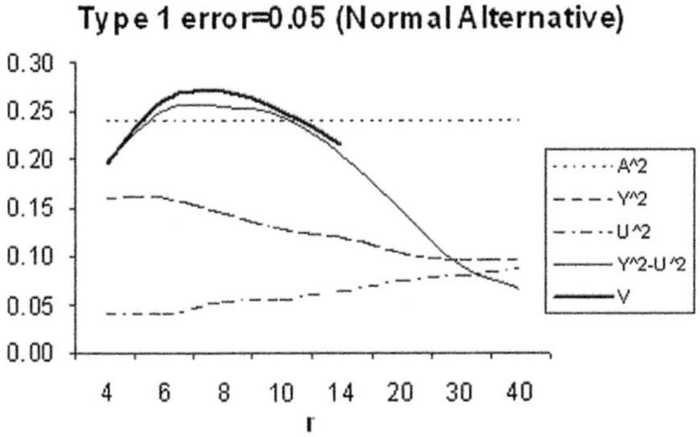

FIGURE 6.3 Power values of $\mathbf{V}_n(V)$, $Y2_n^2(\bar{\boldsymbol{\theta}}_n)(Y^2)$, $U_n^2(\bar{\boldsymbol{\theta}}_n)(U^2)$, $S1_n^2(\bar{\boldsymbol{\theta}}_n) = Y2_n^2(\bar{\boldsymbol{\theta}}_n) - U_n^2(\bar{\boldsymbol{\theta}}_n)(Y^2 - U^2)$, and A_n^2 as functions of the number of cells r.

Suppose we wish to test a composite null hypothesis about the logistic probability distribution against a normal alternative distribution. To compare the power of $\mathbf{V}_n$ test with that of the scalar tests $Y2_n^2(\overline{\boldsymbol{\theta}}_n)$, $U_n^2(\overline{\boldsymbol{\theta}}_n)$, $Y2_n^2(\overline{\boldsymbol{\theta}}_n) - U_n^2(\overline{\boldsymbol{\theta}}_n)$, and the Anderson-Darling test A_n^2, Monte Carlo simulations were used. The statistic A_n^2 was calculated with the use of the MME $\overline{\boldsymbol{\theta}}_n$. Since theoretical and simulated critical values corresponding to a given level of significance α do not differ significantly, we used the theoretical ones (except for the A_n^2 test). We used the simulated critical values for the A_n^2 test since the corresponding theoretical results are fairly limited (see Sinclair and Spurr, 1988). The rejection region for $\mathbf{V}_n$ was the intersection of rejection regions of the components of $\mathbf{V}_n$ (like in Figure 6.2). Since we can select a rejection region of a given level α in infinitely many ways, we tried several possible variants to choose one that gives the highest power.

The results of a simulation study based on random samples of size $n = 200$ and $\alpha = 0.05$ are shown in Figure 6.3. All modified chi-squared tests considered were calculated for r equiprobable fixed intervals (number of runs was 10,000).

From Figure 6.3, we observe that, despite the correlation between the components, the vector-valued statistic for testing the logistic null distribution against the normal alternative distribution possesses not too much power but, nevertheless, higher power than the scalar tests based on individual components under consideration. It should also be noted that the power of Dzhaparidze-Nikulin test $U_n^2(\overline{\boldsymbol{\theta}}_n)$ is quite low when equiprobable cells are used. More details on the power of $U_n^2(\overline{\boldsymbol{\theta}}_n)$ can be found in Voinov et al. (2009).

6.5 COMBINING NONPARAMETRIC TESTS

In this section, we consider a combination of two correlated nonparametric goodness of fit tests based on the empirical distribution function, viz., the

TABLE 6.5 Power values of K_n, W_n^2, and $\mathbf{R}_n$ against triangular alternative distribution for the rejection region of $\mathbf{R}_n$ taken as a union. The statistical error shown corresponds to one standard deviation.

	S_{1U}	S_{2U}
K_n	0.155	0.155
W_n^2	0.141	0.141
$\mathbf{R}_n$	0.143	**0.161±0.004**

Kolmogorov-Smirnov test K_n and the Cramer-von Mises test W_n^2:

$$\mathbf{R}_n = (K_n, W_n^2)^T. \tag{6.5}$$

Consider the problem of testing the simple null hypothesis of the logistic distribution with mean zero and variance one against the triangular alternative distribution with the same mean and variance. In this case, the power of both K_n and W_n^2, implemented individually, is nearly the same.

To compare the power of vector-valued test in (6.5) with union- and intersection-type of rejection regions of its components, we carried out a simulation study. Table 6.5 shows the results of the power for two different rejection regions of the same significance level taken as a union of rejection regions of the components ($n = 200, \alpha = 0.05$, number of runs was 10,000):

$$S_{1U} = (K_n > 0.1206) \cup (W_n^2 > 0.4634),$$
$$S_{2U} = (K_n > 0.1044) \cup (W_n^2 > 0.6884).$$

Numerical values in above formulas were selected in such a way that the power of the vector-valued test $\mathbf{R}_n$ is approximately maximal for the overall level $\alpha = 0.05$.

From Table 6.5, we see that, in the above considered case, the vector-valued test $\mathbf{R}_n$ in (6.5) may have higher power than the scalar tests based on the components of the vector $\mathbf{R}_n$. It is also evident that the power of $\mathbf{R}_n$ definitely depends on the structure of the rejection region used.

Consider the following two rejection regions for the vector-valued test in (6.5):

$$S_{1I} = (K_n > 0.1044) \cap (W_n^2 > 0.2269),$$
$$S_{2I} = (K_n > 0.0883) \cap (W_n^2 > 0.4420).$$

The numerical values in the above S_{1I} and S_{2I} were selected such that the power of the vector-valued test $\mathbf{R}_n$ is nearly maximal for the overall level $\alpha = 0.05$.

The results of the Monte Carlo study with random samples of size $n = 200$ for this case are presented in Table 6.6. From this table, we once again observe

TABLE 6.6 Power values of K_n, W_n^2, and $\mathbf{R}_n$ against triangular alternative distribution for the rejection region of $\mathbf{R}_n$ taken as an intersection. The statistical error shown corresponds to one standard deviation.

	S_{1l}	S_{2l}
K_n	0.155	**0.162±0.004**
W_n^2	0.141	0.146
$\mathbf{R}_n$	**0.160**	0.155

that the power of the vector-valued test may be more or less than that of scalar components depending on the structure of a rejection region. Note that the power has been considered here only for very simple rejection regions (intersection and union). Of course, more complicated regions can also be considered in this regard and their power properties may also vary depending on their form and structure.

6.6 CONCLUDING COMMENTS

Several examples of vector-valued goodness of fit tests have been discussed in the preceding sections. Simulation studies carried out have revealed that, when combining either correlated or uncorrelated nonparametric or parametric tests with approximately the same power, vector-valued tests may gain power when compared with the components of the vector-valued statistic. Examples considered show that the power of vector-valued goodness of fit tests depends on the structure of the rejection region, correlation between the components of the test, and the dimensionality of the vector. A more detailed and thorough examination of these issues in the consideration of vector-valued goodness of fit tests is warranted both theoretically and empirically.

Chapter 7

Applications of Modified Chi-Squared Tests

7.1 POISSON VERSUS BINOMIAL: APPOINTMENT OF JUDGES TO THE US SUPREME COURT

The problem of discriminating between Poisson and binomial probability distributions is a very difficult one due to the close similarity between these two models; see Bol'shev (1963), Vu and Maller (1996), and Voinov and Voinov (2008, 2010). In this regard, we consider here a detailed statistical analysis of the number of US Supreme Court justices' appointments from 1789 to 2004 (Voinov et al., 2010). Proposed modified chi-squared type tests, the likelihood ratio test of Vu and Maller (1996) and some other well-known tests are all examined with these data. The analysis shows that both simple Poisson and simple binomial models are equally appropriate for describing these data. No firm statistical evidence in favor of an exponential Poisson regression model is found. Two important observations made from a simulation study in this connection are that Vu and Maller (1996) test is the most powerful test among those considered when testing for the Poisson versus binomial, and that the classical variance test with an upper tail critical region is a biased test.

7.1.1 Introduction

The problem of describing the frequency with which Presidents of the US were able to appoint the US Supreme Court justices has been addressed by a number of authors including Wallis (1936), Callen and Leidecker (1971), Ulmer (1982), and King (1987, 1988). They made use of the Poisson distribution and

Chi-Squared Goodness of Fit Tests with Applications. http://dx.doi.org/10.1016/B978-0-12-397194-4.00007-7

an exponential Poisson regression model to describe the process of judges' appointments. As Wallis (1936) noted, this description is of importance "to estimate how often a President will be called upon to appoint judges of the Supreme Court during the 4 years of an administration." The problem, in addition to being of political interest, is of great interest from a statistical viewpoint as well. It is well known that the Poisson distribution is a limiting form of the binomial distribution as the number of trials tends to infinity. The question that is of interest in this case is "whether the data on the US Supreme Court judges' appointments follow a Poisson or a binomial distribution?"

In Section 7.1.2, the pertinent data are presented. Section 7.1.3 is devoted to a thorough statistical analysis of these data. Some new tests as well as some well-known statistical tests, including the most powerful test of Vu and Maller (1996), are used to test the hypotheses of interest. In Section 7.1.4, the results of Wallis (1936) and Ulmer (1982) are revisited. Some remarks concerning the exponential Poisson regression model of King (1987, 1988) are made in Section 7.1.5. Finally, some concluding comments are made in Section 7.1.6.

7.1.2 Data to be analyzed

Data on the number of judges appointed during the period 1789–2004 are presented in Table 7.1. These are available on the website of Members of the Supreme Court of the United States (2008).

The mean value of the number of judges appointed per year and the standard error of the mean are found to be 0.5231 and 0.0512, respectively. Before applying our statistical techniques, we first examine whether the data in Table 7.1 can be regarded as a set of realizations of independent and identically distributed random variables. The nonparametric runs test for randomness has a P-value of 0.764, thus providing a strong evidence toward the hypothesis that they form a random set. The sample autocorrelation function (ACF) for the number of appointments per year for the period 1789–2004 in Figure 7.1 looks like the autocorrelation function from a white noise process and, this together with randomness, provides a weak confirmation of independence.

Next, we check for the equality of probabilities of occurrences of events of interest. For this purpose, we divided the period 1789–2004 into six parts (five of 36 years and the last one of 38 years, so as to use the central limit theorem) and produced the two-sample t-statistics for the difference of means for all 15 pairs. The P-values of the tests varied from 0.178 to 1.000, thus confirming the hypothesis of equality of probabilities of occurrence of events at a level of 0.10.

Dividing the period 1789–2004 into parts of 4 years based on the President's governance, the data in Table 7.1 can be summarized as in Table 7.2. The symbol N_i here denotes the number of appointments during the ith 4-year of presidency.

In creating this table, we have omitted the data for the Presidents Taylor (1849–1950), Garfield (1881), Harding (1921–1922), and Kennedy (1961–1963), who served for less than 4 years. For the data in Table 7.2, the runs

TABLE 7.1 Number n_i of appointments in each year during the period 1789–2004.

Year	n_i	Year	n_i	Year	n_i	Year	n_i	Year	n_i	Year	n_i
1789	2	1825	0	1861	0	1897	0	1933	0	1969	1
1790	4	1826	1	1862	3	1898	1	1934	0	1970	1
1791	0	1827	0	1863	1	1899	0	1935	0	1971	0
1792	1	1828	0	1864	1	1900	0	1936	0	1972	2
1793	1	1829	0	1865	0	1901	0	1937	1	1973	0
1794	0	1830	2	1866	0	1902	1	1938	1	1974	0
1795	1	1831	0	1867	0	1903	1	1939	2	1975	1
1796	2	1832	0	1868	0	1904	0	1940	1	1976	0
1797	0	1833	0	1869	0	1905	0	1941	3	1977	0
1798	0	1834	0	1870	2	1906	1	1942	0	1978	0
1799	1	1835	1	1871	0	1907	0	1943	1	1979	0
1800	1	1836	2	1872	0	1908	0	1944	0	1980	0
1801	1	1837	1	1873	1	1909	0	1945	1	1981	1
1802	0	1838	1	1874	1	1910	3	1946	1	1982	0
1803	0	1839	0	1875	0	1911	2	1947	0	1983	0
1804	1	1840	0	1876	0	1912	1	1948	0	1984	0
1805	0	1841	0	1877	1	1913	0	1949	2	1985	0
1806	0	1842	1	1878	0	1914	1	1950	0	1986	2
1807	2	1843	0	1879	0	1915	0	1951	0	1987	0
1808	0	1844	0	1880	0	1916	2	1952	0	1988	1
1809	0	1845	2	1881	2	1917	0	1953	1	1989	0
1810	0	1846	1	1882	2	1918	0	1954	0	1990	1
1811	1	1847	0	1883	0	1919	0	1955	1	1991	1
1812	1	1848	0	1884	0	1920	0	1956	1	1992	0
1813	0	1849	0	1885	0	1921	1	1957	1	1993	1
1814	0	1850	0	1886	0	1922	1	1958	1	1994	1
1815	0	1851	1	1887	0	1923	2	1959	0	1995	0
1816	0	1852	0	1888	2	1924	0	1960	0	1996	0
1817	0	1853	1	1889	0	1925	1	1961	0	1997	0
1818	0	1854	0	1890	1	1926	0	1962	2	1998	0
1819	0	1855	0	1891	1	1927	0	1963	0	1999	0
1820	0	1856	0	1892	1	1928	0	1964	0	2000	0
1821	0	1857	0	1893	1	1929	0	1965	1	2001	0
1922	0	1858	1	1894	1	1930	2	1966	0	2002	0
1923	1	1859	0	1895	0	1931	0	1967	1	2003	0
1924	0	1860	0	1896	1	1932	1	1968	0	2004	0

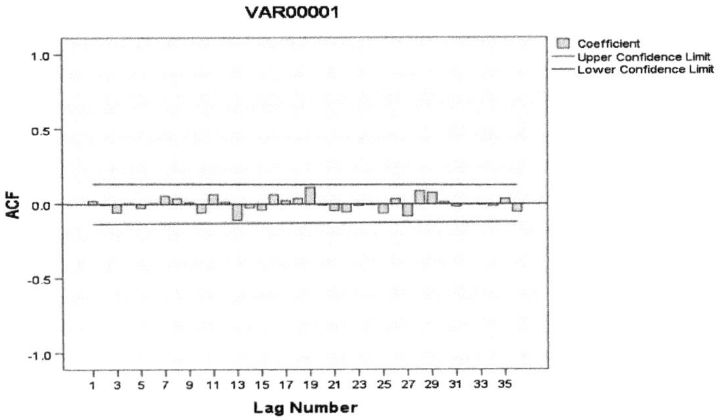

FIGURE 7.1 The sample autocorrelation function (ACF) for the number of appointments per year during the period 1789–2004 (the 95% confidence interval of the absence of the significant correlation is shown as well).

test has a P-value of 0.284, providing evidence for randomness of the data. Next, the sample ACF of the data does not provide evidence for significant correlation between the observations.

7.1.3 Statistical analysis of the data

Let us consider the problem from the viewpoint of discriminating between the Poisson and binomial distributions. To date, many papers have been published using the Poisson model as a base model for inference. We analyze here the data in Table 7.2 by using various chi-squared type tests and the likelihood ratio test of Vu and Maller (1996).

Testing for the Poisson null hypothesis

In this section, we take the null hypothesis as that the data are from a Poisson distribution, and test it against a binomial alternative.

(a) First, the data in Table 7.2 are summarized in Table 7.3 in a frequency count form.

For the Poisson null hypothesis and $r = 5$ intervals in Table 7.3, the values of the test statistics $Y1_n^2(\hat{\theta}_n), U_n^2(\hat{\theta}_n)$, and $S_n^2(\hat{\theta}_n)$, defined by the formulas in (9.2), (3.23), and (3.24), are 3.44, 2.76, and 0.68, respectively. Since critical values of the corresponding chi-squared limiting distributions with 4, 3, and 1 degrees of freedom, for $\alpha = 0.05$ level, are 9.49, 7.81, and 3.84, we conclude that the data are consistent with the Poisson null hypothesis. The simulated power of the tests $Y1_n^2(\hat{\theta}_n), U_n^2(\hat{\theta}_n)$, and $S_n^2(\hat{\theta}_n)$ with respect to the binomial alternative are $0.052 \pm 0.002, 0.051 \pm 0.002$,

TABLE 7.2 Number N_i of appointments during the ith 4-year of presidency for the period 1789–2004.

i	Years	President	N_i	i	Years	President	N_i
1	1789–1792	Washington	7	27	1893–1896	Cleveland	2
2	1793–1796	Washington	4	28	1897–1900	McKinley	1
3	1797–1801	Adams	3	29	1901–1904	Roosevelt	2
4	1801–1804	Jefferson	1	30	1905–1908	Roosevelt	1
5	1805–1808	Jefferson	2	31	1900–1912	Taft	6
6	1809–1812	Madison	2	32	1913–1916	Wilson	3
7	1813–1816	Madison	0	33	1917–1920	Wilson	0
8	1817–1820	Monroe	0	34	1923–1929	Coolidge	1
9	1821–1824	Monroe	1	35	1929–1932	Hoover	3
10	1825–1828	Adams	1	36	1933–1936	Roosevelt	0
11	1829–1832	Jackson	2	37	1937–1940	Roosevelt	5
12	1833–1836	Jackson	3	38	1941–1944	Roosevelt	4
13	1837–1841	Van Buren	2	39	1945–1948	Truman	2
14	1841–1845	Tyler	1	40	1949–1952	Truman	2
15	1845–1848	Polk	2	41	1953–1956	Eisenhower	3
16	1850–1853	Fillmore	1	42	1957–1960	Eisenhower	2
17	1853–1856	Pierce	1	43	1966–1969	Johnson	2
18	1857–1860	Buchanan	1	44	1969–1973	Nixon	4
19	1861–1864	Lincoln	4	45	1974–1977	Ford	1
20	1865–1868	Johnson	0	46	1977–1980	Carter	0
21	1869–1872	Grant	2	47	1981–1984	Reagan	1
22	1873–1876	Grant	2	48	1985–1988	Reagan	3
23	1877–1881	Hayes	2	49	1989–1992	Bush, G.H.W.	2
24	1881–1884	Arthur	2	50	1993–1996	Clinton	2
25	1881–1884	Cleveland	2	51	1997–2000	Clinton	0
26	1889–1892	Harrison	4	52	2001–2004	Bush, G.W.	0

and 0.052 ± 0.002, respectively. These values reveal that these tests are unbiased, but their power for the binomial alternative is very low.

It is known that some tests of continuous models based on two Neyman-Pearson classes possess larger power (Voinov et al., 2009). Let us now apply this idea for the case of discrete Poisson null hypothesis. Define the following two Neyman-Pearson classes: $\Delta_1 = \{x : P_1(X = x) < P(X = x)\}$ and $\Delta_2 = \{x : P_1(X = x) \geqslant P(X = x)\}$, where the probabilities $P(X = x)$ and $P_1(X = x)$ are given by formulas (9.1) and (9.3). In our case, $\Delta_1 = \{0,1\} \cup \{5, \ldots ,41\}$ and $\Delta_2 = \{2,3,4\}$. When the number of intervals r equals two, statistics $Y1_n^2(\hat{\theta}_n)$ and $U_n^2(\hat{\theta}_n)$ are not defined and only $S_n^2(\hat{\theta}_n)$ can be used. For the data in Table 7.3, the value 0.0244 of

TABLE 7.3 A summary of the data in Table 7.2.

k	Interval	Frequency, ν_k
0	1	8
1	2	12
2	3	18
3	4	6
> 3	5	8

TABLE 7.4 Number of years in which specified number of vacancies in the Supreme Court were filled during the period 1789–2004.

Number of vacancies	Total during 1789–2004
0	131
1	64
2	19
3	3
Over 3	1

$S_n^2(\hat{\theta}_n)$, possessing the simulated power of 0.054 ± 0.002, does not reject the null hypothesis.

Next, let us consider the data in a form as in Table 7.4 that was used by Wallis (1936), Callen and Leidecker (1971), and Ulmer (1982). Since the expected frequencies for the number of vacancies being at least 3 are less than 5, the last three totals were combined into one interval, thus leaving $r = 3$ intervals. For the Poisson null hypothesis and $r = 3$ intervals in Table 7.4, the values of $Y1_n^2(\hat{\theta}_n), U_n^2(\hat{\theta}_n)$, and $S_n^2(\hat{\theta}_n)$, defined by formulas (9.2), (3.23), and (3.24), are 0.443, 0.363, and 0.08, respectively. Since the critical values of the corresponding chi-squared limiting distributions with 2 and 1 degrees of freedom, for $\alpha = 0.05$, are 5.991 and 3.841, the conclusion once again is that the data are consistent with the Poisson null hypothesis.

(b) Vu and Maller (1996) presented a likelihood ratio statistic for testing the Poisson null hypothesis against the binomial alternative. For the data in Table 7.3, the value of the test statistic $d_n(N_{\max} = 13)$ of Vu and Maller (1996) (formula 4) is 1.19, which is less than the simulated critical value of 2.833 for $\alpha = 0.05$, and so the null hypothesis of a Poisson distribution is not rejected. For the data in Table 7.4, the value of $d_n(N_{\max} = \infty)$ is 0, which also supports the null hypothesis of a Poisson distribution at level $\alpha = 0.05$. It is of importance to note that the simulated power of the test d_n in the first case is 0.10 ± 0.01 and 0.08 ± 0.01 in the second case (the number of runs used was 1000 in the simulation study). The power of the likelihood ratio test is slightly higher than that of the previous tests. This can be explained by the fact that the test of Vu and Maller (1996) is based on the essential difference between the finite support of a binomial distribution and the infinite support of the Poisson model. Yet, this power is not sufficient to discriminate between the two models with confidence.

(c) A simple test based on Fisher's index of dispersion or variance test (see Berkson, 1940; Cochran, 1954; Selby, 1965; Gbur, 1981) for the Poisson distribution is sometimes suggested (Greenwood and Nikulin, 1996, p. 199). The test statistic in this case is

$$\chi^2 = \sum_{i=1}^{n} \frac{(x_i - \bar{x})^2}{\bar{x}}, \tag{7.1}$$

where $x_i, i = 1, \ldots, n$, are realizations of n i.i.d. random variables. For the Poisson null hypothesis, this statistic approximately follows χ^2_{n-1}, the central chi-square distribution with $n-1$ degrees of freedom. This test was previously used to test the Poisson distribution against compound Poisson, double Poisson, Neyman type A, and negative binomial distributions; see Selby (1965). Berkson (1940) noted that the variance test χ^2 in (7.1) could "distinguish correctly that we are not dealing with a Poisson distribution." He compared the ability of χ^2 to discriminate between the Poisson and binomial distributions and, using simulations, showed that the variance test performs better than the classical Pearson's test. However, no attempt was made to evaluate the power of χ^2 for the binomial alternative. Using 10,000 replications and parameters ($p = 0.11, n = 56$) of Berkson (1940, p. 364), we found the simulated power for the upper tail critical region to be $0.011 \pm 0.001 < 0.05$. This shows that the χ^2 test for the upper tail critical region is biased with respect to the binomial alternative. As Gbur (1981, p. 532) noted, for alternatives like the binomial, when $\sigma^2/\mu < 1$, a lower tail critical region "would be appropriate." Indeed, our simulations showed that, under the same conditions as above, the power for the lower rejection region is 0.130 ± 0.003, implying the unbiasedness of the test.

Using the data in Table 7.2, the MLE $\hat{\theta}_n$ of θ is calculated as $\hat{\theta}_n = 104/52 = 2$. The observed value of $\chi^2 = 60$ corresponds to a P-value of 0.182. The critical value of χ^2_{51} at level $\alpha = 0.05$ is 68.67, and so we do not reject the null hypothesis. For $\theta = 2, n = 50$, and $N = 10{,}000$ replications of χ^2 for the Poisson distribution, the simulated critical value at level $\alpha = 0.05$ turns out to be 68.55, which agrees well with the theoretical value of 68.67. Also, 10,000 Monte Carlo replications of χ^2 with $\theta = 2$ and $n = 50$ gives the probability of falling into the upper tailed rejection region of level $\alpha = 0.05$ to be $0.032 \pm 0.002 < 0.05$ under the binomial alternative.

This means that this χ^2 test is also biased with respect to the binomial alternative. Further, the probability of falling into the lower tailed rejection region of level $\alpha = 0.05$ under the binomial alternative is 0.075 ± 0.003. From this, it follows that, while the Poisson null hypothesis is not rejected, the power of the χ^2 test is not high enough to discriminate between the two specified models with reasonable confidence.

(d) The classical Pearson-Fisher (PF) test statistic in (2.19) for analyzing the data in Table 7.3 with $r = 5$ intervals is given by

$$\chi^2_{PF} = \sum_{k=0}^{3} \frac{(\nu_k - np_k(\theta))^2}{np_k(\theta)} + \frac{\left[\nu_4 - n\left(1 - \sum_{k=0}^{3} p_k(\theta)\right)\right]^2}{n\left(1 - \sum_{k=0}^{3} p_k(\theta)\right)}, \quad (7.2)$$

where $p_k(\theta) = \theta^k e^{-\theta}/k!, \nu_4$ is the frequency for the fifth interval, and the unknown parameter θ can be replaced by the estimate $\tilde{\theta}_n$ that minimizes the sum χ^2_{PF} in (7.2). For the data in Table 7.3, the Microsoft Excel Solver gives $\tilde{\theta}_n = 1.973$ and $\chi^2_{PF} = 2.777$. Since, at level $\alpha = 0.05$, the critical value of the chi-squared distribution with $r - 2 = 3$ degrees of freedom is 7.815, the Poisson null hypothesis is not rejected. 5000 Monte Carlo simulations of χ^2_{PF} with $\theta = 2$ and $n = 50$ gave the probability of falling into the rejection region of level $\alpha = 0.05$ under the binomial alternative to be 0.047 ± 0.003. Low power means that the PF test is very weak in distinguishing between the Poisson and binomial distributions.

The simulated powers of the modified chi-squared test $S_n^2(\hat{\theta}_n)$ for two Neyman-Pearson (NP) classes, Vu and Maller's test, the variance test, and the classical PF test against the binomial alternative are all presented in Table 7.5 along with the estimates of the parameter θ. From this table, it is seen that Vu and Maller's test possesses power that is not very high, but still the highest among all considered tests. Moreover, we observe that the classical Pearson-Fisher test is not applicable.

Testing for the binomial null hypothesis

In this section, we consider the binomial distribution as the null and test it against the Poisson alternative.

(a) In this case, the classical PF test for analyzing the data in Table 7.3 for $r = 5$ intervals is given by (7.2), where

$$p_k(\theta) = \binom{n}{k}\left(\frac{\theta}{n}\right)^k \left(1 - \frac{\theta}{n}\right)^{n-k},$$

and the unknown parameter θ can be replaced by the estimate $\tilde{\theta}_n$ that minimizes the sum χ^2_{PF}. For the data in Table 7.3, the Microsoft Excel

TABLE 7.5 Simulated power values of the modified chi-squared test $S_n^2(\hat{\theta}_n)$, Vu and Maller's test, variance test, and the classical Pearson-Fisher test against the binomial alternative. Estimates of the parameter θ used are given in Column 3. The estimate of the power is shown with the corresponding simulated standard error.

Test statistics	Power	Estimates of θ
$S_n^2(\hat{\theta}_n)$ (NP classes)	0.054 ± 0.002	$\hat{\theta}_n = 2$
Vu and Maller	0.100 ± 0.010	$\hat{\theta}_n = 2$
Variance	0.075 ± 0.003	$\hat{\theta}_n = 2$
Pearson-Fisher	0.047 ± 0.003	$\hat{\theta}_n = 1.973$

Solver gives $\tilde{\theta}_n = 1.969$ and $\chi^2_{PF} = 2.839$. Since the critical value of the chi-squared distribution with $r - 2 = 3$ degrees of freedom is 7.815 at level $\alpha = 0.05$, the binomial null hypothesis is not rejected. To examine the power of this test against the Poisson alternative, 5,000 Monte Carlo replications of χ^2_{PF} with $\theta = 2$ and $n = 50$ gave the probability of falling into the rejection region of level $\alpha = 0.05$ under the Poisson alternative to be 0.054 ± 0.003. The very low power means that, as in the case of the Poisson null hypothesis, the PF test is very weak for discriminating between the binomial and Poisson distributions.

(b) Consider the modified chi-squared tests for the binomial model. For the binomial null hypothesis, the values of $Y1_n^2(\hat{\boldsymbol{\theta}}_n), U_n^2(\hat{\theta}_n)$, and $S_n^2(\hat{\theta}_n)$, computed from the formulas in (9.4), (3.23), and (3.25), are 3.851, 2.822, and 1.029, respectively. We conclude that there is enough evidence for the binomial null hypothesis. It is of importance to note that the simulated power of these tests is not higher than 0.052 ± 0.002 which is very low compared to the level $\alpha = 0.05$. For the same two Neyman-Pearson classes used before (see Section 7.1.3), but for the binomial null hypothesis, the value of $S_n^2(\hat{\theta}_n)$ is 0.0005 with a simulated power of 0.058 ± 0.002. Similarly, for the approaches of Section 7.1.4, but for the binomial null hypothesis, the values of $Y1(\hat{\theta}_n), U(\hat{\theta}_n)$, and $S_n^2(\hat{\theta}_n)$, computed from the formulas in (9.4), (3.23), and (3.25), are 0.477, 0.385, and 0.092, respectively. So, once again, the hypothesis of the binomial distribution is not rejected. As before, the powers of these tests are only slightly higher than the nominal level of $\alpha = 0.05$.

(c) Consider the data in Table 7.6. The original data were grouped mainly by periods of 34 years (136 quarters), with the exception of 14 years (1925–1938) before World War II, 8 years (1939–1946) during World War II, and the last 24 years (1981–2004). These data can be presented in the form of proportions $p_i = x_i/n_i, i = 1, 2, \ldots, M$ $(M = 8)$. Assume that x_i are realizations of independent binomial random variables. To test for the

TABLE 7.6 The number of judges appointed during the corresponding number of quarters for specified periods.

i	Years	Number of quarters, n_i	Number of judges, x_i
1	1789–1822	136	19
2	1823–1856	136	15
3	1857–1890	136	18
4	1891–1924	136	22
5	1925–1938	56	6
6	1939–1946	32	9
7	1947–1980	136	16
8	1981–2004	96	8

binomial null hypothesis, Tarone (1979) suggested the binomial variance test X_ν^2 for homogeneity and two Pitman asymptotically efficient tests, viz., X_c^2, against a correlated binomial alternative, and X_m^2, against alternatives given by the multiplicative generalizations of the binomial model. X_ν^2 is asymptotically chi-square distributed with $M-1$ degrees of freedom, and both X_c^2 and X_m^2 are asymptotically chi-square distributed with one degree of freedom. For the data in Table 7.6, we have $X_\nu^2 = 10.505$. At level 0.05, we do not reject the binomial null hypothesis since $\chi_7^2(0.05) = 14.067$. The same result holds for the other two tests since $X_c^2 = 0.183$ and $X_m^2 = 0.393$ with $\chi_1^2(0.05) = 3.841$. The fact that the X_c^2 test against a correlated binomial alternative does not reject the simple binomial model can also be considered as an argument against King's exponential Poisson regression model (see Section 7.1.5). It is also important to note that the simulated powers of X_ν^2, X_c^2, and X_m^2 tests are all negligible.

The very low power of all these tests (see Table 7.7) can be easily explained by the closeness of the models (Poisson and binomial) under consideration (see Figure 7.2).

7.1.4 Revisiting the analyses of Wallis and Ulmer

Wallis (1936) (also see Brownlee, 1960) analyzed the number of vacancies filled each year from 1790 to 1932. Using the classical Pearson's test, Wallis showed that the data are consistent with a Poisson distribution with a mean of 0.5. But this analysis contains two inaccuracies. The first is that Wallis estimated the population mean by the maximum likelihood estimator based on ungrouped (raw) data. In this case, Pearson's sum does not follow in the limit a chi-square distribution and it depends on the unknown parameter (Chernoff and Lehmann, 1954). The second inaccuracy is that the expected frequency for the fourth

TABLE 7.7 Simulated power values of the modified chi-squared test $S_n^2(\hat{\theta}_n)$, Tarone's tests X_v^2, X_c^2, X_m^2, and the classical PF test for the binomial null hypothesis against the Poisson alternative. Estimates of the parameter θ used are given in column 3. The estimate of the power is shown with the corresponding simulated standard error.

Test statistics	Power	Estimates of θ
$S_n^2(\hat{\theta}_n)$ (NP classes)	0.058 ± 0.002	$\hat{\theta}_n = 2$
X_v^2, X_c^2, X_m^2	0.054 ± 0.003	$\hat{\theta}_n = 2$
Pearson-Fisher	0.054 ± 0.003	$\tilde{\theta}_n = 1.969$

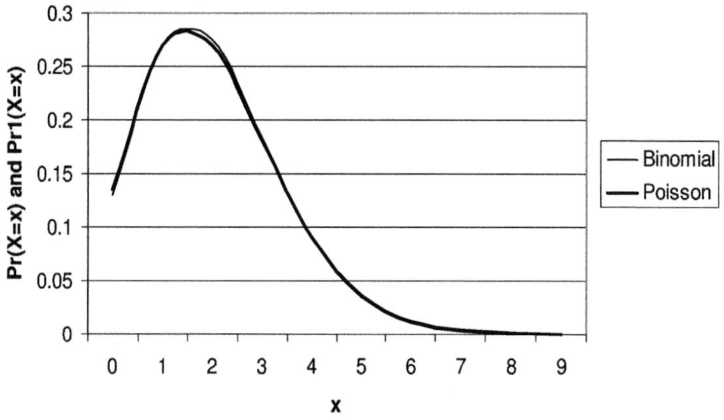

FIGURE 7.2 The smoothed curves for the Poisson $Pr(X = x)$ and the binomial distribution $Pr_1(X = x)$ for the same observed value of the parameter $\theta = 104/52 = 2$.

interval (see Table 2 of Wallis, 1936) is 1.383, which is less than the commonly recommended 5 (Lawal, 1980). To avoid these, we may use the PF test with the minimum chi-square estimate of the mean from grouped data. For the Poisson distribution and the data in Table 7.8, obtained from Table 7.2 of Wallis (1936) by combining the "2" and "over 2" lines, the test would require minimizing the following sum with respect to the parameter θ:

$$X_n^2 = \frac{(59 - 96e^{-\theta})^2}{96e^{-\theta}} + \frac{(27 - 96\theta e^{-\theta})^2}{96\theta e^{-\theta}} + \frac{[(10 - 96(1 - e^{-\theta} - \theta e^{-\theta})]^2}{96(1 - e^{-\theta} - \theta e^{-\theta})}.$$

The Microsoft Excel Solver gives $\tilde{\theta}_n = 0.4668$ as the estimate of θ, and $X_n^2 = 0.0709$ with one degree of freedom. The observed P-value of 0.79 is even higher than the P-value of 0.713 of the test of Wallis. Let the null hypothesis

TABLE 7.8 Observed frequencies for the number of vacancies in the Supreme Court during the period 1837–1932.

Class number	Number of vacancies	Observed frequency
1	0	59
2	1	27
3	>1	10
		Total 96

be the binomial distribution. Then, for the PF test, we have to minimize

$$X_n^2 = \frac{\left[59 - 96\left(1-\theta/96\right)^{96}\right]^2}{96\left(1-\theta/96\right)^{96}} + \frac{\left[27 - 96\theta\left(1-\theta/96\right)^{95}\right]^2}{96\theta\left(1-\theta/96\right)^{95}}$$
$$+\frac{\left\{10 - 96\left[1-\left(1-\theta/96\right)^{96} - \theta\left(1-\theta/96\right)^{95}\right]\right\}^2}{96\left[1-\left(1-\theta/96\right)^{96} - \theta\left(1-\theta/96\right)^{95}\right]},$$

with respect to θ. The Microsoft Excel Solver gives $\tilde{\theta}_n = 0.4647$ as the estimate of θ, and $X_n^2 = 0.07645$ with one degree of freedom. The observed P-value is 0.78, which means that the data of Wallis (1936) do not contradict either the Poisson or the binomial distributions. Doing the same for the data in Table 7.2 of Ulmer (1982) for the Poisson null hypothesis, we obtain $\tilde{\theta}_n = 0.533$, $X_n^2 = 0.0467$, and $P = 0.83$. If the null hypothesis corresponds to the binomial distribution, then $\tilde{\theta}_n = 0.5316, X_n^2 = 0.0516$, and $P = 0.82$. So, the extended data of Ulmer (1982) also do not contradict either the Poisson or binomial distributions. Thus, the use of the correct statistical PF test does not change the conclusions of Ulmer (1982) and Wallis (1936).

7.1.5 Comments about King's exponential Poisson regression model

King (1987) questioned the independence of events as well as the equality of probabilities of those events, in these data, and thus the validity of the Poisson model. He proposed an exponential Poisson regression model instead. This model is of interest, but we may ask the question whether it is needed here. A methodology like the one used in Section 7.1.2 can be made of use as well. King used data on the US Supreme Court Judges' appointment data for the period 1790–1984, and for these data, the runs test has a P-value of 0.697, thus supporting the hypothesis of randomness. The sample ACF shown in Figure 7.3 also supports the independence of occurrences. To check for the equality of probabilities, we divided the period 1790–1984 into six parts (five of 33 years

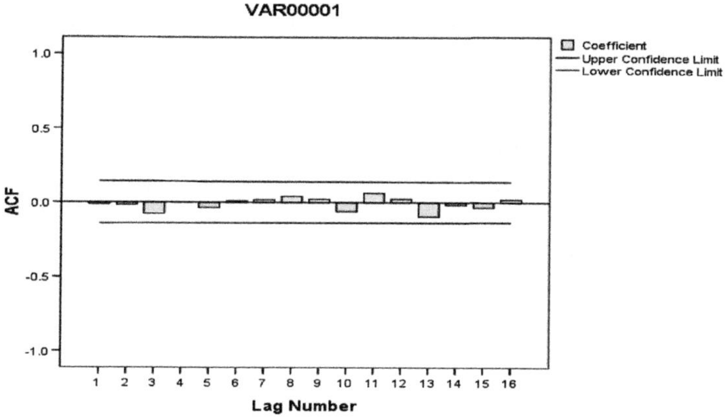

FIGURE 7.3 The sample autocorrelation function (ACF) for the number of appointments per year in the period 1790–1984 (the 95% confidence interval of the absence of the significant correlation is also shown).

and last one of 30 years) and performed the two-sample *t*-test for the difference of means for all 15 pairs. The *P*-values of the tests varied from 0.36 to 1.00, which provides strong evidence to the hypothesis of equality of probabilities. It should be added that binomial goodness of fit tests (see Section 7.1.3) of Tarone (1979), which are asymptotically optimal against correlated binomial alternatives and a multiplicative generalization of the binomial model, do not reject the simple binomial hypothesis as well. Thus, these tests also do not support the criticisms of King, and they seem to provide support to a simple Poisson or simple binomial model for describing the data under consideration.

7.1.6 Concluding remarks

The discussion in this section has been on two main aspects. The first one is purely statistical and it pertains to which test is best to use for discriminating between Poisson and binomial distributions. Most statistical tests implemented (including the modified chi-squared tests) cannot discriminate between the two models with high confidence. This fact does not mean that tests considered are bad and that one needs to search for a better test. It simply means that the models are so close to each other that one can hardly develop a test with larger power.

The second aspect is purely applied and it pertains to which statistical model is best to use for estimating the number of judges of the Supreme Court appointed by the US President. The detailed statistical analysis of the number of appointments of US Supreme Court justices during the period 1789–2004 shows that the binomial model is as appropriate as the Poisson model. No test used can reject the Poisson model in favor of the binomial model or vice versa. Our statistical analysis in Section 7.1.5 does not confirm the arguments of King that there is no "independence of events and equality of probabilities of occurrences of those events," and so there does not seem to be any necessity for the exponential Poisson regression model. Indeed, taking into account the closeness of the Poisson and the binomial models, the independence of events and equality of probabilities are statistically confirmed not only by the analysis in Section 7.1.5, but also by Tarone's tests for the binomial distribution conducted in Section 7.1.3. Thus, the overall conclusion is that there is no firm statistical need for the exponential Poisson regression model since both Poisson and binomial models can be equally used for assessing the data on how many times a President has to appoint judges to the Supreme Court.

7.2 REVISITING RUTHERFORD'S DATA

The data from the classical experiment of Rutherford et al. (1930) on radioactive alpha decay is reanalyzed here (see Voinov and Voinov, 2008, 2010). To perform this analysis, unified versions of the modified chi-squared NRR tests and some other tests for the Poisson, binomial and a "binomial" approximation of the Feller's distribution (which takes into account the dead time of a counter under the assumption that it is not random) are used. At this point, it is worth mentioning that "Feller's" distribution in (9.7), being actually a "binomial" approximation of the incomplete gamma-function, can be used as an approximation of traffic counting distribution when a minimum spacing or headway between units of traffic (cars, airplanes, etc.) is taken into account (see Oliver, 1961). We show that the experimental data of Rutherford et al. (1930) contradict both Poisson and binomial null hypotheses, but support "Feller's" distribution. Modified chi-squared tests for the Poisson, binomial, and "binomial" approximation of Feller's distributions are all described in Chapter 9. The next section is devoted to the reanalysis of the classical Rutherford experimental data, and the concluding remarks are made in Section 7.2.2.

7.2.1 Analysis of the data

Over the years, several authors have analyzed the data of Rutherford et al. (1930), trying to show that the data do not contradict the exponential distribution for mean free paths of alpha particles from a radioactive radium source, from which one could then support the hypothesis that the number of particles registered follow the Poisson distribution; see, for example Rutherford et al. (1930), Cramer (1946), Bol'shev and Mirvaliev (1978), Voinov and Nikulin (1984), and Mirvaliev (2001). Their results seem to confirm the Poisson model, but due to some incorrectly implemented statistical tests, they cannot be considered as a proper statistical confirmation of that now well-known result in physics. This is why, we reanalyze these data here with the methods seen thus far in our discussion.

TABLE 7.9 Rutherford et al. (1930) experimental data in the form of a frequency table.

i	N_i	c, class number	N_c
0	57	1	57
1	203	2	203
2	383	3	383
3	525	4	525
4	532	5	532
5	408	6	408
6	273	7	273
7	139	8	139
8	45	9	45
9	27	10	27
10	10	11 ($i \geq 10$)	16
11	4		
12	0		
13	1		
14	1		
≥ 15	0		
	$\sum = 2608$		$\sum = 2608$

In the experiment of Rutherford et al. (1930), the number of flashes produced by alpha particles from a radioactive radium source, which hit the screen made of sulfide of zinc during non-overlapping time intervals of 7.5 s, were registered. The total number of such intervals was $n = 2608$. Let $X_1, X_2, \ldots, X_n$ be a sample from the probability distribution $P(X = i), i = 1,2, \ldots$, representing the number of scintillations during 7.5 s. Observed frequencies N_i of events $(X_j = i), j = 1,2, \ldots, n$, as recorded by Rutherford et al. (1930), are presented in Table 7.9.

If the exponential law of radioactive decay is true, the random variable X must follow the Poisson distribution. To test this hypothesis, Cramer (1946) used Pearson's sum $X_n^2(\hat{\theta}_n) = \mathbf{V}^{(n)T}(\hat{\theta}_n)\mathbf{V}^{(n)}(\hat{\theta}_n)$ with the MLE $\hat{\theta}_n$ of θ obtained from the raw data. The conclusion drawn by Cramer was that the data do not contradict the Poisson distribution. Later, Chernoff and Lehmann (1954) showed that, for fixed grouping intervals, this test is incorrect since the limiting distribution of $X_n^2(\hat{\theta}_n) = \mathbf{V}^{(n)T}(\hat{\theta}_n)\mathbf{V}^{(n)}(\hat{\theta}_n)$ will not follow χ_{r-1}^2 and may depend on the unknown parameters (see Section 2.5).

Bol'shev and Mirvaliev (1978) developed a distribution-free modified chi-squared test for the Poisson distribution based on minimum variance unbiased estimators of the hypothesized probabilities. Using $r = 11$ grouping classes shown in 3rd and 4th columns of Table 7.9 and their modified chi-squared test, they showed that the data do not contradict the Poisson distribution. Since the power of such a test is very low, Bol'shev and Mirvaliev (1978) implemented their test for two Neyman-Pearson classes (Greenwood and Nikulin, 1996), which maximizes Pearson's measure of distance between the null and alternative hypotheses thus resulting in a more powerful test. These classes are defined as $I_1 = \{c : f(c) \geqslant f_1(c)\}$ and $I_2 = \{c : f(c) < f_1(c)\}$. For the data in Table 7.9, $c \in \{3,4,5,6\}$ for I_1, and $c \in \{0,1,2\} \cup \{7,8,9,10,11\}$ for I_2. For these intervals, Bol'shev and Mirvaliev (1978) test Y (see formula (5.6)) based on the UMVUEs of hypothetical probabilities follows asymptotically the standard normal distribution. Since the calculated $Y = 2.3723$, the null hypothesis of Poisson distribution is rejected at level $\alpha = 0.05$.

By using the MLE $\hat{\theta}_n$ of θ, intervals I_1 and I_2, and the test statistic in (3.24), Mirvaliev (2001) found $S_n^2(\hat{\theta}_n) = 5.6773$. Since the critical value of χ_1^2 is 3.841 at $\alpha = 0.05$, $S_n^2(\hat{\theta}_n) = 5.6773$ results in rejection of the Poisson null hypothesis. Vu and Maller (1931) developed the likelihood ratio test for the Poisson against binomial distributions, as mentioned earlier. For the data in Table 7.9, the "deviance" statistic $d_n = 2.81$ ($\hat{N}_n = 83$) which is larger than the critical value $c_\alpha = 2.71$, and so the Poisson null hypothesis is rejected once again at $\alpha = 0.05$. Spinelli and Stephens (1997) suggested using the statistic A^2 for the Poisson null hypothesis. Strictly speaking, this test, which is not distribution-free, cannot be used in principle for this problem. However, for the data in Table 7.9, the statistic $A^2 = 1.248$. Since all critical values at $\alpha = 0.05$, regardless of the unknown value of θ (see Table 1 of (Spinelli and Stephens, 1997) and Figure 7.4) are less than 1.248, the Poisson null hypothesis is rejected here as well. Greenwood and Nikulin (1996, p. 203) applied Fisher's test based on the index of dispersion in (7.1) and showed that, at level $\alpha = 0.05$, the Poisson null hypothesis is rejected once again.

It is important to mention here that Cramer (1946), Bol'shev and Mirvaliev (1978), Mirvaliev (2001), Cramer (1946), Voinov and Nikulin (1984), and Greenwood and Nikulin (1996) did not control the power of the tests they implemented. As remarked in the preceding section, the simulated power of all these tests against the binomial alternative is only slightly more than the nominal level $\alpha = 0.05$ of the test. This is not surprising since the Poisson and the binomial models are very close to each other (Figure 7.5). Since all considered tests reject the Poisson hypothesis, which is so close to the binomial model, we may naturally expect the binomial alternative to be rejected as well. Basing on their results, Bol'shev and Mirvaliev (1978) and Mirvaliev (2001), however, erroneously concluded that "the binomial approximation can be preferred over the Poisson approximation" (see also Von Mises, 1972). This conclusion is incorrect since the binomial distribution was used only to construct the two Neyman-Pearson grouping intervals and it does not mean that the data do not contradict the binomial null hypothesis. Actually, to do this, we need to test the binomial null hypothesis (using, for example, their own test in (5.6)) to see whether or not the data support the binomial model.

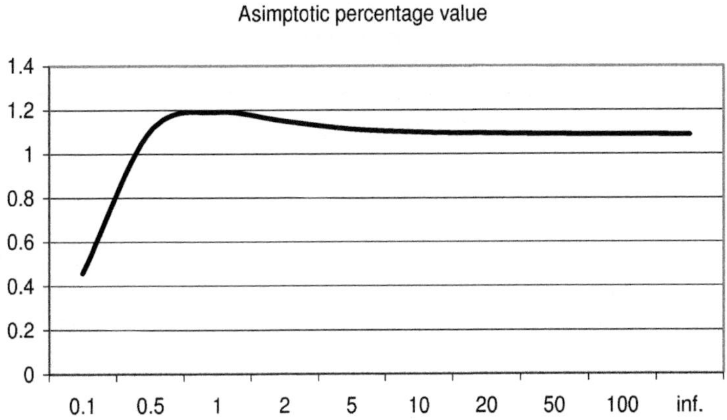

FIGURE 7.4 Asymptotic percentage points for the statistic A^2 as a function of the unknown parameter θ, for $\alpha = 0.05$, constructed by using Table 1 of Spinelli and Stephens (1997).

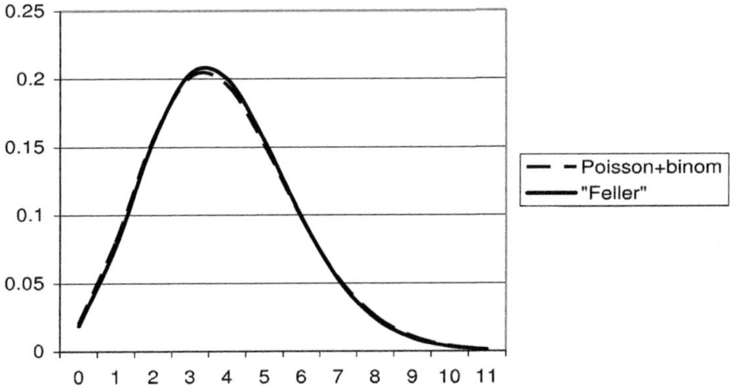

FIGURE 7.5 Smoothed curves of Poisson and binomial distributions with θ replaced by $\hat{\boldsymbol{\theta}}_n$ coincide, and are presented by dashed line. The smooth curve of the "binomial" approximation of Feller's distribution, with $\tilde{n}$ and $\tilde{p}$ replaced by estimates $\bar{\tilde{n}}$ and $\bar{\tilde{p}}$, is the solid line.

We applied the test in (5.6) for the same data, but for the binomial null hypothesis with $n = S$ and $\theta = S/t = 3.87155$. The MVUEs $W_i, i = 0, 1, \ldots, S$ (see formula (2) of Bol'shev and Mirvaliev (1978)) of the binomial probabilities were calculated by formula

$$W_i = \frac{\binom{S}{i}\binom{St - S}{S - i}}{\binom{St}{S}}, \quad i = 0, \ldots, S, \tag{7.3}$$

TABLE 7.10 Values of N_i – observed frequencies and of W_i – MVUEs of the hypothesized binomial probabilities.

i	N_i	$\ln W_i$	W_i	iW_i
0	57	−3.873034	0.020795	0.000000
1	203	−2.518612	0.080571	0.080571
2	383	−1.857536	0.156057	0.312113
3	**525**	−1.602123	**0.201468**	**0.604405**
4	**532**	−1.634589	**0.195032**	**0.780130**
5	**408**	−1.890398	**0.151012**	**0.755058**
6	**273**	−2.328727	**0.097420**	**0.584518**
7	139	−2.921404	0.053858	0.377006
8	45	−3.647811	0.026048	0.208385
9	27	−4.492199	0.011196	0.100764
10	10	−5.442146	0.004330	0.043302
11	4	−6.487602	0.001522	0.016744
12	0	−7.620267	0.000490	0.005885
13	1	−8.833174	0.000146	0.001896
14	1	−10.120390	0.000040	0.000564
15	0	−11.476790	0.000010	0.000156

where $S = 10{,}097$ and $t = 2608$. The logarithms of factorials in the above equation were calculated by the well-known approximation

$$\ln(n!) \approx (n + 1/2)\ln n - n + 1/2\ln(2\pi).$$

The details of calculations are presented in Table 7.10.

For the same two Neyman-Pearson classes used by Bol'shev and Mirvaliev (1978) while testing for the Poisson null hypothesis, we obtain

$$N_1^{(0)} = N_3 + N_4 + N_5 + N_6 = 1738,$$
$$W_1^{(0)} = W_3 + W_4 + W_5 + W_6 = 0.644932,$$
$$W_1^{(1)} = 3W_3 + 4W_4 + 5W_5 + 6W_6 = 2.724112.$$

Substituting these values and $\beta = S/t - S/t^2 = 3.87$ into (5.6), we get $Y = 2.362$. This means that the P-value of this test equals 0.009 which leads to the rejection of the binomial null hypothesis. If the logic of Bol'shev and Mirvaliev (1978) was correct, we would conclude that the data do not contradict the Poisson null hypothesis in this case. The apparent contradiction is evident, and in this case we have to reject both the Poisson and the binomial null hypotheses.

Nevertheless, the method of Bol'shev and Mirvaliev (1978) assists us greatly in having a look at the situation from the other side. Quite interestingly, they mentioned that data may contradict the Poisson distribution because of the dead time ("resolving time") of a human eye (Bol'shev, 1965). If the non-random parameter that characterizes the dead time is small, then the approximation of Feller's model mentioned in (9.6) may be used. Using the MMEs $\tilde{n} = \tau/(2\gamma)$ and $\tilde{p} = 2\mu\gamma/(1+\mu\gamma)$, Bol'shev and Mirvaliev (1978) estimated γ as $\hat{\gamma} = 0.044$ s. Then, by comparing this estimate with $\hat{\gamma} = 0.05$ s of Von Mises (1931) from another experiment, they concluded that the binomial approximation is valid. No attempt has been made, however, to develop tests to verify whether the data are in conformance with models in (9.5)–(9.7). Voinov and Nikulin (1984) revisited the problem and analyzed the data with corrections for frequencies N_c (see Table 7.9) that account for the dead time on average for each cell c, and then by applying Bol'shev and Mirvaliev's test, showed that the data do support Feller's model. Since such a way of data correction cannot be justified well, we reanalyze the data here by using the model in (9.7) that takes into account the dead time individually for each observation.

Using the data in Table 7.9 for the binomial null hypothesis and $r = 11$ grouping intervals, we obtain $Y1_n^2(\hat{\theta}_n) = 13.644, U_n^2(\hat{\theta}_n) = 12.692$, and $S_n^2(\hat{\theta}_n) = 0.952$ (see formulas (9.2), (3.23), and (3.24)). With $\alpha = 0.05$, the critical values of the limiting chi-squared distributions with 10, 9, and 1 degrees of freedom are 18.307, 16.919, and 3.841, respectively. From these values, at level 0.05, the null hypothesis is not rejected. The simulated (from $N = 1000$ runs) as well as the exact value calculated with the help of non-central chi-squared distribution (Abdel-Aty, 1954) values of the power of these tests against the Poisson alternative is only slightly more than the level 0.05. For $r = 2$ Neyman-Pearson classes I_1 and I_2, we find the statistic $Y1_n^2(\hat{\theta}_n) = S_n^2(\hat{\theta}_n) = 5.4882$, and since $5.4882 > 3.841$, the 5% critical value from the χ_1^2 distribution, the hypothesis of the binomial distribution, as well as of the Poisson distribution, are rejected. The simulated as well as the exact power of this test for the Poisson alternative is also negligible. However, it is of interest to mention that the simulated (based on $N = 1000$ runs) power of $S_n^2(\hat{\theta}_n)$ for both the Poisson and binomial alternatives for the "Feller's distribution" in (9.7) are 0.244 and 0.230, respectively. The corresponding non-centrality parameters are 1.648 and 1.546, which gives the estimates of exact power as 0.250 and 0.238, respectively, which agree very well with the simulated values. All these results seem to suggest that Feller's distribution in (9.7) differs noticeably from the Poisson and binomial distributions (see Figure 7.5).

Therefore, contrary to the conclusions of Bol'shev and Mirvaliev (1978) and Mirvaliev (2001), the data of Rutherford et al. (1930) do not conform to either Poisson or binomial distributions.

Moreover, from the above results, we cannot conclude that the null hypothesis of the model in (9.7) is acceptable either. For this reason, we carried out the test $Y2_n^2(\bar{\theta}_n)$ defined in (4.12). It is clear that there is no point in using tests with $r = 11$ intervals due to their low power. For the data in Table 7.9, we find $Y2_n^2(\bar{\theta}_n) = 2.791$, and so the null hypothesis of the model in (9.7) is not rejected at level 0.05. The calculated value of Singh's test in (3.25) for the same two Neyman-Pearson classes turns out to be $Q_s^2(\bar{\theta}_n) = 2.745$, which also does not lead to the rejection of the null hypothesis at the same 5% level of significance. To confirm that Rutherford's data do not contradict "Feller's" model, we may assess further the power of the test statistic $Y2_n^2(\bar{\theta}_n)$ against the Poisson and binomial alternatives which were rejected by the test $Y1_n^2(\hat{\theta}_n)$. For this purpose, we simulated values of $Y2_n^2(\bar{\theta}_n)$ under these alternatives using $N = 500$ runs for each case. For 40% of the simulated samples from the Poisson alternative, we found negative estimates for $\tilde{p}$ which evidently means that those samples contradict "Feller's distribution," and consequently the power of $Y2_n^2(\bar{\theta}_n)$ is at least 0.40. The same is observed for the binomial alternative in 41% of the simulated samples meaning that the power of $Y2_n^2(\bar{\theta}_n)$ is at least 0.41. The calculated values of the non-centrality parameters are $\lambda = 3.79$ for the Poisson alternative and $\lambda = 2.77$ for the binomial alternative. Using the non-central chi-squared distribution with one degree of freedom at these non-centrality parameters, we obtain the corresponding estimates of the powers to be 0.49 and 0.38, respectively, which are close to the simulated results. It is clear that the power of $Y2_n^2(\bar{\theta}_n)$ test is high enough to conclude that Rutherford's data do not contradict "Feller's distribution" which takes into account the counter's dead time. These findings may be regarded as a statistical confirmation of the exponential law of radioactive decay based on the experimental data of Rutherford et al. (1930).

It is useful to mention that the approximate model in (9.7) and the HRM statistic in (4.12) based on moment-type estimators may also be used for testing a traffic counting distribution (Oliver, 1961). This model is quite similar to the Feller's distribution in (9.5), and it is expressed in terms of incomplete gamma-function as well.

7.2.2 Concluding remarks

Contrary to a common belief that only efficient maximum likelihood estimators need be used in constructing efficient parametric modified chi-squared tests, one can use modified chi-squared tests based on non-efficient moment-type estimators as well to produce some efficient goodness of fit tests (see Alloyarova et al., 2007; Voinov et al., 2008a, 2009). This has been demonstrated in the preceding sections.

We have used different statistical validation procedures as well as some formal goodness of fit tests to shed some additional light on statistical analyses of data from Rutherford's experiment. The proposed tests perform well in discriminating between the Poisson, binomial, and the approximate Feller's distributions which are all quite close to each other. The power of the tests is sufficient to conclude that "Feller's model" is most appropriate for Rutherford's experimental data.

7.3 MODIFIED TESTS FOR THE LOGISTIC DISTRIBUTION

The logistic distribution has found important applications in many different fields of science. For a comprehensive review of various developments on the theory, methods and applications of the logistic distribution, one may refer to the book by Balakrishnan (1992).

Now, let us consider the problem of testing the null hypothesis (H_0) that the distribution function of a random variable X belongs to the family of logistic distributions

$$P\{X \leqslant x|H_0\} = F(x,\boldsymbol{\theta}) = G\left(\frac{x-\theta_1}{\theta_2}\right) = \left(1+\exp\left\{-\frac{\pi}{\sqrt{3}}\left(\frac{x-\theta_1}{\theta_2}\right)\right\}\right)^{-1},$$

where $x \in R^1, \theta_1 \in R^1$ and $\theta_2 > 0$. The corresponding pdf of X is

$$\begin{aligned} f(x,\boldsymbol{\theta}) &= G'\left(\frac{x-\theta_1}{\theta_2}\right) = \frac{1}{\theta_2} g\left(\frac{x-\theta_1}{\theta_2}\right) \\ &= \frac{\pi}{\sqrt{3}\theta_2} \frac{\exp\left(-\frac{\pi}{\sqrt{3}\theta_2}(x-\theta_1)\right)}{\left\{1+\exp\left(-\frac{\pi}{\sqrt{3}\theta_2}(x-\theta_1)\right)\right\}^2}. \end{aligned} \tag{7.4}$$

Let $\hat{\boldsymbol{\theta}}_n = (\hat{\theta}_{1n},\hat{\theta}_{2n})^T$ be the MLE of the parameter $\boldsymbol{\theta} = (\theta_1,\theta_2)^T$. Since there is no other sufficient statistic for $\boldsymbol{\theta}$ than the trivial one $\mathbf{X} = (X_1,\ldots,X_n)^T$ in the case of logistic distribution in (7.4), the maximum likelihood estimates have no explicit forms. For this reason, Harter and Moore (1967) (see also Harter and Balakrishnan, 1996) discussed numerical determination of the maximum likelihood estimate of $\boldsymbol{\theta}$ based on complete as well as doubly Type-II censored samples. However, by using some approximations in the resulting likelihood equations, Balakrishnan (1990) produced approximate maximum likelihood estimator of $\boldsymbol{\theta}$ which are simple explicit estimators of θ_1 and θ_2 and are also nearly as efficient as the maximum likelihood estimators. To present the expressions of these estimators, let us assume that from the random sample $X_1,\ldots,X_n$ from the logistic density function in (7.4), we observe a doubly Type-II censored sample of the form $X_{r+1:n} \leqslant \cdots \leqslant X_{n-s:n}$, where the smallest r and the largest s observations have been censored. Then, from the corresponding likelihood function

$$L = \frac{n!}{r!s!}\left(F\left(x_{r+1:n},\boldsymbol{\theta}\right)\right)^r\left(1-F\left(x_{n-s:n},\boldsymbol{\theta}\right)\right)^s \prod_{i=r+1}^{n-s} f(x_{i:n},\boldsymbol{\theta}),$$

we readily obtain the likelihood equations for θ_1 and θ_2 as follows:

$$\begin{aligned}\frac{\partial \ln L}{\partial \theta_1} &= \frac{\pi}{\sqrt{3}\theta_2}\left[-r\frac{f^*(z_{r+1:n})}{F^*(z_{r+1:n})} + s\frac{f^*(z_{n-s:n})}{1-F^*(z_{n-s:n})}\right.\\ &\qquad \left. +(n-r-s) - 2\sum_{i=r+1}^{n-s}\frac{f^*(z_{i:n})}{F^*(z_{i:n})}\right]\\ &= 0\end{aligned}$$

and

$$\begin{aligned}\frac{\partial \ln L}{\partial \theta_2} &= \frac{1}{\sigma}\left[-(n-r-s) - rz_{r+1:n}\frac{f^*(z_{r+1:n})}{F^*(z_{r+1:n})} + sz_{n-s:n}\frac{f^*(z_{n-s:n})}{1-F^*(z_{n-s:n})}\right.\\ &\quad \left. + \sum_{i=r+1}^{n-s} z_{i:n} - 2\sum_{i=r+1}^{n-s} z_{i:n}\frac{f^*(z_{i:n})}{F^*(z_{i:n})}\right]\\ &= 0,\end{aligned}$$

where $z_{i:n} = \pi(x_{i:n} - \theta_1)/(\sqrt{3}\theta_2)$,

$$f^*(z) = \frac{e^{-z}}{(1+e^{-z})^2} \quad \text{and} \quad F^*(z) = \frac{1}{1+e^{-z}} \quad \text{for } z \in R^1.$$

Then, Balakrishnan (1990) suggested expanding the function $F^*(z_{i:n})$ in a Taylor series around the point $F^*(p_i) = \ln(p_i/q_i)$ and then approximating it by

$$F^*(z_{i:n}) \simeq \gamma_i + \delta_i z_{i:n},$$

where

$$p_i = 1 - q_i = \frac{i}{n+1}, \quad \gamma_i = p_i - p_i q_i \ln(p_i/q_i) \quad \text{and} \quad \delta_i = p_i q_i.$$

Upon using this approximation in the above likelihood equations for θ_1 and θ_2, and then solving the resulting simplified equations, the approximate maximum likelihood estimators of θ_1 and θ_2 are derived to be

$$\hat{\theta}_1 = B - \frac{\sqrt{3}}{\pi}C\hat{\theta}_2 \quad \text{and} \quad \hat{\theta}_2 = \frac{\pi}{\sqrt{3}}\left\{\frac{D+\sqrt{D^2+4AE}}{2A}\right\},$$

where

$$A = n - r - s,$$

$$B = \frac{1}{m}\left\{ r\delta_{r+1}x_{r+1:n} + s\delta_{n-s}x_{n-s:n} + 2\sum_{i=r+1}^{n-s}\delta_i x_{i:n} \right\},$$

$$C = \frac{1}{m}\left\{ n - s - r\gamma_{r+1} - s\gamma_{n-s} - 2\sum_{i=r+1}^{n-s}\gamma_i \right\},$$

$$D = \sum_{i=r+1}^{n-s}(2\gamma_i - 1)x_{i:n} - r(1-\gamma_{r+1})x_{r+1:n} + s\gamma_{n-s}x_{n-s:n} + mBC,$$

$$E = r\delta_{r+1}x^2_{r+1:n} + s\delta_{n-s}x^2_{n-s:n} + 2\sum_{i=r+1}^{n-s}\delta_i x^2_{i:n} - mB^2,$$

$$m = r\delta_{r+1} + s\delta_{n-s} + 2\sum_{i=r+1}^{n-s}\delta_i.$$

As shown by Balakrishnan (1990), these estimators are nearly as efficient as the maximum likelihood estimators of θ_1 and θ_2 in the case of complete as well as censored samples. Of course, the approximate maximum likelihood estimators of θ_1 and θ_2 for the complete sample situation are deduced from the above expressions simply by setting $r = s = 0$. Note that, in this case, the following simple expressions are obtained for A,B,C,D,E, and m:

$$A = n, m = 2\sum_{i=1}^{n}\delta_i, \quad B = \frac{2}{m}\sum_{i=1}^{n}\delta_i x_{i:n}, \quad C = \frac{1}{m}\left\{ n - 2\sum_{i=1}^{n}\gamma_i \right\},$$

$$D = \sum_{i=1}^{n}(2\gamma_i - 1)x_{i:n} + mBC, \quad \text{and} \quad E = 2\sum_{i=1}^{n}\delta_i x^2_{i:n} - mB^2.$$

Based on the above estimators, Aguirre and Nikulin (1994) suggested the following modified chi-squared test. The limiting covariance matrix of the random vector $\sqrt{n}(\hat{\boldsymbol{\theta}}_n - \boldsymbol{\theta})$ is $\mathbf{J}^{-1}$, where

$$\mathbf{J} = \frac{1}{9\theta_2^2}\begin{bmatrix} \pi^2 & 0 \\ 0 & \pi^2 + 3 \end{bmatrix}.$$

Let $\mathbf{N}^{(n)} = (N_1^{(n)},\ldots,N_r^{(n)})^T$ be the frequency vector arising from grouping $X_1,\ldots,X_n$ over the random equiprobable intervals $(-\infty,b_1]$, $(b_1,b_2],\ldots,(b_{r-1},+\infty)$, where $b_j = \hat{\theta}_{1n} + \sqrt{3}\hat{\theta}_{2n}\ln(j/(r-j))/\pi$, $j = 1,\ldots,r-1$. Denote by $\mathbf{p} = (p_1,\ldots,p_r)^T$ the vector of probabilities $p_j = \int_{b_{j-1}}^{b_j} f(x,\boldsymbol{\theta})dx$, with θ_1 and θ_2 replaced by $\hat{\theta}_{1n}$ and $\hat{\theta}_{2n}$, respectively.

For the above chosen random intervals, the probabilities of falling into each interval are $p_1 = \cdots = p_r = 1/r$. Let $y_j = \ln(j/(r-j))$, for $j = 1,\ldots,r-1$. Further, let

$$\mathbf{a} = (a_1,\ldots,a_r)^T, \quad \mathbf{b} = (b_1,\ldots,b_r)^T, \quad \mathbf{W}^T = -\frac{1}{\theta_2}[\mathbf{a}\vdots\mathbf{b}],$$

where, for $j = 1,2,\ldots,r$,

$$\begin{aligned} a_j &= g(y_j) - g(y_{j-1}) = \frac{\pi}{r^2\sqrt{3}}(r-2j+1), \\ b_j &= y_j g(y_j) - y_{j-1} g(y_{j-1}) \\ &= \frac{1}{r^2}\left[(j-1)(r-j+1)\ln\left(\frac{r-j+1}{j-1}\right) - j(r-j)\ln\left(\frac{r-j}{j}\right)\right], \end{aligned}$$

$$\alpha(\mathbf{N}^{(n)}) = r\sum_{j=1}^{r} a_j N_j^{(n)} = \frac{\pi}{r\sqrt{3}}\left[(r+1)n - 2\sum_{j=1}^{r} jN_j^{(n)}\right],$$

$$\beta(\mathbf{N}^{(n)}) = r\sum_{j=1}^{r} b_j N_j^{(n)} = \frac{1}{r}\sum_{j=1}^{r-1}(N_{j+1}^{(n)} - N_j^{(n)})j(r-j)\ln\left(\frac{r-j}{j}\right),$$

$$\lambda_1 = J(11) - r\sum_{j=1}^{r} a_j^2 = \frac{\pi^2}{9r^2}, \quad \lambda_2 = J(22) - r\sum_{j=1}^{r} b_j^2.$$

Let $\mathbf{\Sigma} = \mathbf{D} - \mathbf{p}\mathbf{p}^T - \mathbf{W}^T\mathbf{J}^{-1}\mathbf{W}$, where $\mathbf{D}$ is the diagonal matrix with elements $1/r$ on the main diagonal. Then, the matrix $\mathbf{\Sigma}$ does not depend on $\boldsymbol{\theta}$, and its rank is $r-1$, i.e. the matrix $\mathbf{\Sigma}$ is singular. Consider the matrix $\tilde{\mathbf{\Sigma}}$ obtained by deleting the last row and the last column of $\mathbf{\Sigma}$, which is not singular and in fact

$$\tilde{\mathbf{\Sigma}}^{-1} = \mathbf{A} + \mathbf{A}\tilde{\mathbf{W}}^T(\mathbf{J} - \tilde{\mathbf{W}}\mathbf{A}\tilde{\mathbf{W}}^T)^{-1}\tilde{\mathbf{W}}\mathbf{A},$$

where $\mathbf{A} = \tilde{\mathbf{D}}^{-1} + \mathbf{1}\mathbf{1}^T/p_r, \tilde{\mathbf{D}}^{-1}$ is a diagonal matrix with elements $1/p_1,\ldots,1/p_{r-1}$ on the main diagonal, $\mathbf{1} = \mathbf{1}_{r-1}$ is the $r-1$ dimensional vector with all elements being one, and $\tilde{\mathbf{W}}$ is obtained from $\mathbf{W}$ by deleting the last column. The vector $\tilde{\mathbf{N}}^{(n)} = (N_1^{(n)},\ldots,N_{r-1}^{(n)})^T$ is asymptotically normally distributed with mean vector $\mathbf{E}\tilde{\mathbf{N}}^{(n)} = n\tilde{\mathbf{p}}$ and covariance matrix $\mathbf{E}(\tilde{\mathbf{N}}^{(n)} - n\tilde{\mathbf{p}})^T(\tilde{\mathbf{N}}^{(n)} - n\tilde{\mathbf{p}}) = n\tilde{\boldsymbol{\sigma}}$, where $\tilde{\mathbf{p}} = (p_1,\ldots,p_{r-1})^T$. Then, we have the following result.

Theorem 7.1 (Aguirre and Nikulin (1994)) *The statistic*

$$\begin{aligned} Y1_n^2 &= \frac{1}{n}(\tilde{\mathbf{N}}^{(n)} - n\tilde{\mathbf{p}})^T\tilde{\boldsymbol{\sigma}}^{-1}(\tilde{\mathbf{N}}^{(n)} - n\tilde{\mathbf{p}}) \\ &= \sum_{j=1}^{r}\frac{(N_j^{(n)} - np_j)^2}{np_j} + \frac{\lambda_1\beta^2(\mathbf{N}^{(n)}) + \lambda_2\alpha^2(\mathbf{N}^{(n)})}{n\lambda_1\lambda_2} \end{aligned} \tag{7.5}$$

follows in the limit the chi-square distribution with $r-1$ degrees of freedom.

Remark 7.1. *Consider the alternative hypothesis H_η according to which the random variables $X_i, i = 1,\ldots,n$, follow a probability distribution $G((x-\theta_1)/\theta_2,\eta)$, where $G(x,\eta)$ is continuous, $|x| < \infty, \eta \in \mathbf{H} \in R^1$, $G(x,0) = G(x)$, and $\eta = 0$ is a limit point of $\mathbf{H}$. Assume that $\frac{\partial}{\partial x}G(x,\eta) = g(s,\eta)$ and $\frac{\partial}{\partial \eta}g(x,\eta)|_{\eta=0} = \psi(x)$, where $g(x,0) = g(x) = G'(x)$. If $\frac{\partial^2 g(x,\eta)}{\partial \eta^2}$ exists and is continuous for all x in the neighborhood of $\eta = 0$, then for $b_j = \hat{\theta}_1 + \frac{\sqrt{3}\hat{\theta}_2}{\pi}\ln\left(\frac{j}{j-1}\right), j = 1,\ldots,r-1$, we have $P\{b_{j-1} < X \leqslant b_j | H_\eta\} = p_j + \eta c_j + o(\eta)$, where $c_j = \int_{y_{r-1}}^{y_j}\psi(x)dx, j = 1,\ldots,r$. In the limit, as $n \to \infty$, the test statistic in (7.5) will follow the non-central chi-square distribution with $r-1$ degrees of freedom and the non-centrality parameter*

$$\lambda = \sum_{j=1}^{r}\frac{c_j^2}{p_j} + \frac{\lambda_2\alpha^2(\mathbf{c}) + \lambda_1\beta^2(\mathbf{c})}{\lambda_1\lambda_2}, \quad \text{with } \mathbf{c} = (c_1,\ldots,c_r)^T.$$

Remark 7.2. Voinov et al. (2003) *(see also Voinov et al., 2009) constructed the HRM test $Y2_n^2(\bar{\boldsymbol{\theta}}_n)$ in (4.9) for logistic distribution as the null hypothesis and also examined its power properties. Analytical expressions for the elements of matrices $\mathbf{B},\mathbf{C},\mathbf{K},\mathbf{V}$ needed to evaluate the test statistic $Y2_n^2(\bar{\boldsymbol{\theta}}_n)$ are presented in Section 9.5. As it is difficult to obtain the distribution of the test statistic $Y2_n^2(\bar{\boldsymbol{\theta}}_n)$ under alternative hypotheses, a Monte Carlo experiment was used to estimate the power, and the results of this simulation study are all detailed in Sections 4.4.2 and 4.5.2.*

7.4 MODIFIED CHI-SQUARED TESTS FOR THE INVERSE GAUSSIAN DISTRIBUTION

7.4.1 Introduction

Schrödinger (1915) was the first to derive the probability distribution of the first passage time in Brownian motion. Tweedie (1957) investigated the basic characteristics of this distribution and proposed the name inverse Gaussian, and it is also known as Wald's distribution. Folks and Chhikara (1978) gave a review of this distribution (denoted here by IGD) and mentioned numerous applications of it. An application of the IGD as a life time model is possibly the most appealing one; see Chhikara and Folks (1989), Gunes et al. (1997), and Seshadri (1993).

Let $X_1,\ldots,X_n$ be i.i.d. random variables that follow the IGD with probability density function

$$f(x;\boldsymbol{\theta}) = \sqrt{\frac{\lambda}{2\pi x^3}}\exp\left\{-\frac{\lambda(x-\mu)^2}{2\mu^2 x}\right\}, \quad x \geqslant 0,$$
$$\boldsymbol{\theta} = (\mu,\lambda)^T \in R_+^1 \times R_+^1 \subset R^2,$$

where μ is the location (in fact, the mean) and λ is the shape parameter. The hazard rate function of the IGD is

$$h(x,\boldsymbol{\theta}) = \sqrt{\frac{\lambda}{2\pi x^3}} \exp\left[-\frac{\lambda(x-\mu)^2}{2\mu^2 x}\right]\left\{\Phi\left[-\sqrt{\frac{\lambda}{x}}\left(\frac{x}{\mu}-1\right)\right]\right.$$
$$\left.-\exp\frac{2\lambda}{\mu}\Phi\left[-\sqrt{\frac{\lambda}{x}}\left(\frac{x}{\mu}+1\right)\right]\right\}^{-1}, \quad x \geqslant 0.$$

The complete sufficient statistics for parameters μ and λ are (Folks and Chhikara, 1978, Seshadri, 1993) $S = \overline{X} = \frac{1}{n}\sum_{i=1}^{n} X_i$ and $T = \sum_{i=1}^{n}\left(\frac{1}{X_i} - \frac{1}{\overline{X}}\right)$, respectively. Moreover, the MLEs of μ and λ are $\hat{\mu} = S$ and $\hat{\lambda} = n/T$.

Gunes et al. (1997) (see also Lemeshko et al., 2010) used Monte Carlo simulations to investigate the power properties of the following EDF tests: Kolmogorov-Smirnov, Kuiper, Cramer-von Mises, Watson and Anderson-Darling. Gunes et al. (1997) showed that, among all the non-parametric tests investigated, the Watson test based on the statistic

$$W = \frac{1}{12n} + \sum_{i=1}^{n}\left[F(X_i) - \frac{2i-1}{2n}\right]^2 - n\left[\left(\sum_{i=1}^{n}\frac{F(X_i)}{n}\right) - \frac{1}{2}\right]^2,$$

where $F(x)$ is the cumulative distribution function of the IGD, possesses the highest power.

7.4.2 The NRR, DN and McCulloch tests

The NRR test in (3.8) for IGD has been investigated by Nikulin and Saaidia (2009), Lemeshko (2010), and Saaidia and Tahir (2012). This test is based on the MLE of the Fisher information matrix $\mathbf{J}(\boldsymbol{\theta})$ for one observation from IGD given by

$$\mathbf{J}(\boldsymbol{\theta}) = \begin{pmatrix} \frac{\lambda}{\mu^3} & 0 \\ 0 & \frac{1}{2\lambda^2} \end{pmatrix},$$

and the matrix

$$\mathbf{B}(\boldsymbol{\theta}) = \begin{pmatrix} b_{11}\, b_{12} \\ \cdots\cdots \\ b_{r1}\, b_{r2} \end{pmatrix},$$

where

$$b_{i1} = \frac{1}{\sqrt{p_i(\boldsymbol{\theta})}}\frac{\partial p_i(\boldsymbol{\theta})}{\partial \mu}, \quad b_{i2} = \frac{1}{\sqrt{p_i(\boldsymbol{\theta})}}\frac{\partial p_i(\boldsymbol{\theta})}{\partial \lambda}, \qquad i = 1,\ldots,r,$$

r is the number of grouping cells, and $p_i(\boldsymbol{\theta})$ is the probability of falling into the ith cell.

For r equiprobable grouping intervals with random boundaries (when $p_i(\boldsymbol{\theta}) = 1/r$), Saaidia and Tahir (2012) investigated the power of the NRR test in (3.8), the DN test in (3.23), and the McCulloch test in (3.24) with respect to the following alternatives:

$$f(x;\,\mu,\sigma) = \frac{1}{x\sigma\sqrt{2\pi}}\exp\left\{-(\ln x-\mu)^2/2\sigma^2\right\},\quad x,\sigma>0,\ \mu\in R^1;$$

the loglogistic

$$f(x;\,\alpha,\beta) = \frac{(\beta/\alpha)(x/\alpha)^{\beta-1}}{\{1+(x/\alpha)^{\beta}\}^2},\quad x,\alpha,\beta>0;$$

the generalized Weibull

$$F(x) = 1-\exp\left\{1-\left[1+(x/\sigma)^{\nu}\right]^{1/\gamma}\right\},\quad x,\sigma,\nu,\gamma>0;$$

and the exponentiated Weibull

$$F(x) = \left\{1-\exp\left(1-x/\alpha\right)^{\beta}\right\}^{\gamma},\quad x,\alpha,\beta,\gamma>0.$$

They showed that, for equiprobable random cells for a sample of size $n = 200$ and $\alpha = 0.05$, the McCulloch test (as in Sections 3.5 and 4.4) is the most powerful among all modified chi-squared tests considered in their study.

Chapter 8

Probability Distributions of Interest

In this chapter, we describe various discrete and continuous probability distributions that are most pertinent to the discussion in preceding chapters, and provide details on their forms and key distributional properties. For more elaborate discussions on all univariate and multivariate distributions described here, interested readers may refer to the books by Johnson et al. (1994, 1995, 2005), Kotz et al. (2000), Balakrishnan and Nevzorov (2003), and Balakrishnan and Lai (2009).

8.1 DISCRETE PROBABILITY DISTRIBUTIONS

8.1.1 Binomial, geometric, and negative binomial distributions

Consider a sequence of identical and independent trials indexed by $k = 1,2,\ldots$. Suppose the outcome of each trial is one of two possible events E_1 and $E_2 = \overline{E}_1$, one of which, for example E_1, is referred to as a "*success*" and the other one, E_2, as a "*failure*." These independent trials, in which the probability of success $p = \mathbf{P}(E_1)$ stays fixed, are called Bernoulli trials. Let $q = 1 - p = \mathbf{P}(E_2)$ be the probability of failure. Corresponding to each trial, we define the Bernoulli random variable X_i (for $i = 1,\ldots,n$) with parameter p. Then, the random variables $\{X_i : i = 1,2,\ldots\}$ form a sequence of independent random variables, following the same law of Bernoulli with parameter $p, 0 < p < 1$.

Chi-Squared Goodness of Fit Tests with Applications. http://dx.doi.org/10.1016/B978-0-12-397194-4.00008-9

With different stopping rules for these trials in a Bernoulli experiment, different probability models are obtained.

Let n ($\geqslant 2$) be a fixed positive integer. Then, consider the statistic

$$\mu_n = \sum_{i=1}^{n} X_i$$

which represents the number of successes among the first n trials. The statistic μ_n follows the binomial law $B(n,p)$ with parameters n and $p, 0 \leqslant p \leqslant 1$, which we denote by $\mu_n \sim B(n,p)$. It is easy to show that

$$\mathbf{P}\{\mu_n = k|n,p\} = \binom{n}{k} p^k (1-p)^{n-k}, \quad k \in \{0,1,\ldots,n\},$$

$$\mathbf{E}\mu_n = np, \quad \mathbf{Var}\mu_n = np(1-p) = npq < \mathbf{E}\mu_n.$$

The cumulative distribution function of μ_n is the step function given by the formula

$$\begin{aligned}
&\mathbf{P}\{\mu_n \leqslant m|n,p\} \\
&\quad = \sum_{k=0}^{m} \binom{n}{k} p^k (1-p)^{n-k} \\
&\quad = P_0 + P_1 + \cdots + P_m \\
&\quad = 1 - I_p(m+1, n-m) = I_{1-p}(n-m, m+1), \quad m = 0,1,\ldots,n,
\end{aligned}$$

where

$$I_x(a,b) = \frac{1}{B(a,b)} \int_0^x u^{a-1}(1-u)^{b-1} du, \quad 0 < u < 1,$$

is the incomplete beta ratio ($a > 0, b > 0$), and

$$B(a,b) = \int_0^1 u^{a-1}(1-u)^{b-1} du$$

is the complete beta function. Note that

$$B(a,b) = \frac{\Gamma(a)\Gamma(b)}{\Gamma(a+b)},$$

where

$$\Gamma(f) = \int_0^\infty t^{f-1} e^{-t} dt, \quad f > 0,$$

is the complete gamma function.

Let Z_1 be the number of Bernoulli trials performed until the occurrence of the first success. Then, the possible values of the statistic Z_1 are $1,2,3,\ldots$.

In this case, we say that the statistic Z_1 follows the geometric distribution with parameter $p, 0 < p < 1$, and its probability mass function is

$$\mathbf{P}\{Z_1 = k|p\} = p(1-p)^{k-1}, \quad k = 1,2,\ldots$$

We shall denote it by $Z_1 \sim G(p)$. In this case, it can be easily shown that

$$\mathbf{E}Z_1 = \sum_{k=1}^{\infty} kp(1-p)^{k-1} = \frac{1}{p}$$

and

$$\mathbf{Var}Z_1 = \mathbf{E}{Z_1}^2 - (\mathbf{E}(Z_1))^2 = \frac{1-p}{p^2}.$$

Note that

$$\mathbf{Var}Z_1 < \mathbf{E}Z_1$$

if and only if $p \in \left(\frac{1}{2}, 1\right]$. The distribution function of Z_1 is

$$F(x) = \mathbf{P}\{Z_1 \leqslant x|p\} = \sum_{k=1}^{[x]} p(1-p)^{k-1} = 1-(1-p)^{[x]}, \quad x \geqslant 1.$$

Next, let Z_r be the number of Bernoulli trials performed until the r-th success. Then, the possible values of Z_r are $r, r+1, \ldots$ The event $Z_r = k$ would occur if and only if during the first $k-1$ trials $r-1$ successes were observed, and in the kth trial a success is observed, which readily yields

$$\mathbf{P}(Z_r = k|p) = \binom{k-1}{r-1} p^{r-1}(1-p)^{k-1-(r-1)} p = \binom{k-1}{r-1} p^r (1-p)^{k-r}.$$

The probability distribution of the statistic Z_r is called the negative binomial distribution with parameters r and $p, r \in \{1,2,\ldots\}, 0 < p < 1$, and we shall denote it by $Z_r \sim NB(r,p)$. We then observe the following:

1. If $r = 1$, then we have the geometric distribution with parameter p, that is, $NB(1,p) = G(p)$;
2. Consider r is independent and identically distributed random variables $Z_{1,1}, Z_{1,2}, \ldots, Z_{1,r}$, where $Z_{1,i} \sim G(p)$. Then, from the definition of the negative binomial distribution, it follows that

$$Z_r \stackrel{d}{=} \sum_{i=1}^{r} Z_{1,i},$$

where $Z_r \sim NB(r,p)$. So, we immediately have

$$\mathbf{E}Z_r = \frac{r}{p} \quad \text{and} \quad \mathbf{Var}Z_r = \frac{r(1-p)}{p^2};$$

3. In a similar manner, we can show that if $Z_{r_1}, Z_{r_2}, \ldots, Z_{r_n}$ are independent random variables such that $Z_{r_i} \sim NB(r_i, p)$, then the statistic

$$U_n = \sum_{i=1}^{n} Z_{r_i}$$

follows the negative binomial distribution with parameters $r = \sum_{i=1}^{n} r_i$ and p, that is, $U_n \sim NB(r,p)$;

4. $\mathbf{Var} Z_r < \mathbf{E} Z_r$ if and only if $p \in \left(\frac{1}{2}, 1\right]$.

Note that $\mathbf{P}(Z_r > n) = \mathbf{P}(\mu_n < r)$, from which it follows that

$$\begin{aligned}\mathbf{P}\{Z_r \leqslant n|p\} &= 1 - \mathbf{P}\{Z_r > n|p\} = 1 - \mathbf{P}\{\mu_n < r|r+n-1, p\}\\ &= 1 - \mathbf{P}\{\mu_n \leqslant r-1|r+n-1, p\} = I_p(r, n-r+1)\\ &= 1 - I_{1-p}(n-r+1, r)\end{aligned}$$

and so the distribution function of the statistic Z_r is

$$F_{Z_r}(x) = I_p(r, [x]-r+1) = 1 - I_{1-p}([x]-r+1, r), \quad x \geqslant r. \tag{8.1}$$

Suppose the Bernoulli trials are continued until the rth success. Then, the statistic $T_r = Z_r - r$ shows the number of failures observed in these trials. Then, the statistic T_r takes on values $0, 1, \ldots,$ and

$$\mathbf{P}\{T_r = k|p\} = \mathbf{P}\{Z_r = k + r|p\} = \binom{k+r-1}{r-1} p^r (1-p)^k.$$

We say that the statistic T_r follows the Pascal distribution with parameters r and p.

The mean and variance of the statistic T_r are evidently

$$\mathbf{E}T_r = \frac{r(1-p)}{p}, \quad \mathbf{Var}T_r = \frac{r(1-p)}{p^2},$$

and its distribution function is

$$F_{T_r}(x) = F_{Z_r}(x+r) = I_p(r, [x]+1) = 1 - I_{1-p}([x]+1, r), \quad x \geqslant 0. \tag{8.2}$$

From (8.1) and (8.2), it follows that

$$\mathbf{P}\{T_r \geqslant n|p\} = \mathbf{P}\{\mu_n \leqslant r-1|r+n-1, p\} = 1 - I_p(r,n) = I_{1-p}(n,r).$$

Note that

$$\frac{\mathbf{Var}T_r}{\mathbf{E}T_r} = \frac{1}{p} > 1$$

for $p \in [0,1]$.

8.1.2 Multinomial distribution

Consider a sequence of n independent trials, and suppose the outcome of each trial is one of k mutually exclusive events $E_1, E_2, \ldots, E_k$ with positive probabilities

$$p_1 = \mathbf{P}(E_1), p_2 = \mathbf{P}(E_2), \ldots, p_k = \mathbf{P}(E_k),$$

where $p_1 + \cdots + p_k = 1$.

Let $\mathbf{p} = (p_1, \ldots, p_k)^T$ and $\boldsymbol{\nu} = (\nu_1, \ldots, \nu_k)^T$, where ν_i is the frequency of the event E_i in the sequence of trials ($i = 1, \ldots, k$). It is then evident that the possible values of ν_i are non-negative integers n_i, $0 \leqslant n_i \leqslant n$, and

$$n_1 + n_2 + \cdots + n_k = n. \tag{8.3}$$

The statistic $\boldsymbol{\nu}$ follows the multinomial probability distribution with parameters n and $\mathbf{p}$, and its probability mass function is

$$\mathbf{P}\{\nu_1 = n_1, \ldots, \nu_k = n_k\} = \frac{n!}{n_1! \ldots n_k!} p_1^{n_1} p_2^{n_2} \ldots p_k^{n_k},$$

for any integers $n_1, \ldots, n_k$ satisfying (8.3). We shall denote it by $\boldsymbol{\nu} \sim M_k(n, \mathbf{p})$.

By direct computation, one can see that the mean vector and the variance–covariance matrix of $\boldsymbol{\nu}$ are given by

$$\mathbf{E}\boldsymbol{\nu} = n\mathbf{p} \quad \text{and} \quad \mathbf{E}(\boldsymbol{\nu} - n\mathbf{p})(\boldsymbol{\nu} - n\mathbf{p})^T = n\boldsymbol{\Sigma} = n(\mathbf{P} - \mathbf{p}\mathbf{p}^T),$$

where $\mathbf{P}$ is the diagonal matrix with $p_1, \ldots, p_k$ on its main diagonal. It can be shown that, due to the condition in (8.3), the rank of $\boldsymbol{\Sigma} = k - 1$.

8.1.3 Poisson distribution

The discrete random variable X follows the Poisson distribution with parameter $\lambda > 0$, if

$$\mathbf{P}\{X = k\} = \frac{\lambda^k}{k!} e^{-\lambda}, \quad k = 0, 1, \ldots,$$

and we shall denote it by $X \sim P(\lambda)$. It is easy to show that

$$\mathbf{E}X = \mathbf{Var}X = \lambda,$$

and so

$$\frac{\mathbf{Var}X}{\mathbf{E}X} = 1.$$

The distribution function of X is

$$\mathbf{P}\{X \leqslant m\} = \sum_{k=0}^{m} \frac{\lambda^k}{k!} e^{-\lambda} = 1 - I_\lambda(m+1), \tag{8.4}$$

where

$$I_x(f) = \frac{1}{\Gamma(f)} \int_0^x t^{f-1} e^{-t} dt, \quad x > 0$$

is the incomplete gamma function. Often, for large values of λ, to compute (8.4), we can use a normal approximation

$$\mathbf{P}\{X \leqslant m\} = \Phi\left(\frac{m + 0.5 - \lambda}{\sqrt{\lambda}}\right) + O\left(\frac{1}{\sqrt{\lambda}}\right), \quad \lambda \to \infty.$$

Let $\{X_n\}_{n=1}^{\infty}$ be a sequence of independent and identically distributed random variables following the same Bernoulli distribution with parameter $p, 0 < p < 1$, with

$$\mathbf{P}\{X_i = 1\} = p, \quad \mathbf{P}\{X_i = 0\} = q = 1 - p.$$

Let

$$\mu_n = X_1 + \cdots + X_n, \quad F_n(x) = \mathbf{P}\left\{\frac{\mu_n - np}{\sqrt{npq}} \leqslant x\right\}, \quad x \in \boldsymbol{R}^1.$$

Then, uniformly for $x \in \boldsymbol{R}^1$, we have

$$F_n(x) \to \Phi(x) = \frac{1}{\sqrt{2\pi}} \int_{-\infty}^x e^{-t^2/2} dt, \quad n \to \infty.$$

From this result, it follows that for large values of n,

$$\mathbf{P}\left\{\frac{\mu_n - np}{\sqrt{npq}} \leqslant x\right\} \approx \Phi(x).$$

Often this approximation is used with the so-called continuity correction given by

$$\mathbf{P}\left\{\frac{\mu_n - np + 0.5}{\sqrt{npq}} \leqslant x\right\} \approx \Phi(x).$$

We shall now describe the Poisson approximation to the binomial distribution. Let $\{\mu_n\}$ be a sequence of binomial random variables, $\mu_n \sim B(n, p_n)$, $0 < p_n < 1$, such that

$$np_n \to \lambda \quad \text{if} \quad n \to \infty \quad \text{and} \quad \lambda > 0.$$

Then,

$$\lim_{n \to \infty} \mathbf{P}\{\mu_n = m | n, p_n\} = \frac{\lambda^m}{m!} e^{-\lambda}.$$

In practice, this means that for "large" values of n and "small" values of p, we may approximate the binomial distribution $B(n, p)$ by the Poisson distribution with parameter $\lambda = np$, that is,

$$\mathbf{P}\{\mu_n = m | n, p\} \approx \frac{\lambda^m}{m!} e^{-\lambda}.$$

It is of interest to note that (Hodges and Le Cam, 1960)

$$\sup_x \left| \sum_{m=0}^{x} \binom{n}{m} p^m (1-p)^{n-m} - \sum_{m=0}^{x} \frac{\lambda^m}{m!} e^{-\lambda} \right| \leqslant \frac{C}{\sqrt{n}}, \quad \text{where } C \leqslant 3\sqrt{\lambda}.$$

Hence, if the probability of success in Bernoulli trials is small, and the number of trials is large, then the number of observed successes in the trials can be regarded as a random variable following the Poisson distribution.

8.2 CONTINUOUS PROBABILITY DISTRIBUTIONS

A random variable X is said to be continuous if its distribution function

$$F(x) = \mathbf{P}(X \leqslant x), \quad x \in \boldsymbol{R}^1$$

is absolutely continuous, i.e., it can be presented as an integral of a probability density function $f(x)$ in the form

$$F(x) = \int_{-\infty}^{x} f(y)dy, \quad x \in \boldsymbol{R}^1.$$

The probability that X takes a value $x \in \boldsymbol{R}^1$ is equal to 0, i.e., $\mathbf{P}(X = x) = 0$. But, we may speak about the probability that X takes a value in an interval, and it is given by

$$\mathbf{P}(X \in [a,b]) = \int_a^b f(x)dx.$$

If the distribution function is differentiable at the point x, then

$$f(x) = \lim_{h \downarrow 0} \frac{\mathbf{P}(x < X \leqslant x+h)}{h} = F'(x).$$

The quantity

$$x_p = \inf\{x : F(x) \geqslant p\}, \quad 0 < p < 1$$

is called the p-quantile of the random variable X.

If the support $\mathcal{X} = \{x : f(x) > 0\}$ of the density (the set on which the density is strictly positive) is an interval, then the distribution function is continuous and strictly increasing on this interval, and consequently

$$x_p = F^{-1}(p), \quad 0 < p < 1,$$

where F^{-1} is the inverse function of $F(\cdot)$.

In the case when a random variable is continuous and positive, it can be regarded as a lifetime variable. In this case, we often denote it by T. The value of this random variable can be considered as the failure time of a system or a unit. The law of a lifetime T is often specified in terms of the survival or reliability function, given by

$$S(t) = \mathbf{P}\{T > t\} = 1 - F(t), \quad t \geqslant 0.$$

For any $t > 0, S(t) = 1 - F(t)$ gives the probability that the lifetime of the considered system (unit) will exceed t, or the probability not to have the failure of the system during the time interval $[0,t]$. It is evident that $F(t)$ is then the probability of failure of the system in the interval $[0,t]$.

The value $f(t)$ of the density $f(\cdot)$ represents the probability of failure in a small interval after the moment t.

In survival analysis and reliability theory, it is more natural to formulate models in terms of *failure rate function* or *hazard function* of the lifetime T defined by

$$\lambda(t) = \lim_{h \downarrow 0} \frac{\mathbf{P}(t < T \leqslant t + h | T > t)}{h} = \frac{f(t)}{S(t)}.$$

For any positive t, the value $\lambda(t)$ represents the probability of failure in a small interval after the moment t given that the system is still functioning at the moment t. Thus, it is the *instantaneous failure rate* of a system which has survived t units of time. Next, we introduce the important notion of Mean Time to Failure (MTTF). It is the average length of time until failure of a system, and is given by

$$\begin{aligned}\mathbf{E}T &= \int_0^\infty t f(t) dt = \int_0^\infty t dF(t) = -\int_0^\infty t d[1 - F(t)] \\ &= -\lim_{t \to \infty} t[1 - F(t)] + \int_0^\infty [1 - F(t)] dt = \int_0^{+\infty} S(t) dt,\end{aligned}$$

provided it exists. In the same way, we can see that if $\mathbf{Var}T$ exists, then it can be expressed as

$$\mathbf{Var}T = 2 \int_0^\infty t S(t) dt - (\mathbf{E}T)^2.$$

8.2.1 Exponential distribution

Let $\boldsymbol{X} = (X_1, \ldots, X_n)^T$ be a sample from an exponential distribution $\mathcal{E}(\mu,\sigma)$, i.e.

$$X_i \sim f(x; \boldsymbol{\theta}), \quad \boldsymbol{\theta} \in \boldsymbol{\Theta} = \{\boldsymbol{\theta} = (\mu,\sigma)^T : |\mu| < \infty, \sigma > 0\},$$

where

$$f(x; \boldsymbol{\theta}) = \begin{cases} \frac{1}{\sigma} \exp\left(-\frac{x-\mu}{\sigma}\right), & x \geqslant \mu, \\ 0, & \text{otherwise.} \end{cases}$$

We can alternatively write

$$f(x;\boldsymbol{\theta}) = \frac{1}{\sigma}\exp\left(-\frac{x-\mu}{\sigma}\right)H(x-\mu), \tag{8.5}$$

where

$$H(x) = \begin{cases} 1, \text{ if } & x \geqslant 0, \\ 0, \text{ if } & x < 0. \end{cases}$$

It is easy to show that

$$\mathbf{E}X_i = \mu + \sigma \quad \text{and} \quad \mathbf{Var}X_i = \sigma^2. \tag{8.6}$$

Let $\boldsymbol{X}^{(n)} = (X_{(1)},X_{(2)},\ldots,X_{(n)})^T$ denote the vector of order statistics obtained by arranging the vector of observations $\boldsymbol{X}$, with

$$\mathbf{P}\{X_{(1)} < X_{(2)} < \cdots < X_{(n)}\} = 1. \tag{8.7}$$

Then, it can be shown that $\mathbf{T} = (X_{(1)},S)^T$ is the minimal sufficient statistic for the parameter $\boldsymbol{\theta}$, where

$$X_{(1)} = \min(X_1,X_2,\ldots,X_n) \quad \text{and} \quad S = \sum_{i=2}^{n}(X_{(i)} - X_{(1)}). \tag{8.8}$$

Indeed, the likelihood function of $\boldsymbol{X}$ is

$$\begin{aligned} L(\boldsymbol{X};\theta) &= \prod_{i=1}^{n} f(X_i;\theta) = \frac{1}{\sigma^n}\exp\left\{-\frac{1}{\sigma}\sum_{i=1}^{n}(X_i-\mu)\right\}H(X_{(1)}-\mu) \\ &= \frac{1}{\sigma^n}\exp\left\{-\frac{1}{\sigma}\sum_{i=1}^{n}(X_{(i)}-\mu)\right\}H(X_{(1)}-\mu). \end{aligned} \tag{8.9}$$

Since

$$\sum_{i=1}^{n} X_i = \sum_{i=1}^{n} X_{(i)} = \sum_{i=2}^{n}(X_{(i)} - X_{(1)}) + nX_{(1)} = \sum_{i=2}^{n}(X_{(i)} - X_{(1)}) + nX_{(1)},$$

we readily have $\mathbf{T} = (X_{(1)},S)^T$ to be the minimal sufficient statistic for $\boldsymbol{\theta} = (\mu,\sigma)^T$. Then, the statistic vector

$$\boldsymbol{U} = \left(X_{(1)}, \sum_{i=2}^{n} X_{(i)}\right)^T$$

is also minimal sufficient for $\boldsymbol{\theta}$. It is easy to show that the density of $X_{(1)}$ is given by

$$\frac{n}{\sigma}\exp\left\{-\frac{n}{\sigma}(x_{(1)}-\mu)\right\}H(x_{(1)}-\mu), \tag{8.10}$$

i.e. $X_{(1)}$ follows the exponential law $\mathcal{E}(\mu,\sigma/n)$, and consequently

$$\mathbf{E}X_{(1)} = \mu + \frac{\sigma}{n} \quad \text{and} \quad \mathbf{Var}X_{(1)} = \frac{\sigma^2}{n^2}. \tag{8.11}$$

Thus, we see that $nX_{(1)} \sim \mathcal{E}(n\mu,\sigma)$, and so from (8.11) it readily follows that

$$\mathbf{E}\{nX_{(1)}\} = n\mu + \sigma \quad \text{and} \quad \mathbf{Var}\{nX_{(1)}\} = \sigma^2. \tag{8.12}$$

Now, we shall show that $X_{(1)}$ and S are independent. First, we note that the density of $\boldsymbol{X}^{(\cdot)}$ is

$$\begin{aligned} g(\boldsymbol{x}^{(\cdot)};\boldsymbol{\theta}) &= n!\prod_{i=1}^{n} f(x_{(i)};\theta) \\ &= \frac{n!}{\sigma^n}\exp\left\{-\frac{1}{\sigma}\sum_{i=1}^{n}(x_{(i)}-\mu)\right\}H(x_{(1)}-\mu) \\ &= \frac{n}{\sigma}\exp\left\{-\frac{n}{\sigma}(x_{(1)}-\mu)\right\}H(x_{(1)}-\mu)\frac{(n-1)!}{\sigma^{n-1}} \\ &\quad \times \exp\left\{-\frac{1}{\sigma}\sum_{i=2}^{n}(x_{(i)}-x_{(1)})\right\}H(x_{(2)}-x_{(1)}), \end{aligned} \tag{8.13}$$

where

$$\begin{aligned} \boldsymbol{x}^{(\cdot)} &= (x_{(1)},\ldots,x_{(n)})^T \in \boldsymbol{B}_\mu \\ &= \{\boldsymbol{x} \in \boldsymbol{R}^n : \mu \leqslant x_1 \leqslant x_2 \leqslant \cdots \leqslant x_n\}. \end{aligned} \tag{8.14}$$

From (8.13), it follows that

$$\frac{(n-1)!}{\sigma^{n-1}}\exp\left\{-\frac{1}{\sigma}\sum_{i=2}^{n}(x_{(i)}-x_{(1)})\right\}, \quad x_{(1)} \leqslant x_{(2)} \leqslant \cdots \leqslant x_{(n)} \tag{8.15}$$

represents the conditional joint density function of the vector $(X_{(2)}, X_{(3)},\ldots,X_{(n)})^T$, given $X_{(1)} = x_{(1)}$. One can see that this conditional law does not depend on μ. Moreover, from (8.8) and (8.13), it follows that for a given value of the statistic $X_{(1)}$, the vector $(X_{(2)},X_{(3)},\ldots,X_{(n)})^T$ represents the vector of order statistics associated with a sample of dimension $n-1$ from an exponential distribution shifted by $x_{(1)}$ (instead of μ). Also, the elements of this sample follow the same exponential law

$$\frac{1}{\sigma}\exp\left\{-\frac{x-x_{(1)}}{\sigma}\right\}H(x-x_{(1)}).$$

Now, we shall obtain the joint density

$$q(\boldsymbol{y};\boldsymbol{\theta}), \quad \boldsymbol{y} = (y_1,\ldots,y_n)^T \in \boldsymbol{B}_\mu = \{\boldsymbol{x} \in \boldsymbol{R}^n : \mu \leqslant y_1, 0 \leqslant y_2 \leqslant \cdots \leqslant y_n\},$$

of the statistics

$$X_{(1)} \text{ and } (X_{(2)} - X_{(1)}, \ldots, X_{(n)} - X_{(1)})^T,$$

i.e., the density of the statistic

$$\boldsymbol{Y} = (Y_1, Y_2, \ldots, Y_n)^T,$$

where

$$Y_1 = X_{(1)}, \quad Y_j = X_{(j)} - X_{(1)}, \quad j = 2, \ldots, n. \tag{8.16}$$

Note that the statistic $\boldsymbol{Y}$ is a linear transformation of the statistic $\boldsymbol{X}^{(n)}$ expressed as

$$\boldsymbol{Y} = \boldsymbol{B}\boldsymbol{X}^{(n)},$$

where

$$\boldsymbol{B} = \begin{Vmatrix} 1 & 0 & 0 & \ldots & 0 \\ -1 & 1 & 0 & \ldots & 0 \\ -1 & 0 & 1 & \ldots & 0 \\ \vdots & & & & \\ -1 & 0 & 0 & \ldots & 1 \end{Vmatrix},$$

and so

$$\boldsymbol{X}^{(n)} = \boldsymbol{B}^{-1}\boldsymbol{Y},$$

where

$$\boldsymbol{B}^{-1} = \begin{Vmatrix} 1 & 0 & 0 & \ldots & 0 \\ 1 & 1 & 0 & \ldots & 0 \\ 1 & 0 & 1 & \ldots & 0 \\ \vdots & & & & \\ 1 & 0 & 0 & \ldots & 1 \end{Vmatrix}.$$

Since $\det \boldsymbol{B} = 1$, it follows from (8.13) that

$$\begin{aligned} q(\boldsymbol{y}; \boldsymbol{\theta}) &= g(\boldsymbol{B}^{-1}\boldsymbol{y}; \boldsymbol{\theta}) |\det \boldsymbol{B}^{-1}| = g(y_1, y_1 + y_2, \ldots, y_1 + y_n; \boldsymbol{\theta}) \\ &= \frac{n}{\sigma} \exp\left\{-\frac{n}{\sigma}(y_1 - \mu)\right\} H(y_1 - \mu) \frac{(n-1)!}{\sigma^{n-1}} \left\{-\frac{1}{\sigma} \sum_{i=2}^{n} y_i\right\}, \\ &\qquad \boldsymbol{y} \in \boldsymbol{B}_\mu \subset \boldsymbol{R}^n. \end{aligned} \tag{8.17}$$

From (8.17), we observe that the joint density of $X_{(1)}$ and $(X_{(2)} - X_{(1)}, \ldots, X_{(n)} - X_{(1)})^T$ is the product of two densities of the statistics $X_{(1)}$ and $(X_{(2)} - X_{(1)}, \ldots, X_{(n)} - X_{(1)})^T$, which implies that the statistics $X_{(1)}$ and $\sum_{i=2}^{n} (X_{(i)} - X_{(1)})$ are independent. Moreover, from (8.17), it also follows that $\sum_{i=2}^{n} (X_{(i)} - X_{(1)})$ follows the gamma distribution with density

$$\frac{1}{\sigma^{n-1}\Gamma(n-1)} y^{n-2} e^{-y/\sigma} H(y).$$

Since

$$\frac{(n-1)!}{\sigma^{n-1}}\exp\left\{-\frac{1}{\sigma}\sum_{i=2}^{n}y_i\right\},\quad 0 \leqslant y_2 \leqslant y_3 \leqslant \cdots \leqslant y_n,$$

represents the joint density of the $(n-1)$-dimensional statistic associated with the exponential law

$$\frac{1}{\sigma}\exp\left\{-\frac{1}{\sigma}y\right\}H(y),$$

it follows that the statistic

$$\frac{1}{\sigma}\sum_{i=2}^{n}Y_i = \frac{1}{\sigma}\sum_{i=2}^{n}(X_{(i)} - X_{(1)}) = \gamma_{n-1}$$

is distributed as the sum of $(n-1)$ independent random variables following the standard exponential distribution $\mathcal{E}(0,1)$. This means that S follows the gamma distribution with shape parameter $(n-1)$ and scale parameter σ, i.e.,

$$S = \sum_{i=2}^{n}Y_i = \sum_{i=2}^{n}(X_{(i)} - X_{(1)}) = \sigma\gamma_{n-1}, \tag{8.18}$$

and so

$$\mathbf{E}S = \mathbf{E}\{\sigma\gamma_{n-1}\} = (n-1)\sigma,\quad \mathbf{Var}S = \mathbf{Var}\{\sigma\gamma_{n-1}\} = \sigma^2(n-1). \tag{8.19}$$

In this case, the statistic

$$\bar{\sigma}_n = \frac{1}{n-1}\sum_{i=2}^{n}(X_{(i)} - X_{(1)}) = \frac{n}{n-1}(\bar{X}_n - X_{(1)}) \tag{8.20}$$

is the minimum variance unbiased estimator (MVUE) of σ. From (8.19), it then follows that

$$\mathbf{Var}\bar{\sigma}_n = \frac{\sigma^2}{n-1}.$$

Note that, by using (8.11) and (8.20), we may construct the MVUE $\bar{\mu}_n$ of μ as

$$\bar{\mu}_n = X_{(1)} - \frac{\bar{\sigma}_n}{n} = X_{(1)} - \frac{1}{n(n-1)}\sum_{i=2}^{n}(X_{(i)} - X_{(1)}) = X_{(1)} - \frac{1}{n-1}(\bar{X}_n - X_{(1)}).$$

Since the statistics $X_{(1)}$ and S are independent, the statistics $X_{(1)}$ and $\bar{\sigma}_n$ are also independent, and so

$$\mathbf{Var}\bar{\mu}_n = \mathbf{Var}X_{(1)} + \frac{1}{n^2}\mathbf{Var}\bar{\sigma}_n = \frac{\sigma^2}{n^2} + \frac{\sigma^2}{(n-1)n^2} = \frac{\sigma^2}{n(n-1)}. \tag{8.21}$$

Remark 8.1. *Since*

$$\sum_{i=2}^{n}(X_{(i)}-X_{(1)})=\sum_{i=2}^{n}Y_i=\sum_{i=2}^{n}(n-i+1)(X_{(i)}-X_{(i-1)}), \tag{8.22}$$

from (8.13) *and* (8.16), *it follows that the statistics*

$n(X_{(1)}-\mu),(n-1)(X_{(2)}-X_{(1)}),\ldots,(n-i+1)(X_{(i)}-X_{(i-1)}),\ldots,X_{(n)}-X_{(n-1)}$

are independent, and

$$nX_{(1)}\sim\mathcal{E}(n\mu,\sigma),\quad i.e.\quad n(X_{(1)}-\mu)\sim\mathcal{E}(0,\sigma), \tag{8.23}$$

$$(n-i+1)(X_{(i)}-X_{(i-1)})\sim\mathcal{E}(0,\sigma),\quad i=2,3,\ldots,n. \tag{8.24}$$

It is evident that these nice properties of the exponential distribution are consequences of the lack of memory of the exponential distribution.

Remark 8.2. *From* (8.5), *it follows that the likelihood function is*

$$L(\boldsymbol{X};\boldsymbol{\theta})=L(\boldsymbol{X};\mu,\sigma)=\frac{1}{\sigma^n}\exp\left\{-\frac{1}{\sigma}\sum_{i=1}^{n}(X_i-\mu)\right\}H(X_{(1)}-\mu),$$

and so the maximum likelihood estimator of μ is simply

$$\hat{\mu}_n=X_{(1)}.$$

Since

$$\frac{\partial\ln L(\boldsymbol{X};\boldsymbol{\theta})}{\partial\sigma}=-\frac{n}{\sigma}+\frac{1}{\sigma^2}\sum_{i=1}^{n}(X_i-\mu),$$

we see that the maximum likelihood estimator of σ ($\hat{\sigma}_n$) is the solution of the equation

$$-\frac{n}{\sigma}+\frac{1}{\sigma^2}\sum_{i=1}^{n}(X_i-X_{(1)})=0,$$

i.e.

$$\hat{\sigma}_n=\frac{1}{n}\sum_{i=1}^{n}(X_i-X_{(1)})=\overline{X}_n-X_{(1)}.$$

Since

$$\left.\frac{\partial^2\ln L(\boldsymbol{X};\boldsymbol{\theta})}{\partial\sigma^2}\right|_{\hat{\sigma}_n,\hat{\mu}_n}=-\frac{n}{(\bar{X}_n-X_{(1)})^2}<0,$$

we have the maximum likelihood estimator (MLE) $\hat{\theta}_n$ of $\boldsymbol{\theta}$ as

$$\hat{\boldsymbol{\theta}}_n=(\hat{\mu}_n,\hat{\sigma}_n)^T=\left(X_{(1)},\overline{X}_n-\overline{X}_{(1)}\right)^T.$$

It is important to mention here that analogous results can be developed for different forms of censored data as well under the exponential distribution. For a detailed review of all these results, one may refer to the book by Balakrishnan and Basu (1995).

8.2.2 Uniform distribution

Definition 8.1. *A random variable U follows the uniform distribution on the interval* $[a,b]$ *if the density function of U is given by*

$$f(x;a,b) = \frac{1}{b-a}\mathbf{1}_{[a,b]}(x), \quad x \in \boldsymbol{R}^1.$$

The distribution function of U is

$$F(x;a,b) = \mathbf{P}\{U \leqslant x\} = \frac{x-a}{b-a}\mathbf{1}_{[a,b]}(x) + \mathbf{1}_{]b,+\infty[}(x), \quad x \in \boldsymbol{R}^1.$$

It is easy to verify that

$$\mathbf{E}U = \frac{a+b}{2}, \quad \mathbf{Var}U = \frac{(b-a)^2}{12}.$$

Remark 8.3. *Let X be a continuous random variable, with* $F(x)$ *as the distribution function of X. It is then easy to show that the statistic* $U = F(X)$ *follows the uniform probability distribution on* $[0,1]$. *This transformation is often referred to as the probability integral transformation.*

8.2.3 Triangular distribution

Let X and Y be independent random variables uniformly distributed over the interval $[\frac{a}{2},\frac{b}{2}]$. Consider the statistic $Z = X + Y$. The probability density function of Z is given by

$$f(z;a,b) = \left\{\frac{2}{b-a} - \frac{2|a+b-2z|}{(b-a)^2}\right\}\mathbf{1}_{[a,b]}(z).$$

Since this function has the form of a triangle on $[a,b]$ with the maximum value $2/(a+b)$ at the point $\frac{a+b}{2}$, this distribution is known as the *triangular distribution*. It is known also as *Simpson's distribution*. It is easy to verify that

$$\mathbf{E}Z = \frac{a+b}{2} \quad \text{and} \quad \mathbf{Var}Z = \frac{(b-a)^2}{24}.$$

Triangular distribution is a suitable substitute for the beta distribution in some analyses; see Johnson (1997). The symmetric triangular distribution is commonly used in audio dithering, for example.

8.2.4 Pareto model

The Pareto model has found key applications in many fields including economics, actuarial science, and reliability. Its probability distribution function is given by

$$F(x,\theta) = 1 - \left(\frac{\theta}{x}\right)^{\alpha}\mathbf{1}_{[\theta,+\infty[}(x), \theta,\alpha > 0, x \in \boldsymbol{R}^1. \tag{8.25}$$

Correspondingly, the survival function and the density function are

$$S(x;\theta,\alpha) = \left(\frac{\theta}{x}\right)^{\alpha} \mathbf{1}_{[\theta,+\infty[}(x)$$

and

$$f(x;\theta,\alpha) = \frac{\alpha\theta^{\alpha}}{x^{\alpha+1}}, \quad x \geqslant \theta.$$

The hazard rate function $\lambda(x) = \alpha/x, x \geqslant \theta$, is clearly a decreasing function. It is easy to show that the population mean and variance for the model in (8.25) are

$$\mathbf{E}X = \frac{\alpha\theta}{\alpha - 1}, \quad \alpha > 1, \quad \text{and} \quad \mathbf{Var}X = \frac{\theta^2\alpha}{(\alpha-1)(\alpha-2)}, \quad \alpha > 2.$$

Let $X_1, \ldots, X_n$ be i.i.d random variables from the Pareto distribution in (8.25). Then, the MLEs $\hat{\theta}_n$ and $\hat{\alpha}_n$ of the parameters θ and α are

$$\hat{\theta}_n = X_{(1)}, \hat{\alpha}_n = n/\sum_{i=1}^{n}\left(\ln X_i - \ln X_{(1)}\right).$$

8.2.5 Normal distribution

Definition 8.2. *A random variable Z follows the standard normal distribution, denoted by $N(0,1)$, if the density function $\varphi(x)$ of Z is*

$$\varphi(x) = \frac{1}{\sqrt{2\pi}} e^{-x^2/2}, \quad x \in \boldsymbol{R}^1. \tag{8.26}$$

The distribution function Φ of the standard normal law is well known and is given by

$$\Phi(x) = \mathbf{P}\{Z \leqslant x\} = \frac{1}{\sqrt{2\pi}} \int_{-\infty}^{x} e^{-z^2/2} dz, \quad x \in \boldsymbol{R}^1. \tag{8.27}$$

From (8.27), it follows that

$$\Phi(x) + \Phi(-x) \equiv 1, \quad x \in \boldsymbol{R}^1. \tag{8.28}$$

Let x be fixed, and

$$p = \Phi(x), \quad 0 < p < 1. \tag{8.29}$$

Also, let $\Psi(\cdot) = \Phi^{-1}(\cdot)$ be the inverse function for $y = \Phi(x), 0 < y < 1$. Then, from (8.28) and (8.29), it follows that

$$\Phi[\Psi(p)] \equiv p \quad \text{and} \quad \Phi[\Psi(1-p)] \equiv 1 - p \tag{8.30}$$

for any $p, 0 < p < 1$. Moreover,

$$\Phi(-x) = 1 - \Phi(x) = 1 - p \quad \text{and} \quad -x = \Psi(1-p),$$

where $x = z_p = \Psi(p)$ is the p-quantile of the standard normal distribution, and so

$$\Psi(p) + \Psi(1-p) \equiv 0, \quad 0 < p < 1, \tag{8.31}$$

i.e. $z_p = -z_{1-p}$. This also can be noted readily from the symmetry of the standard normal distribution around 0. It can be verified that

$$\mathbf{E}Z = 0, \quad \mathbf{Var}Z = 1.$$

Let $X = \sigma Z + \mu$, where $Z \sim N(0,1), |\mu| < \infty$ and $\sigma > 0$. In this case, we say that X follows the normal distribution, denoted by $N(\mu,\sigma^2)$, with parameters

$$\mu = \mathbf{E}X \quad \text{and} \quad \sigma^2 = \mathbf{Var}X. \tag{8.32}$$

Its density function is

$$\frac{1}{\sigma}\varphi\left(\frac{x-\mu}{\sigma}\right) = \frac{1}{\sqrt{2\pi}\sigma}\exp\left\{-\frac{(x-\mu)^2}{2\sigma^2}\right\}, \quad x \in \boldsymbol{R}^1, \tag{8.33}$$

distribution function is

$$\mathbf{P}\{X \leqslant x\} = \Phi\left(\frac{x-\mu}{\sigma}\right), \quad x \in \boldsymbol{R}^1, \tag{8.34}$$

and the p-quantile is

$$x_p = \mu + \sigma z_p = \mu + \sigma\Phi^{-1}(p).$$

8.2.6 Multivariate normal distribution

Let $\boldsymbol{X} = (X_1, \ldots, X_p)^T$ be a p-dimensional random vector. If, for any vector $\mathbf{z} \in R^p$ with $\mathbf{z} \neq \mathbf{0}$, the scalar random variable

$$\mathbf{z}^T\boldsymbol{X} = \sum_{i=1}^{p} z_i X_i$$

has a normal distribution, then we say that the vector $\boldsymbol{X}$ has a p-dimensional normal distribution $N_p(\mathbf{a}, \boldsymbol{\Sigma})$ in R^p with parameters $\mathbf{a}$ and $\boldsymbol{\Sigma}$, where

$$\mathbf{a} = \mathbf{E}\boldsymbol{X} \quad \text{and} \quad \boldsymbol{\Sigma} = \mathbf{Var}\boldsymbol{X} = \mathbf{E}(\boldsymbol{X} - \mathbf{a})(\boldsymbol{X} - \mathbf{a})^T,$$

where $\mathbf{a} \in R^p, \boldsymbol{\Sigma} = |\sigma_{ij}|_{p\times p}$; we shall denote it by $\boldsymbol{X} \sim N_p(\mathbf{a}, \boldsymbol{\Sigma})$.

Let $k = rank\,\boldsymbol{\Sigma}$. If $p = k$, then there exists the inverse matrix $\boldsymbol{\Sigma}^{-1}$, and in this case we say that the distribution of $\boldsymbol{X}$ is non-degenerate, and $\boldsymbol{X}$ then has its density function as

$$f_X(\boldsymbol{x}) = (2\pi)^{-p/2}\left(\det\boldsymbol{\Sigma}\right)^{-1/2}\exp\left\{-\frac{1}{2}(\boldsymbol{x} - \mathbf{a})^T\boldsymbol{\Sigma}^{-1}(\boldsymbol{x} - \mathbf{a})\right\} \tag{8.35}$$

for any $\boldsymbol{x} \in \boldsymbol{R}^p$.

Consider one particular and important case when

$$\mathbf{a} = \mathbf{0_p} \quad \text{and} \quad \mathbf{\Sigma} = \mathbf{I}_p,$$

where $\mathbf{I}_p$ is the identity matrix of rank p. Let $\mathbf{Z} = (Z_1, \ldots, Z_p)^T$ be a random vector with

$$\mathbf{Z} \sim N_p(\mathbf{0}_p, \mathbf{I}_p).$$

Then, we say that the random vector $\mathbf{Z}$ follows the p-dimensional standard normal distribution in R^p. Since the components Z_i of $\mathbf{Z}$ are independent and follow marginally the same standard normal distribution $N(0,1)$, the density function of $\mathbf{Z}$ is

$$f_{\mathbf{Z}}(\mathbf{z}) = \frac{1}{(2\pi)^{p/2}} e^{-\frac{1}{2}\sum_{i=1}^{p} z_i^2} = \frac{1}{(2\pi)^{p/2}} e^{-\frac{1}{2}\mathbf{z}^T\mathbf{z}}, \tag{8.36}$$

for any $\mathbf{z} = (z_1, z_2, \ldots, z_p)^T \in R^p$, where

$$\mathbf{z}^T\mathbf{z} = \sum_{i=1}^{p} z_i^2.$$

Now, consider a linear transformation $Y = \mathbf{AZ} + \mathbf{a}$ of the vector $\mathbf{Z} \sim N_p(\mathbf{0}_p, \mathbf{I}_p)$, where $\mathbf{A}$ is a matrix of order p, the *rank* $\mathbf{A} = p$, and $\mathbf{a} \in R^p$. Then, the mean and covariance matrix of Y are

$$\mathbf{EY} = \mathbf{E}(\mathbf{AZ} + \mathbf{a}) = \mathbf{AEZ} + \mathbf{a} = \mathbf{a}$$

and

$$\mathbf{E}(Y - \mathbf{a})(Y - \mathbf{a})^T = \mathbf{EAZ}(\mathbf{AZ})^T = \mathbf{AEZZ}^T\mathbf{A}^T = \mathbf{AA}^T = \mathbf{\Sigma}.$$

Evidently, we have $Y \sim N_p(\mathbf{a}, \mathbf{\Sigma})$, and so the density of the vector $Y = \mathbf{AZ} + \mathbf{a}$ is

$$f_Y(y) = (2\pi)^{-p/2} \left(\det \mathbf{\Sigma}\right)^{-1/2} \exp\left\{-\frac{1}{2}(y - \mathbf{a})^T \mathbf{\Sigma}^{-1}(y - \mathbf{a})\right\}, \quad \mathbf{y} \in \mathbf{R}^p,$$

where

$$\mathbf{\Sigma} = \mathbf{AA}^T.$$

Note that

$$|\det \mathbf{A}^{-1}| = (\det \mathbf{\Sigma})^{-1/2} \quad \text{and} \quad (\mathbf{A}^{-1})^T\mathbf{A}^{-1} = \mathbf{\Sigma}^{-1}.$$

Remark 8.4. *In the special case of $p = 2$, i.e., the bivariate case, let us denote*

$$\sigma_i^2 = \mathbf{Var}(X_i), \quad i = 1,2, \quad \text{and} \quad \rho = \frac{\mathbf{Cov}(X_1, X_2)}{\sigma_1\sigma_2},$$

where ρ is the correlation coefficient. We then have

$$\boldsymbol{\Sigma} = \begin{pmatrix} \sigma_1^2 & \rho\sigma_1\sigma_2 \\ \rho\sigma_1\sigma_2 & \sigma_2^2 \end{pmatrix}, \quad \det \boldsymbol{\Sigma} = \sigma_1^2\sigma_2^2(1-\rho^2),$$

$$\boldsymbol{\Sigma}^{-1} = \frac{1}{1-\rho^2}\begin{pmatrix} \frac{1}{\sigma_1^2} & \frac{-\rho}{\sigma_1\sigma_2} \\ \frac{-\rho}{\sigma_1\sigma_2} & \frac{1}{\sigma_2^2} \end{pmatrix},$$

$$(\mathbf{y}-\mathbf{a})^T\boldsymbol{\Sigma}^{-1}(\mathbf{y}-\mathbf{a}) = \frac{1}{1-\rho^2}\left\{\frac{(y_1-a_1)^2}{\sigma_1^2} - \frac{2\rho(y_1-a_1)(y_2-a_2)}{\sigma_1\sigma_2} + \frac{(y_2-a_2)^2}{\sigma_2^2}\right\},$$

using which we find the density function as

$$f_{Y_1,Y_2}(y_1,y_2) = \frac{1}{2\pi\sigma_1\sigma_2\sqrt{1-\rho^2}}\exp\left\{-\frac{1}{2(1-\rho^2)}\left[\frac{(y_1-a_1)^2}{\sigma_1^2} - \frac{2\rho(y_1-a_1)(y_2-a_2)}{\sigma_1\sigma_2} + \frac{(y_2-a_2)^2}{\sigma_2^2}\right]\right\}$$

for any $(y_1,y_2) \in \boldsymbol{R}^2$.

If $k < p$, then we say that $\boldsymbol{X}$ has a degenerate p-dimensional normal distribution, which is concentrated on a k-dimensional subspace R^k of R^p.

Consider a k-dimensional vector of frequencies $\boldsymbol{\nu} = (\nu_1,\ldots,\nu_k)^T$ that follows the multinomial probability distribution $M_k(n,\mathbf{p})$ with parameters n and $\mathbf{p}$ and probability mass function

$$\mathbf{P}\{\nu_1 = n_1,\ldots,\nu_k = n_k\} = \frac{n!}{n_1!\cdots n_k!}p_1^{n_1}p_2^{n_2}\cdots p_k^{n_k},$$

for all $n_1,\ldots,n_k$ such that $n_i = 0,1,\ldots,n$ and $n_1+n_2 = \cdots = n_k = n$. Then, as mentioned earlier in Section 8.1.2, we have

$$\mathbf{E}\boldsymbol{\nu} = n\mathbf{p} \quad \text{and} \quad \mathbf{E}(\boldsymbol{\nu}-n\mathbf{p})(\boldsymbol{\nu}-n\mathbf{p})^T = n\boldsymbol{\Sigma} = n(\mathbf{P}-\mathbf{p}\mathbf{p}^T),$$

where

$$\boldsymbol{\Sigma} = \boldsymbol{\Sigma}_k = \mathbf{P}-\mathbf{p}\mathbf{p}^T,$$

$\mathbf{P} = \mathbf{P}_k$ is the diagonal matrix with elements $p_1,\ldots,p_k$ on the main diagonal, and $\mathbf{p} = \mathbf{p}_k = (p_1,\ldots,p_k)^T$. It is easy to verify that *rank* $\boldsymbol{\Sigma} = k-1$. Then, the central limit theorem implies that, as $n \to \infty$, we have

$$\frac{1}{\sqrt{n}}(\boldsymbol{\nu}-n\mathbf{p}) \sim AN_k(\mathbf{0}_k,\boldsymbol{\Sigma}).$$

Since the matrix $\mathbf{\Sigma}$ is degenerate, we also obtain the k-dimensional asymptotic normal distribution $N_k(\mathbf{0}_k, \mathbf{\Sigma})$ to be degenerate and so, with probability 1, it is concentrated in $(k-1)$-dimensional subspace $R^{k-1} \subset R^k$. We need to find the inverse of the matrix $\mathbf{\Sigma}$. Since $\mathbf{\Sigma}$ is degenerate, it does not have the ordinary inverse and for this reason we shall construct the so-called generalized inverse $\mathbf{\Sigma}^-$ of $\mathbf{\Sigma}$. For this, we shall use the following definition of the generalized inverse.

Let $\mathbf{A} = |a_{ij}|_{n\times m}$ be an arbitrary matrix, *rank* $\mathbf{A} = f$, and $f \leqslant \min(m,n)$. We shall say that the matrix $\mathbf{A}^- = \mathbf{A}^-_{m\times n}$ is a generalized inverse of $\mathbf{A}$ if

$$\mathbf{A}\mathbf{A}^-\mathbf{A} = \mathbf{A}.$$

Note that the matrix $\mathbf{A}^-$ is not unique in general.

Let us construct a generalized inverse Σ^- of Σ, for example, in the following way (see Rao, 1965). Consider the square matrix $(k-1)\times(k-1)$ of the following form:

$$\tilde{\mathbf{\Sigma}} = \mathbf{\Sigma}_{k-1} = \begin{pmatrix} p_1(1-p_1) & -p_1p_2 & \cdots & -p_1p_{k-1} \\ -p_2p_1 & p_2(1-p_2) & \cdots & -p_2p_{k-1} \\ \cdots & \cdots & \cdots & \cdots \\ -p_{k-1}p_1 & -p_{k-1}p_2 & \cdots & p_{k-1}(1-p_{k-1}) \end{pmatrix}_{(k-1)\times(k-1)}$$

$$= \mathbf{P}_{k-1} - \mathbf{p}_{k-1}\mathbf{p}_{k-1}^T = \tilde{\mathbf{P}} - \tilde{\mathbf{p}}\tilde{\mathbf{p}}^T,$$

which is obtained from $\mathbf{\Sigma}$ simply by deleting the last row and the last column. Evidently, we have

$$\mathbf{E}\tilde{\boldsymbol{\nu}} = n\tilde{\mathbf{p}} \quad \text{and} \quad \mathbf{E}(\tilde{\boldsymbol{\nu}} - n\tilde{\mathbf{p}})(\tilde{\boldsymbol{\nu}} - n\tilde{\mathbf{p}})^T = n\tilde{\Sigma} = n(\tilde{\mathbf{P}} - \tilde{\mathbf{p}}\tilde{\mathbf{p}}^T),$$

$\tilde{\mathbf{\Sigma}} = \mathbf{\Sigma}_{k-1}$ is the covariance matrix of the statistic $\tilde{\boldsymbol{\nu}} = (\nu_1, \ldots, \nu_{k-1})^T$, and $\tilde{\mathbf{p}} = (p_1, \ldots, p_{k-1})^T$. Since *rank* $\tilde{\mathbf{\Sigma}} = k-1$, then there exists the inverse $\tilde{\mathbf{\Sigma}}^{-1}$. It is easy to verify that

$$\tilde{\mathbf{\Sigma}}^{-1} = \begin{pmatrix} \frac{1}{p_1} + \frac{1}{p_k} & \frac{1}{p_k} & \cdots & \frac{1}{p_k} \\ \frac{1}{p_k} & \frac{1}{p_2} + \frac{1}{p_k} & \cdots & \frac{1}{p_k} \\ \cdots & \cdots & \cdots & \cdots \\ \frac{1}{p_k} & \frac{1}{p_k} & \cdots & \frac{1}{p_{k-1}} + \frac{1}{p_k} \end{pmatrix}_{(k-1)\times(k-1)}$$

$$= \mathbf{P}_{k-1}^{-1} + \frac{1}{p_k}\mathbf{1}_{k-1}\mathbf{1}_{k-1}^T = \tilde{\mathbf{P}}^{-1} + \frac{1}{p_k}\tilde{\mathbf{1}}\tilde{\mathbf{1}}^T.$$

It can be verified that

$$\tilde{\mathbf{\Sigma}}\tilde{\mathbf{\Sigma}}^{-1} = \mathbf{I}_{k-1},$$

where $\mathbf{I}_{k-1}$ is the identity matrix of order $k-1$. It is also easy to check that

$$\mathbf{\Sigma}^- = (\mathbf{P} - \mathbf{p}\mathbf{p}^T)^- = \mathbf{P}^{-1} + \frac{1}{p_k}\mathbf{1}\mathbf{1}^T$$

is the generalized inverse of $\mathbf{\Sigma}$, where $\mathbf{1} = \mathbf{1}_k$ is the unit vector $(1,1,\ldots,1)^T \in R^k$ (Nikulin, 1973a). Another suitable way of calculating generalized inverse was also suggested by Rao (1965). Indeed,

$$\mathbf{\Sigma}^- = \begin{pmatrix} \tilde{\mathbf{\Sigma}}^{-1} & \mathbf{0} \\ \mathbf{0} & 0 \end{pmatrix}$$

$$= \begin{pmatrix} \frac{1}{p_1} + \frac{1}{p_k} & \frac{1}{p_k} & \cdots & \frac{1}{p_k} & 0 \\ \frac{1}{p_k} & \frac{1}{p_2} + \frac{1}{p_k} & \cdots & \frac{1}{p_k} & 0 \\ \cdots & \cdots & \cdots & \cdots & 0 \\ \frac{1}{p_k} & \frac{1}{p_k} & \cdots & \frac{1}{p_{k-1}} + \frac{1}{p_k} & 0 \\ 0 & 0 & 0 & 0 & 0 \end{pmatrix}_{k\times k}.$$

This generalized inverse was used by Hsuan and Robson (1976), Moore (1977), and Mirvaliev (2001) in the development of their test statistics.

Lemma 8.1. *Let*

$$\mathbf{Z} \sim N_r(\mathbf{a},\mathbf{A}) \quad with\ rank\ \mathbf{A} = r, and\ \mathbf{a} \in R^r,$$

be a r-dimensional normal random vector with a non-degenerate matrix $\mathbf{A}$, *so that* $\mathbf{A}^{-1}$ *exists. In this case, for any vector* $\mathbf{a}_0 \in R^r$, *the quadratic form*

$$(\mathbf{Z} - \mathbf{a}_0)^T \mathbf{A}^{-1} (\mathbf{Z} - \mathbf{a}_0) = \chi_r^2(\lambda)$$

is a random variable that follows a non-central chi-square distribution, $\chi_r^2(\lambda)$, *with r degrees of freedom and non-centrality parameter as*

$$\lambda = (\mathbf{a} - \mathbf{a}_0)^T \mathbf{A}^{-1} (\mathbf{a} - \mathbf{a}_0).$$

Consider now the case of degenerate distributions, which we often encounter while constructing the modified chi-square tests. In particular, let us consider the vector of frequencies $\boldsymbol{\nu}$ that possesses the asymptotic property

$$\frac{1}{\sqrt{n}}(\boldsymbol{\nu} - n\mathbf{p}) \sim AN_k(\mathbf{0}_k, \mathbf{\Sigma}), \quad \text{with } rank\ \mathbf{\Sigma} = k-1.$$

Since the distribution of the vector $\boldsymbol{\nu}$ is degenerate, its limiting distribution will also be degenerate. But at the same time, we know that the distribution of its subvector $\tilde{\boldsymbol{\nu}}$ is not degenerate in R^{k-1}, and that

$$\frac{1}{\sqrt{n}}(\tilde{\boldsymbol{\nu}} - n\tilde{\mathbf{p}}) \sim AN_{k-1}(\mathbf{0}_{k-1}, \tilde{\mathbf{\Sigma}}), \quad \text{with } rank \tilde{\mathbf{\Sigma}} = k-1.$$

This implies that, for any vector $\mathbf{p}_0 \in R^k$ such that $\mathbf{p}_0^T \mathbf{1}_k = 1$, we have

$$\frac{1}{\sqrt{n}}(\tilde{\boldsymbol{\nu}} - n\tilde{\mathbf{p}}_0) \sim AN_{k-1}(\sqrt{n}(\tilde{\mathbf{p}} - \tilde{\mathbf{p}}_0), \tilde{\mathbf{\Sigma}}), \quad \text{with } rank \tilde{\mathbf{\Sigma}} = k-1.$$

Hence, from Lemma 8.1, it follows that the quadratic form

$$X_n^2 = \frac{1}{n}(\tilde{\mathbf{v}} - n\tilde{\mathbf{p}}_0)^T \left(\tilde{\mathbf{P}}^{-1} + \frac{1}{p_k}\tilde{\mathbf{1}}\tilde{\mathbf{1}}^T\right)(\tilde{\mathbf{v}} - n\tilde{\mathbf{p}}_0)$$

is asymptotically distributed as $\chi^2_{k-1}(\lambda)$, where

$$\lambda = n(\tilde{\mathbf{p}} - \tilde{\mathbf{p}}_0)^T \left(\tilde{\mathbf{P}}^{-1} + \frac{1}{p_k}\tilde{\mathbf{1}}\tilde{\mathbf{1}}^T\right)(\tilde{\mathbf{p}} - \tilde{\mathbf{p}}_0).$$

Theorem 8.1. *Let*

$$\mathbf{Z} \sim N_r(\mathbf{0}_r, \mathbf{A}), \quad with\ rank\ \mathbf{A} = f, \quad f \leqslant r,$$

i.e. the distribution of the vector $\mathbf{Z}$ *may be degenerate, and let* $\boldsymbol{B} = |b_{ij}|_{r\times r}$ *be an arbitrary square matrix of order r. Then:*

1. *The statistic* $\mathbf{Z}^T\boldsymbol{B}\mathbf{Z}$ *follows the chi-square distribution if and only if*

$$\boldsymbol{B}\mathbf{A}\boldsymbol{B} = \boldsymbol{B};$$

2. *The statistic* $\mathbf{Z}^T\boldsymbol{B}\mathbf{Z}$ *follows the chi-square distribution with f degrees of freedom, i.e.,*

$$\mathbf{Z}^T\mathbf{B}\mathbf{Z} = \chi^2_f,$$

   if and only if

$$\boldsymbol{B} = \mathbf{A}^-;$$

3. *The statistic* $\mathbf{Z}^T\mathbf{A}^-\mathbf{Z}$ *is invariant with respect to the choice of the generalized inverse* $\mathbf{A}^-$ *(Moore, 1977).*

For a detailed discussion on the parameter estimation for the multivariate normal distribution, interested readers may refer to Voinov and Nikulin (1996) and Kotz et al. (2000).

8.2.7 Chi-square distribution

Definition 8.3. *We say that a random variable* χ^2_f *follows the chi-square distribution with* $f\,(>0)$ *degrees of freedom if its density is given by*

$$q_f(x) = \frac{1}{2^{\frac{f}{2}}\Gamma\left(\frac{f}{2}\right)} x^{\frac{f}{2}-1} e^{-x/2} \mathbf{1}_{]0,\infty[}(x), \quad x \in \boldsymbol{R}^1, \tag{8.37}$$

where

$$\Gamma(a) = \int_0^\infty t^{a-1}e^{-t}dt, \quad a > 0,$$

is the complete gamma function.

Let $Q_f(x) = \mathbf{P}\{\chi_f^2 \leqslant x\}$ denote the distribution function of χ_f^2. It can be easily shown

$$\mathbf{E}\chi_f^2 = f \quad \text{and} \quad \mathbf{Var}\chi_f^2 = 2f. \tag{8.38}$$

This definition of the chi-square law in not constructive. To construct a random variable $\chi_n^2, n \in N^*$, one may take n independent random variables $Z_1, \ldots, Z_n$, following the same standard normal $N(0,1)$ distribution, and then consider the statistic

$$Z_1^2 + \cdots + Z_n^2, \quad \text{for } n = 1,2,\ldots$$

It is easily seen that $\mathbf{P}\{Z_1^2 + \cdots + Z_n^2 \leqslant x\} = Q_n(x)$, i.e.

$$Z_1^2 + \cdots + Z_n^2 \sim \chi_n^2. \tag{8.39}$$

Quite often, (8.39) is used for the definition of a chi-square random variable χ_n^2, and here we shall also follow this tradition.

From the central limit theorem, it follows that if n is sufficiently large, then the following normal approximation is valid:

$$\mathbf{P}\left\{\frac{\chi_n^2 - n}{\sqrt{2n}} \leqslant x\right\} = \Phi(x) + O\left(\frac{1}{\sqrt{n}}\right).$$

This approximation implies the so-called Fisher's approximation, according to which

$$\mathbf{P}\left\{\sqrt{2\chi_n^2} - \sqrt{2n-1} \leqslant x\right\} = \Phi(x) + O\left(\frac{1}{\sqrt{n}}\right), \quad n \to \infty.$$

The best normal approximation of the chi-square distribution is the Wilson–Hilferty approximation given by

$$\mathbf{P}\{\chi_n^2 \leqslant x\} = \Phi\left[\left(\sqrt[3]{\frac{x}{n}} - 1 + \frac{2}{9n}\right)\sqrt{\frac{9n}{2}}\right] + O\left(\frac{1}{n}\right), \quad n \to \infty.$$

8.2.8 Non-central chi-square distribution

Let $X_1, \ldots, X_n$ be independent random variables, with $X_i \sim N(\mu_i, 1), i = 1, \ldots, n$. Then, the distribution of the statistic

$$\chi_n^2(\delta) = X_1^2 + \cdots + X_n^2$$

depends on two parameters, n and $\delta = \sum_{i=1}^n \mu_i^2$. The probability distribution $\mathbf{P}[\chi_n^2(\delta) \leqslant x]$ of the statistic $\chi_n^2(\delta)$ is known as the non-central chi-square distribution with n degrees of freedom and the parameter of non-centrality δ. The mean and the variance of $\chi_n^2(\delta)$ can be shown to be

$$\mathbf{E}\chi_n^2(\delta) = n + \delta \quad \text{and} \quad \mathbf{Var}\chi_n^2(\delta) = 2(n + 2\delta).$$

If $\mu_1 = \cdots = \mu_k = 0$, then $\delta = 0$, and in this case we have $\chi_n^2(0) = \chi_n^2$. One can verify that

$$\chi_n^2(\delta) = \chi_n^2 + \rho(\delta),$$

where $\rho(\delta)$ is independent of χ_n^2, and

$$\mathbf{E}e^{it\rho(\delta)} = e^{it\delta/(1-2it)},$$

from which it follows that

$$\mathbf{E}\rho(\delta) = \delta \quad \text{and} \quad \mathbf{Var}\rho(\delta) = 4\delta.$$

A random variable $\gamma_f = \gamma_{1,f}$ follows the gamma distribution with f degrees of freedom ($f > 0$) if, for any $x > 0$, we have

$$\mathbf{P}\{\gamma_f \leqslant x\} = I_x(f),$$

where

$$I_x(f) = \frac{1}{\Gamma(f)} \int_0^x t^{f-1} e^{-t} dt$$

is the incomplete gamma ratio. Then, it is easy to verify that

$$\frac{1}{2}\chi_{2f}^2 \overset{d}{=} \gamma_f. \tag{8.40}$$

Indeed, $\forall x > 0$, we have

$$\mathbf{P}\left\{\frac{1}{2}\chi_{2f}^2 \leqslant x\right\} = \mathbf{P}\{\chi_{2f}^2 \leqslant 2x\} = Q_{2f}(2x) = \frac{1}{2^f\Gamma(f)} \int_0^{2x} t^{f-1} e^{-t/2} dt.$$

By performing the change of variable $t = 2u$, we find

$$\mathbf{P}\left\{\frac{1}{2}\chi_{2f}^2 \leqslant x\right\} = \frac{1}{\Gamma(f)} \int_0^x u^{f-1} e^{-u} du = I_x(f) = \mathbf{P}\{\gamma_f \leqslant x\},$$

where γ_f is a random variable having the gamma distribution with f degrees of freedom. Using the relation in (8.40), we readily obtain

$$\mathbf{E}\gamma_f = \mathbf{E}\left(\frac{1}{2}\chi_{2f}^2\right) = f, \quad \mathbf{Var}\gamma_f = \mathbf{Var}\left(\frac{1}{2}\chi_{2f}^2\right) = \frac{1}{4}\mathbf{Var}\chi_{2f}^2 = f.$$

The random variable $\gamma = \gamma_{\theta,\nu}$ follows the gamma distribution $\Gamma(\theta,\nu)$, with parameters $\theta > 0$ and $\nu > 0$, if its density function is

$$f(t;\theta,\nu) = \frac{1}{\theta^\nu \Gamma(\nu)} t^{\nu-1} e^{-\frac{t}{\theta}}, \quad t \geqslant 0.$$

One can see that if $\theta = 1$, then $\gamma_{1,\nu} = \gamma_\nu$. Note that $G(\theta,1) = \mathcal{E}(\theta)$, i.e. the exponential distribution is a particular case of the gamma distribution.

The distribution function of $\gamma_{\theta,\nu}$ can be presented in terms of the incomplete gamma ratio as

$$F(t;\theta,\nu) = \frac{1}{\Gamma(\nu)} \int_0^{t/\theta} u^{\nu-1} e^{-u} du = I_{t/\theta}(\nu).$$

Remark 8.5. *It can be easily shown that*

$$\frac{1}{2}\chi^2_{2\nu} \stackrel{d}{=} \gamma_{1,\nu}.$$

Indeed, for any $x > 0$, we have

$$\mathbf{P}\left\{\frac{1}{2}\chi^2_{2\nu} \leqslant x\right\} = \mathbf{P}\{\chi^2_{2\nu} \leqslant 2x\} = \frac{1}{2^\nu \Gamma(\nu)} \int_0^{2x} t^{\nu-1} e^{-t/2} dt.$$

By changing the variable $t = 2u$, we find that

$$\mathbf{P}\left\{\frac{1}{2}\chi^2_{2\nu} \leqslant x\right\} = \frac{1}{\Gamma(\nu)} \int_0^{x} u^{\nu-1} e^{-u} du = I_x(\nu) = \mathbf{P}\{\gamma_{1,\nu} \leqslant x\}.$$

If $X_1,\ldots,X_n$ are independent random variables following the same exponential distribution, i.e., $X_i \sim \mathcal{E}(\theta)$, then the statistic

$$S_n = X_1 + \cdots + X_n$$

follows the gamma distribution $\Gamma(\theta,n)$. If $f = 1$, then

$$\mathbf{P}\{\gamma_1 \leqslant x\} = \int_0^x e^{-t} dt = 1 - e^{-x}, \quad x > 0, \tag{8.41}$$

i.e. the random variable γ_1 follows the standard exponential distribution. From this result and from the relation in (8.40), it follows that $\frac{1}{2}\chi^2_2$ also follows the standard exponential distribution.

Theorem 8.2. *If independent random variables $X_1,\ldots,X_n$ follow the exponential distribution in (8.41), then their sum follows the gamma distribution with shape parameter n, i.e. $X_1 + \cdots + X_n \stackrel{d}{=} \gamma_n$.*

Remark 8.6. *Let X be a random variable that follows the Poisson distribution $\mathbf{P}(\lambda)$ with parameter $\lambda > 0$. It is easy to show that, for any $m \in \mathbf{N}$,*

$$\begin{aligned}\mathbf{P}\{X \leqslant m\} &= \mathbf{P}\{\gamma_{m+1} \geqslant \lambda\} = \mathbf{P}\{\chi^2_{2m+2} \geqslant 2\lambda\} \\ &= 1 - \mathbf{P}\{\chi^2_{2m+2} \leqslant 2\lambda\}.\end{aligned}$$

Remark 8.7. *Several approximations for the non-central χ^2 cumulative distribution* $\mathbf{P}[\chi_n^2(\delta) \leqslant x]$ *are known; see, for example,* Johnson et al. (1995). *However, we can easily produce codes (for example, in Microsoft Excel) by using the following well-known approximation* (Abdel-Aty, 1954, Cohen, 1988, Ding, 1992):

$$\mathbf{P}[\chi_n^2(\delta) \leqslant x] = \sum_{i=0}^{\infty} \frac{\exp(-\delta/2)(\delta/2)^i}{i!} \mathbf{P}(\chi_{n+2i}^2 \leqslant x), \qquad (8.42)$$

where χ_{n+2i}^2 are central chi-square random variables.

8.2.9 Weibull distribution

Let a random variable X follow the Weibull probability distribution

$$W(x;\theta,\nu) = 1 - \exp\left\{1 - \left(\frac{x}{\theta}\right)^{\nu}\right\}, \quad x > 0,$$

with parameters $\theta > 0$ and $\nu > 0$. The corresponding density function is

$$f(x;\theta,\nu) = \frac{\nu}{\theta^{\nu}} x^{\nu-1} e^{-(x/\theta)^{\nu}}, \quad x \geqslant 0.$$

The applicability of the Weibull family can be justified in many cases by the following fact.

Remark 8.8. *Let $X_1, \ldots, X_n$ be a set of i.i.d. random variables such that*

$$\mathbf{P}(X_i \leqslant x) = G(x;\theta,\nu), \quad i = 1,2,\ldots,n, \quad \theta,\nu,x > 0,$$

where $G(x;\theta,\nu)$ is a distribution function such that $\lim_{x\downarrow 0} G(x;\theta,\nu) = \theta^{-\nu}x^{\nu}$ *and $G(x;\theta,\nu) = 0$ if $x \leqslant 0$ for all fixed θ and ν. Then, $n^{1/\nu}X_{(1)}$, where $X_{(1)} = \min(X_1,\ldots,X_n)$ is the first-order statistic, converges in probability to $W(x;\theta,\nu)$.*

The Weibull probability distribution and its generalized forms are often used as lifetime models; see Barlow and Proshan (1991), Harter (1991), and Bagdonavičius and Nikulin (2002).

The survival function and the hazard rate of this distribution are $S(x;\theta,\nu) = e^{-(x/\theta)^{\nu}}, x \geqslant 0$ and $\lambda(x;\theta,\nu) = \nu x^{\nu-1}/\theta^{\nu}$, respectively. Note that $W(\theta,1) = \mathcal{E}(\theta)$, i.e., the exponential distribution is a particular case of the Weibull distribution if $\nu = 1$

The expected value and the variance of X can be easily shown to be

$$\mathbf{E}X = \theta\Gamma(1+1/\nu) \text{ and } \mathbf{Var}X = \theta^2\left\{\Gamma(1+2/\nu) - \Gamma^2(1+1/\nu)\right\}.$$

Note that the coefficient of variation of X is

$$\frac{\sqrt{\mathbf{Var}X}}{\mathbf{E}X} = \sqrt{\frac{\Gamma\left(1+\frac{2}{\nu}\right)}{\Gamma^2\left(1+\frac{1}{\nu}\right)} - 1} = \frac{\pi}{\nu\sqrt{6}} + O\left(\frac{1}{\nu^2}\right), \quad \nu \to \infty.$$

Remark 8.9. *Following* Efron (1988), *we can consider the so-called Exponentiated Weibull distribution as its hazard rate function possesses very interesting properties. These properties were studied in detail by* Mudholkar et al. (1995).

The random variable X follows the Exponentiated Weibull distribution $EW(\theta,\nu,\gamma)$ with parameters $\theta > 0, \nu > 0$, and $\gamma > 0$ if

$$F_{EW}(x) = \left\{1 - \exp\left[-\left(\frac{x}{\alpha}\right)^{\beta}\right]\right\}^{\gamma}, \quad x,\alpha,\beta,\gamma > 0. \tag{8.43}$$

All the moments of this distribution are finite. Its survival function is

$$S_{EW}(t) = S(t;\alpha,\beta,\gamma) = 1 - \left\{1 - \exp\left[-\left(\frac{t}{\alpha}\right)^{\beta}\right]\right\}^{\gamma}, \quad t \geqslant 0.$$

Evidently, $EW(\alpha,\beta,1) = W(\alpha,\beta)$ and $EW(\alpha,1,1) = \mathcal{E}(\alpha)$.

Depending on the values of the parameters, the hazard rate can be constant, monotone (increasing or decreasing), $\bigcap$-shaped, and $\bigcup$-shaped. More specifically, the hazard rate

$$\lambda(t;\alpha,\beta,\gamma) = \frac{\beta\gamma\left\{1 - \exp\left[-\left(\frac{t}{\alpha}\right)^{\beta}\right]\right\}^{\gamma-1}\exp\left[-\left(\frac{t}{\alpha}\right)^{\beta}\right]\left(\frac{t}{\alpha}\right)^{\beta-1}}{\alpha\left\{1 - \left(1 - \exp\left[-\left(\frac{t}{\alpha}\right)^{\beta}\right]\right)^{\gamma}\right\}}$$

possesses the following properties:

If $\beta > 1, \beta \geqslant 1/\gamma$, then the hazard rate is increasing from 0 to ∞;
If $\beta = 1, \gamma \geqslant 1$, then the hazard rate is increasing from $(\alpha/\gamma)^{-1}$ to ∞;
If $0 < \beta < 1, \beta < 1/\gamma$, then the hazard rate is decreasing from ∞ to 0;
If $0 < \beta < 1, \beta = 1/\gamma$, then the hazard rate is decreasing from α^{-1} to 0;
If $1/\gamma < \beta < 1$, then the hazard rate is increasing from 0 to its maximum value and then is decreasing to 0, i.e. it is $\cap$-shaped;
If $1/\gamma > \beta > 1$, then the hazard rate decreases from ∞ to its minimum value and then it increases to ∞, i.e. it is $\bigcup$-shaped.
The p-quantile is seen to be

$$t_p = \alpha[-\ln(1 - p^{1/\gamma})]^{1/\beta}, \quad 0 < p < 1.$$

Remark 8.10. *The three-parameter Weibull distribution has the following probability density function*:

$$f(x;\theta,\mu,p) = \frac{p}{\theta}\left(\frac{x-\mu}{\theta}\right)^{p-1} \exp\left\{-\left(\frac{x-\mu}{\theta}\right)^{p}\right\},$$
$$x > \mu,\theta,p > 0,\mu \in R^1. \qquad (8.44)$$

The population mean, median, mode, and variance of this distribution are as follows:

$$\mathbf{E}X = \mu + \theta\Gamma\left(\frac{1}{p}+1\right),$$
$$\mu + \theta(\ln 2)^{1/2},$$
$$\mu + \theta\left(1-\frac{1}{p}\right)^{1/p},$$
$$\mathbf{Var}X = \theta^2\left[\Gamma\left(\frac{2}{p}+1\right) - \Gamma^2\left(\frac{1}{p}+1\right)\right],$$

respectively. The failure rate function of the distribution in (8.44) *is*

$$\lambda(x) = \frac{p}{\theta}\left(\frac{x-\mu}{\theta}\right)^{p-1}.$$

The corresponding parameter esitmation has been discussed in Section 4.6.1.

Remark 8.11. Mudholkar et al. (1996) *presented an extension of the Weibull family called "generalized Weibull" that contains distributions with unimodal and bathtub hazard rates and yields a broader class of monotone failure rates. The probability distribution function of this model is given by*

$$F_{GW}(x) = 1 - \left\{1 - \lambda\left(\frac{x}{\sigma}\right)^{1/\alpha}\right\}^{1/\lambda}. \qquad (8.45)$$

The range of the generalized Weibull random variable X is $(0,\infty)$ *for* $\lambda \leqslant 0$ *and* $(0,\sigma/\lambda^\alpha)$ *for* $\lambda > 0$. *Evidently, the hazard function for the family in* (8.45) *is*

$$h(x) = \frac{(x/\sigma)^{(1/\alpha)-1}}{\alpha\sigma(1-\lambda(x/\sigma)^{1/\alpha})}.$$

This hazard function is bathtub shaped if $\alpha > 1$ *and* $\lambda > 0$, *monotone decreasing if* $\alpha \geqslant 1$ *and* $\lambda \leqslant 0$, *unimodal if* $\alpha < 1$ *and* $\lambda < 0$, *monotone increasing if* $\alpha \leqslant 1$ *and* $\lambda \geqslant 0$, *and is constant if* $\alpha = 1$ *and* $\lambda = 0$.

8.2.10 Generalized Power Weibull distribution

In accelerated life studies, the Generalized Power Weibull (GPW) family with distribution function

$$F(t;\theta,\nu,\gamma) = 1 - \exp\left\{1 - \left[1 + \left(\frac{t}{\theta}\right)^{\nu}\right]^{1/\gamma}\right\}, \quad t,\theta,\nu,\gamma > 0 \qquad (8.46)$$

proves to be very useful; see Bagdonavičius and Nikulin (2002). All moments of this distribution are finite. The corresponding probability density function is

$$f(t;\theta,\nu,\gamma) = \frac{\nu}{\theta\gamma}\left(\frac{t}{\theta}\right)^{\nu-1}\left[1 + \left(\frac{t}{\theta}\right)^{\nu}\right]^{\frac{1}{\gamma}-1}\exp\left\{1 - \left[1 + \left(\frac{t}{\theta}\right)^{\nu}\right]^{1/\gamma}\right\}.$$

The survival function and the hazard rate function are given by

$$S(t;\theta,\nu,\gamma) = \exp\left\{1 - \left(1 + \left(\frac{t}{\theta}\right)^{\nu}\right)^{1/\gamma}\right\}, \quad t \geqslant 0,$$

$$\lambda(t;\theta,\nu,\gamma) = \frac{\nu}{\gamma\theta^{\nu}}t^{\nu-1}\left\{1 + \left(\frac{t}{\theta}\right)^{\nu}\right\}^{1/\gamma-1}, \quad t \geqslant 0.$$

Here again, depending on the values of the parameters, the hazard rate can be constant, monotone increasing and decreasing, cap-shaped, and cup-shaped. More specifically, we have the following behaviors in this case:

If $\nu > 1, \nu > \gamma$, then the hazard rate is increasing from 0 to ∞;
If $\nu = 1, \gamma < 1$, then the hazard rate is increasing from $(\gamma\theta)^{-1}$ to ∞;
If $0 < \nu < 1, \nu < \gamma$, then the hazard rate is decreasing from ∞ to 0;
If $0 < \nu < 1, \nu = \gamma$, then the hazards rate is decreasing from ∞ to θ^{-1};
If $\gamma > \nu > 1$, then the hazard rate is increasing from 0 to its maximum value and then is decreasing to 0, i.e. it is $\cap$-shaped;
If $0 < \gamma < \nu < 1$, then the hazard rate decreases from ∞ to its minimal value and then it decreases to ∞, i.e. it is $\cup$-shaped.

8.2.11 Birnbaum-Saunders distribution

The Birnbaum-Saunders (BS) family of distributions was proposed by Birnbaum and Saunders (1969a) to model the length of cracks on surfaces. In fact, it is a two-parameter distribution for a fatigue life with unimodal hazard rate function. Considerable amount of work has been done on this distribution. The cumulative distribution function of the two-parameter Birnbaum–Saunders distribution, denoted by $\text{BS}(\alpha,\beta)$, is

$$F(t;\alpha,\beta) = \Phi\left[\frac{1}{\alpha}\left\{\left(\frac{t}{\beta}\right)^{\frac{1}{2}} - \left(\frac{\beta}{t}\right)^{\frac{1}{2}}\right\}\right], \quad 0 < t < \infty, \quad \alpha,\beta > 0,$$

where α is the shape parameter, β is the scale parameter, and $\Phi(x)$ is the standard normal distribution function. The corresponding probability density function is

$$f(t;\alpha,\beta) = \frac{1}{2\sqrt{2\pi}\alpha\beta}\left\{\left(\frac{\beta}{t}\right)^{\frac{1}{2}} + \left(\frac{\beta}{t}\right)^{\frac{3}{2}}\right\}\exp\left[-\frac{1}{2\alpha^2}\left(\frac{t}{\beta}+\frac{\beta}{t}-2\right)\right],$$
$$0 < t < \infty, \quad \alpha,\beta > 0.$$

It is easy to show that

$$\mathbf{E}(T) = \beta\left(1+\frac{\alpha^2}{2}\right) \quad \text{and} \quad \mathbf{Var}(T) = (\alpha\beta)^2\left(1+\frac{5}{4}\alpha^2\right).$$

The hazard function is

$$\lambda(t,\alpha,\beta) = \frac{f(t,\alpha,\beta)}{1-F(t,\alpha,\beta)},$$

which increases from 0 to its maximum value and then decreases to $1/2\alpha\beta^2$, i.e. it is ∩-shaped; see Kundu et al. (2008). For comprehensive reviews on various developments concerning the BS distribution, one may refer to Johnson et al. (2005), Leiva et al. (2008), and Sanhueza et al. (2008).

8.2.12 Logistic distribution

Let Z be a random variable with probability density function

$$f(z) = \frac{e^{-z}}{(1+e^{-z})^2}, \quad z \in \boldsymbol{R}^1,$$

and cumulative distribution function

$$F(z) = \frac{1}{1+e^{-z}}, \quad z \in \boldsymbol{R}^1.$$

In this case, Z is said to have a logistic distribution. From the above two expressions, it is readily seen that the hazard rate

$$\frac{f(z)}{1-F(Z)} = \frac{1}{1+e^{-z}}, \quad z \in \boldsymbol{R}^1,$$

is a monotone increasing function. Furthermore, it can be shown that the moment generating function of Z is

$$Ee^{tz} = B(1+t,1-t) = \Gamma(1+t)\Gamma(1-t) \quad \text{for } t < 1,$$

where $B(\cdot,\cdot)$ is the complete beta function. From this expression, it can be shown that

$$\mathbf{E}Z = 0 \quad \text{and} \quad \mathbf{Var}Z = 2\left\{\Gamma''(1) - (\Gamma'(1))^2\right\} = \pi^2/3.$$

For this reason, one can consider the following standardized logistic distribution, denoted by $L(0,1)$, with probability density function

$$f(z) = \frac{\pi}{\sqrt{3}} \frac{e^{-\pi z/\sqrt{3}}}{(1+e^{-\pi z/\sqrt{3}})^2}, \quad z \in \boldsymbol{R}^1,$$

and cumulative distribution function

$$F(z) = \frac{1}{1+e^{-\pi z/\sqrt{3}}}, \quad z \in \boldsymbol{R}^1.$$

Evidently, the mean and variance of this form are 0 and 1, respectively.

Now, by using the linear transformation $X = \mu + \sigma Z$, we can introduce the logistic $L(\mu,\sigma)$ distribution with probability density function

$$\frac{\pi}{\sqrt{3}\sigma} \frac{e^{-\frac{\pi(x-\mu)}{\sqrt{3}\sigma}}}{\left(1+e^{-\frac{\pi(x-\mu)}{\sqrt{3}\sigma}}\right)^2}, \quad x \in \boldsymbol{R}^1,$$

where $\mu \in \boldsymbol{R}^1$ is the mean and $\sigma > 0$ is the standard deviation. For elaborate discussion on properties, various inferential methods and diverse applications of the logistic distribution, one may refer to the book by Balakrishnan (1992).

Chapter 9

Chi-Squared Tests for Specific Distributions

9.1 TESTS FOR POISSON, BINOMIAL, AND "BINOMIAL" APPROXIMATION OF FELLER'S DISTRIBUTION

Let $X_1,X_2,\ldots,X_n$ be independent and identically distributed random variables. Consider the problem of testing the composite hypothesis H_0 according to which the distribution of X_i is a member of a parametric family with the Poisson distribution

$$\mathbf{P}(X=x)=f(x;\theta)=\frac{\theta^x}{x!}e^{-\theta},\quad x=0,1,\ldots,\theta>0. \tag{9.1}$$

Let $N_j^{(n)}$ be the observed frequencies meaning the number of realized values of X_i that fall into a specific class or interval $\Delta_j, p_j(\theta)=\sum_{x\in\Delta_j} f(x;\theta), j=1,\ldots,r$, where the fixed integer grouping classes Δ_j are such that $\Delta_1\cup\cdots\cup\Delta_r=\{0,1,\ldots\}$. As before, let $\mathbf{V}^{(n)}(\theta)$ be a column vector of standardized grouped frequencies with its components as

$$v_j^{(n)}(\theta)=[np_j(\theta)]^{-\frac{1}{2}}(N_j^{(n)}-np_j(\theta)),\quad j=1,\ldots,r.$$

If the nuisance parameter θ is estimated effectively from grouped data by $\tilde{\theta}_n$, then the standard Pearson's sum $X_n^2(\tilde{\theta}_n)=\mathbf{V}^{(n)T}(\tilde{\theta}_n)\mathbf{V}^{(n)}(\tilde{\theta}_n)$ will follow in the limit the chi-square distribution with $r-2$ degrees of freedom. If, on the other hand, the parameter θ is estimated from raw (ungrouped) data, for example, by the maximum likelihood estimate (MLE), then the standard Pearson test must

Chi-Squared Goodness of Fit Tests with Applications. http://dx.doi.org/10.1016/B978-0-12-397194-4.00009-0

be modified. Let $\mathbf{B}$ be a $r \times s$ (s being the dimensionality of the parameter space) matrix with its elements as

$$\frac{1}{\sqrt{p_j(\theta)}} \sum_{x \in \Delta_j} \frac{\partial f(x;\theta)}{\partial \theta_k}, \quad j = 1,\ldots,r, \; k = 1,\ldots,s.$$

For the null hypothesis in (9.1), the $r \times 1$ matrix $\mathbf{B}$ possesses its elements as

$$B_j = \frac{1}{\sqrt{p_j(\theta)}} \sum_{x \in \Delta_j} f(x;\theta)\left(\frac{x}{\theta} - 1\right), \quad j = 1,\ldots,r.$$

Let $\hat{\theta}_n$ be the MLE of θ based on the raw data. Then, the well-known Nikulin-Rao–Robson (NRR) modified chi-squared test statistic, distributed in the limit as χ^2_{r-1}, can be expressed as

$$Y1^2_n(\hat{\theta}_n) = \mathbf{V}^{(n)T}(\hat{\theta}_n)(\mathbf{I} - \mathbf{B}_n\mathbf{J}_n^{-1}\mathbf{B}_n^T)^{-1}\mathbf{V}^{(n)}(\hat{\theta}_n), \tag{9.2}$$

where $\mathbf{J}_n = 1/\hat{\theta}_n$ and $\mathbf{B}_n = \mathbf{B}(\hat{\theta}_n)$ are the estimators of the Fisher information matrix (scalar) and of the column vector $\mathbf{B}$, respectively.

Next, let us consider the binomial null hypothesis. Let the probability mass function be specified as

$$\mathbf{P}_1(X = x) = f_1(x;\theta) = \binom{n}{x}\left(\frac{\theta}{n}\right)^x\left(1 - \frac{\theta}{n}\right)^{n-x}, \tag{9.3}$$

where $x = 0,1,\ldots,n$ and $\theta > 0$. In this case, the Fisher information matrix (scalar) is $\mathbf{J}_1 = n/\{\theta(n-\theta)\}$ and the elements of the matrix $\mathbf{B_1}$ are

$$B_{1j} = \frac{1}{\sqrt{p_j(\theta)}} \sum_{x \in \Delta_j} f_1(x;\theta)\left(\frac{n(x-\theta)}{\theta(n-\theta)}\right), \quad j = 1,2,\ldots,r.$$

Let $\hat{\theta}_n$ be the MLE estimator of θ. Then, the modified chi-squared test statistic, distributed in the limit as χ^2_{r-1}, can be expressed as

$$Y1^2_n(\hat{\theta}_n) = \mathbf{V}^{(n)T}(\hat{\theta}_n)(\mathbf{I} - \mathbf{B}_{1n}\mathbf{J}_{1n}^{-1}\mathbf{B}_{1n}^T)^{-1}\mathbf{V}^{(n)}(\hat{\theta}_n), \tag{9.4}$$

where $\mathbf{B}_{1n} = \mathbf{B}_1(\hat{\theta})$ and $\mathbf{J}_{1n} = \mathbf{J}_1(\hat{\theta})$.

Now, let the probability distribution of the null hypothesis follow Feller's (1948, pp. 105–115) discrete distribution with cumulative distribution function

$$\begin{aligned}\mathbf{P}_2(X \leqslant x) &= \frac{\Gamma(x+1,\mu(\tau - x\gamma))}{\Gamma(x+1)} \\ &= \sum_{k=0}^{x} \frac{[\mu(\tau - x\gamma)]^k}{k!} e^{-\mu(\tau - x\gamma)}, \quad \tau > \gamma, \; 0 \leqslant x < \frac{\tau}{\gamma},\end{aligned} \tag{9.5}$$

where $\Gamma(a,b)$ is the complement gamma function.

If the parameter $\tau = \text{const.}$ and $\gamma \to 0$, then the distribution function in (9.5) can be approximated as

$$\mathbf{P}_2(X \leqslant x) = \sum_{k=0}^{x} \frac{(\mu\tau)^k}{k!} e^{-\mu\tau} + O(\gamma).$$

Using the results of Bol'shev (1963) a more accurate approximation of (9.5) can be obtained as

$$\mathbf{P}_2(X \leqslant x) = \sum_{k=0}^{x} \frac{\Gamma(\tilde{n}+1)}{k!\Gamma(\tilde{n}-k+1)} \tilde{p}^k (1-\tilde{p})^{\tilde{n}-k} + O(\gamma^2), \tag{9.6}$$

where $\tilde{n} = \tau/(2\gamma), \tilde{p} = 2\mu\gamma/(1+\mu\gamma)$ and $x = 0,1,\ldots,[\tilde{n}]$. Note that (9.6) looks like the binomial distribution function with the exception that the parameter $\tilde{n}$ can be any real positive number. Consider the probability mass function of (9.6) given by

$$\mathbf{P}_2(X = x) = f_2(x;\boldsymbol{\theta}) = \frac{\Gamma(\tilde{n}+1)}{x!\Gamma(\tilde{n}-x+1)} \tilde{p}^x (1-\tilde{p})^{\tilde{n}-x}, \quad x = 0,1,\ldots,[\tilde{n}], \tag{9.7}$$

where the parameter $\boldsymbol{\theta} = (\tilde{n},\tilde{p})^T$. In this case, there are three possibilities to construct a modified chi-squared test for testing a composite null hypothesis about the distribution in (9.6). First one is to use MLEs of $\tilde{n}$ and $\tilde{p}$ and the NRR statistic $Y1_n^2(\hat{\theta}_n)$. Since the MLEs of $\tilde{n}$ and $\tilde{p}$ cannot be derived easily, the modified test $Y2_n^2(\bar{\theta}_n)$ based on MMEs (see Eq. (4.9)) or Singh's $Q_s^2(\tilde{\boldsymbol{\theta}}_n)$ (see Eq. (3.25)) can be used.

For the model in (9.7) and $r > 2$, the DN statistic $U_n^2(\bar{\boldsymbol{\theta}}_n)$ will follow in the limit χ^2_{r-3} and $S1_n^2(\bar{\boldsymbol{\theta}}_n) \sim \chi_2^2$. If $r = 2$, then, as before (see Eq. (4.12)), we will have

$$\begin{aligned} Y2_n^2(\bar{\boldsymbol{\theta}}_n) &= U_n^2(\bar{\boldsymbol{\theta}}_n) + S1_n^2(\bar{\boldsymbol{\theta}}_n) \\ &= U_n^2(\bar{\boldsymbol{\theta}}_n) + W_n^2(\bar{\boldsymbol{\theta}}_n) + R_n^2(\bar{\boldsymbol{\theta}}_n) - Q_n^2(\bar{\boldsymbol{\theta}}_n) \\ &= X_n^2(\bar{\boldsymbol{\theta}}_n) + R_n^2(\bar{\boldsymbol{\theta}}_n) - Q_n^2(\bar{\boldsymbol{\theta}}_n). \end{aligned}$$

In this case, $\mathbf{B}_n(\mathbf{B}_n^T\mathbf{B}_n)^{-1}\mathbf{B}_n^T = \mathbf{I}, W_n^2(\bar{\boldsymbol{\theta}}_n) = X_n^2(\bar{\boldsymbol{\theta}}_n), U_n^2(\bar{\boldsymbol{\theta}}_n) = 0$, and

$$Y2_n^2(\bar{\boldsymbol{\theta}}_n) = S1_n^2(\bar{\boldsymbol{\theta}}_n) \sim \chi^2_{r-1} = \chi^2_1.$$

To specify the above tests, we of course will need explicit expressions of all the matrices involved.

The elements of the Fisher information matrix **J** for the model in (9.7) are

$$J_{11} = \sum_{x=0}^{[\tilde{n}]} f_2(x,\boldsymbol{\theta})[\psi(\tilde{n}+1) - \psi(\tilde{n}-x+1) + \ln(1-\tilde{p})]^2,$$

$$J_{12} = J_{21} = \sum_{x=0}^{[\tilde{n}]} f_2(x,\boldsymbol{\theta}) \frac{(x-\tilde{n}\tilde{p})}{\tilde{p}(1-\tilde{p})}[\psi(\tilde{n}+1) - \psi(\tilde{n}-x+1) + \ln(1-\tilde{p})],$$

$$J_{22} = \sum_{x=0}^{[\tilde{n}]} f_2(x,\boldsymbol{\theta}) \frac{(x-\tilde{n}\tilde{p})^2}{\tilde{p}^2(1-\tilde{p})^2},$$

where $[\tilde{n}]$ is the largest integer contained in $\tilde{n}$ and $\psi(x)$ is the psi-function. It is known that series expansions for the psi-function converge very slowly. But, for integer values of x, a recurrence $\psi(z-x) = \psi(z) - \sum_{k=1}^{x} 1/(\tilde{n}-k+1)$ can be used, from which it follows that

$$\psi(\tilde{n}+1) - \psi(\tilde{n}-x+1) = \sum_{k=1}^{x} \frac{1}{\tilde{n}-k+1}.$$

This result permits us to calculate all expressions containing $\psi(\tilde{n}+1) - \psi(\tilde{n}-x+1)$ with a very high accuracy.

The elements of the matrix **B** are

$$B_{j1} = \frac{1}{\sqrt{p_j(\boldsymbol{\theta})}} \sum_{x\in\Delta_j} f_2(x,\boldsymbol{\theta})[\psi(\tilde{n}+1) - \psi(\tilde{n}-x+1) + \ln(1-\tilde{p})],$$

$$B_{j2} = \frac{1}{\sqrt{p_j(\boldsymbol{\theta})}} \sum_{x\in\Delta_j} f_2(x,\boldsymbol{\theta}) \left(\frac{x-\tilde{n}\tilde{p}}{\tilde{p}(1-\tilde{p})}\right), \quad j = 1,\ldots,r.$$

The elements of the matrix **V** are

$$V_{11} = \tilde{n}\tilde{p}(1-\tilde{p}),$$
$$V_{12} = V_{21} = \tilde{n}\tilde{p} + \tilde{n}(2\tilde{n}-3)\tilde{p}^2 - 2\tilde{n}(\tilde{n}-1)\tilde{p}^3,$$
$$V_{22} = \tilde{n}\tilde{p} + \tilde{n}(6\tilde{n}-7)\tilde{p}^2 - 4\tilde{n}(\tilde{n}-1)(\tilde{n}-3)\tilde{p}^3 + 2\tilde{n}(\tilde{n}-1)(3-2\tilde{n})\tilde{p}^4.$$

The elements of the matrix **K** are

$$K_{11} = \sum_{x=0}^{[\tilde{n}]} x f_2(x,\boldsymbol{\theta})[\psi(\tilde{n}+1) - \psi(\tilde{n}-x+1) + \ln(1-\tilde{p})],$$

$$K_{12} = \sum_{x=0}^{[\tilde{n}]} x^2 f_2(x,\boldsymbol{\theta}) \left(\frac{x-\tilde{n}\tilde{p}}{\tilde{p}(1-\tilde{p})}\right),$$

$$K_{21} = \sum_{x=0}^{[\tilde{n}]} x^2 f_2(x,\boldsymbol{\theta})[\psi(\tilde{n}+1) - \psi(\tilde{n}-x+1) + \ln(1-\tilde{p})],$$

$$K_{22} = \sum_{x=0}^{[\tilde{n}]} x^2 f_2(x,\boldsymbol{\theta}) \left(\frac{x - \tilde{n}\tilde{p}}{\tilde{p}(1-\tilde{p})} \right).$$

The components $W_1(\bar{\boldsymbol{\theta}}_n)$ and $W_2(\bar{\boldsymbol{\theta}}_n)$ of the vector $\mathbf{W}_n(\bar{\boldsymbol{\theta}}_n) = (W_1(\bar{\boldsymbol{\theta}}_n), W_2(\bar{\boldsymbol{\theta}}_n))^T$ for Singh's test $Q_s^2(\tilde{\boldsymbol{\theta}}_n)$ for the model in (9.7) are

$$W_1(\tilde{\theta}_n) = \frac{1}{\sqrt{n}} \sum_{i=1}^{n} [\psi(\tilde{n}+1) + \psi(\tilde{n} - X_i + 1) + \ln(1-\tilde{p})]$$

and

$$W_2(\tilde{\theta}_n) = \frac{1}{\sqrt{n}} \sum_{i=1}^{n} \frac{(X_i - \tilde{n}\tilde{p})}{\tilde{p}(1-\tilde{p})}.$$

It has to be noted that the test in (3.25) is computationally much more complicated than the statistic $Y2_n^2(\bar{\boldsymbol{\theta}}_n)$ for large samples.

For the model in (9.7), $\mathbf{E}_{\boldsymbol{\theta}} X = \tilde{n}\tilde{p}$ and $\mathbf{E}_{\boldsymbol{\theta}} X^2 = \tilde{n}\tilde{p} + \tilde{n}(\tilde{n}-1)\tilde{p}^2$. Denoting the first two sample moments $\frac{1}{n}\sum_{i=1}^{n} X_i = a$ and $\frac{1}{n}\sum_{i=1}^{n} X_i^2 = b$ and then equating them to population moments, the MMEs of $\tilde{n}$ and $\tilde{p}$ are obtained as

$$\bar{\tilde{n}} = \frac{a^2}{a^2 - b + a} \quad \text{and} \quad \bar{\tilde{p}} = \frac{a}{\bar{\tilde{n}}}. \tag{9.8}$$

From (9.8), we see that negative values of $\bar{\tilde{n}}$ and $\bar{\tilde{p}}$ are possible, but the proportion of such estimates will be almost negligible for samples of size $n > 1000$. It seems that $Y2_n^2(\bar{\boldsymbol{\theta}}_n)$ test can be used for analyzing Rutherford's data, but the question about $\sqrt{n}$-consistency of the MMEs in (9.8) is still open.

To examine the rate of convergence of estimators $\bar{\tilde{n}}$ and $\bar{\tilde{p}}$ for sample sizes $n = 1000(500)3500$, we simulated 3,000 estimates of $\bar{\tilde{n}}$ and $\bar{\tilde{p}}$ assuming that $\tilde{n} = 84.79045$ and $\tilde{p} = 0.04566$, values that correspond to Rutherford's data. The power curve fit of $\langle \bar{\tilde{n}} \rangle$, the average value of estimates $\bar{\tilde{n}}$ for 3000 runs, in Figure 9.1 shows that $\langle \bar{\tilde{n}} \rangle \sim n^{-1.039}$ and $R^2 = 0.978$. The power curve fit of $\langle \bar{\tilde{p}} \rangle$ in Figure 9.2 gives $\langle \bar{\tilde{p}} \rangle \sim n^{-0.403}$ and $R^2 = 0.997$. To check for the distribution of the statistic $S1_n^2(\bar{\boldsymbol{\theta}}_n)$ under the null "Feller's" distribution ($\tilde{n} = 84.79045$ and $\tilde{p} = 0.04566$), we simulated $N = 1000$ values of $S1_n^2(\bar{\boldsymbol{\theta}}_n)$. The histogram of these values is well described by the χ_1^2 distribution (see Figure 9.3). The average value $\langle S1_n^2(\bar{\boldsymbol{\theta}}_n) \rangle = 1.016 \pm 0.051$ also does not contradict the assumption that the statistic $S1_n^2(\bar{\boldsymbol{\theta}}_n)$ follows in the limit the chi-squared distribution with one degree of freedom. Another important property of any test statistic is its independence from the unknown parameters. To check for this feature of the test $S1_n^2(\bar{\boldsymbol{\theta}}_n)$, we simulated $N = 1000$ values of $S1_n^2(\bar{\boldsymbol{\theta}}_n)$ assuming that $\tilde{n} = 42.3952$ (two times less than for the null hypothesis H_0) and $\tilde{p} = 0.091132$ (two times more than for the null hypothesis H_0). The results

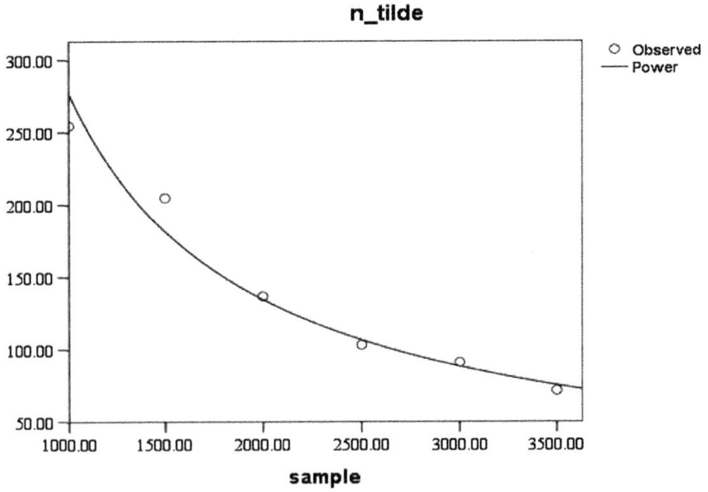

FIGURE 9.1 Simulated average value of $\langle\bar{\tilde{n}}\rangle$ (circles) and the power function fit (solid line) as function of the sample size n.

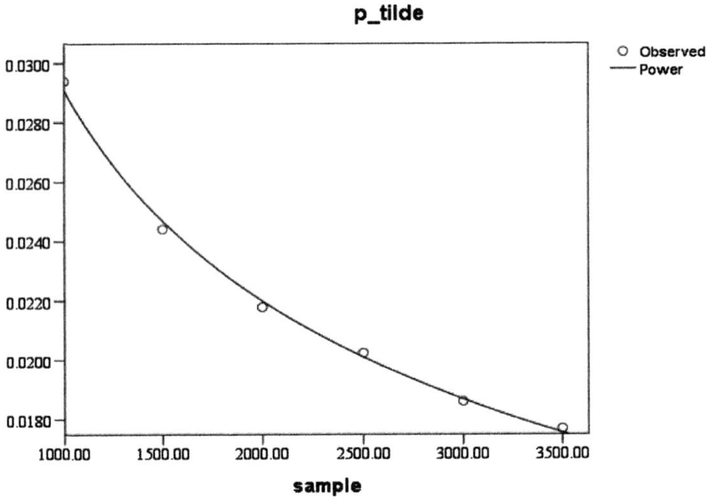

FIGURE 9.2 Simulated average value of $\langle\bar{\tilde{p}}\rangle$ (circles) and the power function fit (solid line) as function of the sample size n.

(Figure 9.4) show that the simulated values do not contradict the independence, because the histogram is again well described by χ_1^2 distribution.

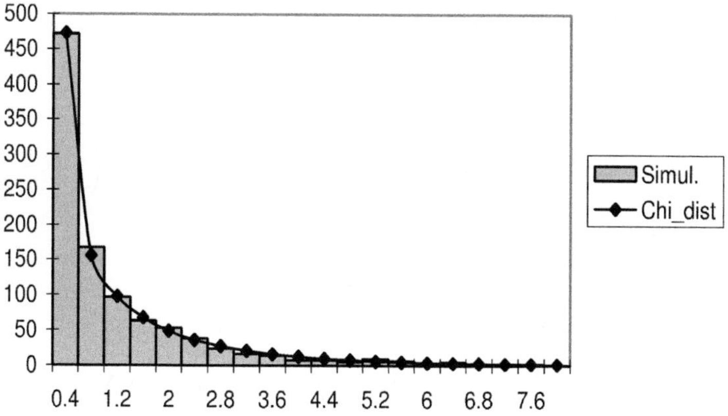

FIGURE 9.3 The histogram of the 1000 simulated values of $S1_n^2(\bar{\boldsymbol{\theta}}_n)$ for the null hypothesis ($\tilde{n} = 84.79045$ and $\tilde{p} = 0.04566$) and the χ_1^2 distribution (solid line).

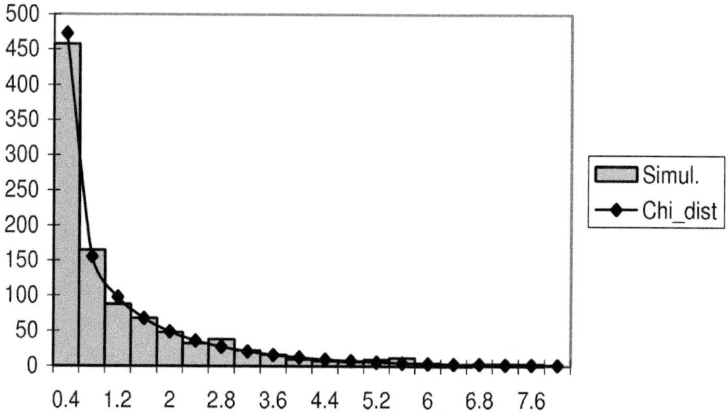

FIGURE 9.4 The histogram of the 1000 simulated values of $S1_n^2(\bar{\boldsymbol{\theta}}_n)$ for $\tilde{n} = 42.3952, \tilde{p} = 0.091132$ and the χ_1^2 distribution (solid line).

The above results evidently allow us to use the HRM statistic $S1_n^2(\bar{\boldsymbol{\theta}}_n)$ for Rutherford's data analysis.

9.2 ELEMENTS OF MATRICES K, B, C, AND V FOR THE THREE-PARAMETER WEIBULL DISTRIBUTION

For r equiprobable cells of the model in (4.25), the borders of equiprobable intervals are defined as:

$$x_j = \mu + \theta \left[-\ln\left(1 - \frac{j}{r}\right)\right]^{1/p}, \quad j = 0,1,\ldots,r.$$

Then, the elements of the matrix **K** are as follows:

$$K_{11} = 1, \quad K_{12} = \frac{1}{p}\Gamma\left(\frac{1}{p}\right), \quad K_{13} = -\frac{\theta}{p^2}\Gamma'\left(1+\frac{1}{p}\right),$$

$$K_{21} = 2\mu + 2\theta\Gamma\left(1+\frac{1}{p}\right), \quad K_{22} = 2\mu\Gamma\left(1+\frac{1}{p}\right) + 2\theta\Gamma\left(1+\frac{1}{p}\right),$$

$$K_{23} = -\frac{2\mu\theta}{p^2}\Gamma'\left(1+\frac{1}{p}\right) - \frac{2\theta^2}{p^2}\Gamma'\left(1+\frac{2}{p}\right),$$

$$K_{31} = 3\mu^2 + 6\mu\theta\Gamma\left(1+\frac{1}{p}\right) + 3\theta^2\Gamma\left(1+\frac{2}{p}\right),$$

$$K_{32} = 3\mu^2\Gamma\left(1+\frac{1}{p}\right) + 6\mu\theta\Gamma\left(1+\frac{2}{p}\right) + 3\theta^2\Gamma\left(1+\frac{3}{p}\right),$$

$$K_{33} = -\frac{3\mu^2\theta}{p^2}\Gamma'\left(1+\frac{1}{p}\right) - \frac{6\mu\theta^2}{p^2}\Gamma'\left(1+\frac{2}{p}\right) - \frac{3\theta^3}{p^2}\Gamma'\left(1+\frac{3}{p}\right),$$

where $\Gamma'(x) = \Gamma(x)\psi(x)$ and $\psi(x)$ is the psi-function. For the required calculation of $\psi(x)$, we used the series expansion

$$\psi(a) = -C + (a-1)\sum_{k=0}^{\infty}\frac{1}{(k+1)(k+a)},$$

where $C = 0.57721566\ldots$ is the Euler's constant.

Similarly, the elements of the matrices **B** and **C** are as follows:

$$B_{j1} = \frac{p}{\theta\sqrt{p_j}}\left(\frac{x_{j-1}-\mu}{\theta}\right)^{p-1}\exp\left\{-\left(\frac{x_{j-1}-\mu}{\theta}\right)^{p}\right\}$$
$$-\frac{p}{\theta\sqrt{p_j}}\left(\frac{x_{j}-\mu}{\theta}\right)^{p-1}\exp\left\{-\left(\frac{x_{j}-\mu}{\theta}\right)^{p}\right\},$$

$$B_{j2} = \frac{p}{\theta\sqrt{p_j}}\left(\frac{x_{j-1}-\mu}{\theta}\right)^{p}\exp\left\{-\left(\frac{x_{j-1}-\mu}{\theta}\right)^{p}\right\}$$
$$-\frac{p}{\theta\sqrt{p_j}}\left(\frac{x_{j}-\mu}{\theta}\right)^{p}\exp\left\{-\left(\frac{x_{j}-\mu}{\theta}\right)^{p}\right\},$$

$$B_{j3} = -\frac{1}{\sqrt{p_j}}\left(\frac{x_{j-1}-\mu}{\theta}\right)^{p}\ln\left(\frac{x_{j-1}-\mu}{\theta}\right)\exp\left\{-\left(\frac{x_{j-1}-\mu}{\theta}\right)^{p}\right\}$$
$$+\frac{1}{\sqrt{p_j}}\left(\frac{x_{j}-\mu}{\theta}\right)^{p}\ln\left(\frac{x_{j}-\mu}{\theta}\right)$$
$$\times\exp\left\{-\left(\frac{x_{j}-\mu}{\theta}\right)^{p}\right\}, \quad j = 1,\ldots,r.$$

$$C_{j1} = \frac{\mu}{\sqrt{p_j}}\exp\left\{-\left(\frac{x_{j-1}-\mu}{\theta}\right)^{p}\right\} - \frac{\mu}{\sqrt{p_j}}\exp\left\{-\left(\frac{x_{j}-\mu}{\theta}\right)^{p}\right\}$$

$$-\sqrt{p_j}m_1 + \frac{\theta\gamma}{\sqrt{p_j}}\gamma\left[\left(1+\frac{1}{p}\right),\left(\frac{x_j-\mu}{\theta}\right)^p\right]$$
$$-\frac{\theta\gamma}{\sqrt{p_j}}\gamma\left[\left(1+\frac{1}{p}\right),\left(\frac{x_{j-1}-\mu}{\theta}\right)^p\right],$$
$$C_{j2} = \frac{\mu^2}{\sqrt{p_j}}\exp\left\{-\left(\frac{x_{j-1}-\mu}{\theta}\right)^p\right\} - \frac{\mu^2}{\sqrt{p_j}}\exp\left\{-\left(\frac{x_j-\mu}{\theta}\right)^p\right\}$$
$$-\sqrt{p_j}m_2 + \frac{2\mu\theta}{\sqrt{p_j}}\gamma\left[\left(1+\frac{1}{p}\right),\left(\frac{x_j-\mu}{\theta}\right)^p\right]$$
$$-\frac{2\mu\theta}{\sqrt{p_j}}\gamma\left[\left(1+\frac{1}{p}\right),\left(\frac{x_{j-1}-\mu}{\theta}\right)^p\right]$$
$$+\frac{\theta^2}{\sqrt{p_j}}\gamma\left[\left(1+\frac{2}{p}\right),\left(\frac{x_j-\mu}{\theta}\right)^p\right]$$
$$-\frac{\theta^2}{\sqrt{p_j}}\gamma\left[\left(1+\frac{2}{p}\right),\left(\frac{x_{j-1}-\mu}{\theta}\right)^p\right],$$
$$C_{j3} = \frac{\mu^3}{\sqrt{p_j}}\exp\left\{-\left(\frac{x_{j-1}-\mu}{\theta}\right)^p\right\} - \frac{\mu^3}{\sqrt{p_j}}\exp\left\{-\left(\frac{x_j-\mu}{\theta}\right)^p\right\}$$
$$-\sqrt{p_j}m_3 + \frac{3\mu^2\theta}{\sqrt{p_j}}\gamma\left[\left(1+\frac{1}{p}\right),\left(\frac{x_j-\mu}{\theta}\right)^p\right]$$
$$-\frac{3\mu^2\theta}{\sqrt{p_j}}\gamma\left[\left(1+\frac{1}{p}\right),\left(\frac{x_{j-1}-\mu}{\theta}\right)^p\right]$$
$$+\frac{3\mu\theta^2}{\sqrt{p_j}}\gamma\left[\left(1+\frac{2}{p}\right),\left(\frac{x_j-\mu}{\theta}\right)^p\right]$$
$$-\frac{3\mu\theta^2}{\sqrt{p_j}}\gamma\left[\left(1+\frac{2}{p}\right),\left(\frac{x_{j-1}-\mu}{\theta}\right)^p\right]$$
$$+\frac{\theta^3}{\sqrt{p_j}}\gamma\left[\left(1+\frac{3}{p}\right),\left(\frac{x_j-\mu}{\theta}\right)^p\right]$$
$$-\frac{\theta^3}{\sqrt{p_j}}\gamma\left[\left(1+\frac{3}{p}\right),\left(\frac{x_{j-1}-\mu}{\theta}\right)^p\right], \quad j=1,\ldots,r,$$

where the population moments are

$$m_i = \sum_{l=0}^{i}\binom{i}{l}\theta^{i-l}\mu^l\Gamma\left(1+\frac{i-l}{p}\right), \quad i=1,2,3,$$

and $\gamma(a,x) = \int_0^x t^{a-1}e^{-t}\,dt$ is the incomplete gamma function. For the required calculation of $\gamma(a,x)$, we used the following series expansion

(Prudnikov et al., 1981, p. 705):

$$\gamma(a,x) = x^a \sum_{n=0}^{\infty} \frac{(-1)^n x^n}{n!(a+n)}.$$

Finally, the elements of the matrix **V** are as follows:

$$\begin{aligned} V_{ij} &= m_{i+j}(\boldsymbol{\theta}) - m_i(\boldsymbol{\theta})m_j(\boldsymbol{\theta}) \\ &= \sum_{n=0}^{i+j} \binom{i+j}{n} \theta^{i+j-n}\mu^n \Gamma\left(1 + \frac{i+j-n}{p}\right) \\ &\quad - \sum_{n=0}^{i} \binom{i}{n} \theta^{i-n}\mu^n \Gamma\left(1 + \frac{i-n}{p}\right) \\ &\quad \times \sum_{n=0}^{j} \binom{j}{n} \theta^{j-n}\mu^n \Gamma\left(1 + \frac{j-n}{p}\right), \quad i,j = 1,2,3. \end{aligned}$$

9.3 ELEMENTS OF MATRICES J AND B FOR THE GENERALIZED POWER WEIBULL DISTRIBUTION

Elements J_{ij} of the Fisher information matrix **J** are as follows:

$$J_{11} = \frac{\nu^2}{\gamma^2\theta^2} \int_1^{\infty} \left[\frac{(y^\gamma - 1)(y-1) - \gamma}{y^\gamma} \right]^2 \exp(1-y)dy,$$

$$\begin{aligned} J_{12} = J_{21} &= \frac{\nu}{\gamma\theta} \int_1^{\infty} \left[\frac{(y^\gamma - 1)(y-1) - \gamma}{y^\gamma} \right] \\ &\quad \times \left[\frac{(y-1)\ln y^\gamma}{\gamma^2} - \frac{1}{\gamma} \right] \exp(1-y)dy, \end{aligned}$$

$$\begin{aligned} J_{13} = J_{31} &= \frac{\nu}{\gamma\theta} \int_1^{\infty} \left[\frac{(y^\gamma - 1)(y-1) - \gamma}{y^\gamma} \right] \left[\frac{1}{\nu} + \ln((y^\gamma - 1)^{1/\nu}) \right. \\ &\quad \left. + \frac{(y^\gamma - 1)\ln((y^\gamma - 1)^{1/\nu})(1 - \gamma - y)}{\gamma y^\gamma} \right] \exp(1-y)dy, \end{aligned}$$

$$J_{22} = \int_1^{\infty} \left[\frac{(y-1)\ln(y^\gamma)}{\gamma^2} - \frac{1}{\gamma} \right]^2 \exp(1-y)dy,$$

$$J_{23} = J_{32} = \int_1^{\infty} \left[\frac{(y-1)\ln(y^\gamma)}{\gamma^2} - \frac{1}{\gamma} \right] \left[\frac{1}{\nu} + \ln((y^\gamma - 1)^{1/\nu}) \right.$$

$$+\frac{(y^{\gamma}-1)\ln((y^{\gamma}-1)^{1/\nu})(1-y-\gamma)}{\gamma y^{\gamma}}\Bigg]\exp(1-y)dy,$$

$$J_{33}=\int_1^{\infty}\Bigg[\frac{1}{\nu}+\ln((y^{\gamma}-1)^{1/\nu})$$
$$+\frac{(y^{\gamma}-1)\ln((y^{\gamma}-1)^{1/\nu})(1-y-\gamma)}{\gamma y^{\gamma}}\Bigg]^2\exp(1-y)dy.$$

Next, the elements of matrix **B** are as follows:

$$B_{11}=-\frac{f(a_1(\boldsymbol{\theta}))}{\sqrt{p_1}}\left\{\left[1-\ln\left(1-\frac{1}{r}\right)\right]^{\gamma}-1\right\}^{1/\nu},$$

$$B_{12}=-\frac{\theta f(a_1(\boldsymbol{\theta}))}{\sqrt{p_1}\nu}\left\{\left[1-\ln\left(1-\frac{1}{r}\right)\right]^{\gamma}-1\right\}^{\frac{1}{\nu}-1}$$
$$\times\left[1-\ln\left(1-\frac{1}{r}\right)\right]^{\gamma}\ln\left[1-\ln\left(1-\frac{1}{r}\right)\right],$$

$$B_{13}=\frac{\theta f(a_1(\boldsymbol{\theta}))}{\sqrt{p_1}\nu^2}\left\{\left[1-\ln\left(1-\frac{1}{r}\right)\right]^{\gamma}-1\right\}^{\frac{1}{\nu}}\ln\left\{\left[1-\ln\left(1-\frac{1}{r}\right)\right]^{\gamma}-1\right\},$$

$$B_{j1}=\frac{1}{\sqrt{p_j}}\Bigg\{-f(a_j(\boldsymbol{\theta}))\left\{\left[1-\ln\left(1-\frac{j}{r}\right)\right]^{\gamma}-1\right\}^{1/\nu}$$
$$+f(a_{j-1}(\boldsymbol{\theta}))\left\{\left[1-\ln\left(1-\frac{j-1}{r}\right)\right]^{\gamma}-1\right\}^{1/\nu},\quad j=2,\ldots,r-1,$$

$$B_{j2}=\frac{\theta}{\sqrt{p_j}\nu}\Bigg\{-f(a_j(\boldsymbol{\theta}))\left\{\left[1-\ln\left(1-\frac{j}{r}\right)\right]^{\gamma}-1\right\}^{\frac{1}{\nu}-1}$$
$$\times\left[1-\ln\left(1-\frac{j}{r}\right)\right]^{\gamma}\ln\left[1-\ln\left(1-\frac{j}{r}\right)\right]$$
$$+f(a_{j-1}(\boldsymbol{\theta}))\left\{\left[1-\ln\left(1-\frac{j-1}{r}\right)\right]^{\gamma}-1\right\}^{\frac{1}{\nu}-1}$$
$$\times\left[1-\ln\left(1-\frac{j-1}{r}\right)\right]^{\gamma}$$
$$\times\ln\left[1-\ln\left(1-\frac{j-1}{r}\right)\right]\Bigg\},\quad j=2,\ldots,r-1,$$

$$B_{j3}=\frac{\theta}{\sqrt{p_j}\nu^2}\Bigg\{f(a_j(\boldsymbol{\theta}))\left\{\left[1-\ln\left(1-\frac{j}{r}\right)\right]^{\gamma}-1\right\}^{\frac{1}{\nu}}$$
$$\times\ln\left\{\left[1-\ln\left(1-\frac{j}{r}\right)\right]^{\gamma}-1\right\}$$

$$
-f(a_{j-1}(\boldsymbol{\theta}))\left\{\left[1-\ln\left(1-\frac{j-1}{r}\right)\right]^{\gamma}-1\right\}^{\frac{1}{\nu}}
$$

$$
\times\ln\left\{\left[1-\ln\left(1-\frac{j-1}{r}\right)\right]^{\gamma}-1\right\}\Bigg\},\quad j=2,\ldots,r-1,
$$

$$
B_{r1}=\frac{1}{\sqrt{p_r}}f(a_{r-1}(\boldsymbol{\theta}))\left\{\left[1-\ln\left(\frac{1}{r}\right)\right]^{\gamma}-1\right\}^{1/\nu},
$$

$$
B_{r2}=\frac{\theta}{\sqrt{p_r}\nu}f(a_{r-1}(\boldsymbol{\theta}))\left\{\left[1-\ln\left(\frac{1}{r}\right)\right]^{\gamma}-1\right\}^{\frac{1}{\nu}-1}
$$

$$
\times\left[1-\ln\left(\frac{1}{r}\right)\right]^{\gamma}\ln\left[1-\ln\left(\frac{1}{r}\right)\right],
$$

$$
B_{r3}=-\frac{\theta}{\sqrt{p_r}\nu^2}f(a_{r-1}(\boldsymbol{\theta}))\left\{\left[1-\ln\left(\frac{1}{r}\right)\right]^{\gamma}-1\right\}^{\frac{1}{\nu}}
$$

$$
\times\ln\left\{\left[1-\ln\left(\frac{1}{r}\right)\right]^{\gamma}-1\right\}.
$$

9.4 ELEMENTS OF MATRICES J AND B FOR THE TWO-PARAMETER EXPONENTIAL DISTRIBUTION

Consider the two-parameter exponential distribution with cumulative distribution function

$$
F(x,\boldsymbol{\theta})=1-\exp\left\{-\frac{x-\mu}{\theta}\right\},\quad x\geqslant\mu,\ \theta>0,\ \mu\in R^1, \tag{9.9}
$$

where the unknown parameter $\boldsymbol{\theta}=(\theta,\mu)^T$. It is easily verified that the matrix $\mathbf{J}$ for the model in (9.9) is

$$
\mathbf{J}=\begin{pmatrix}\frac{1}{\theta^2} & 0\\ 0 & \frac{1}{\theta^2}\end{pmatrix}. \tag{9.10}
$$

Based on the set of n i.i.d. random variables $X_1,\ldots,X_n$, the MLE $\hat{\boldsymbol{\theta}}_n$ of the parameter $\boldsymbol{\theta}$ equals $\hat{\boldsymbol{\theta}}_n=(\hat{\theta}_n,\hat{\mu}_n)^T$, where

$$
\hat{\theta}_n=\frac{1}{n-1}\sum_{i=1}^{n}(X_i-X_{(1)})\quad\text{and}\quad\hat{\mu}_n=X_{(1)}. \tag{9.11}
$$

Consider r disjoint equiprobable intervals

$$
\Delta_j=\left\{\mu-\theta\ln\left(1-\frac{j-1}{r}\right),\mu-\theta\ln\left(1-\frac{j}{r}\right)\right\},\quad j=1,\ldots,r.
$$

For these intervals, the elements of the matrix **B** (see Eq. (3.4)) are

$$B_{j1} = \frac{\sqrt{r}}{\theta}\left\{\left(1-\frac{j}{r}\right)\ln\left(1-\frac{j}{r}\right)-\left(1-\frac{j-1}{r}\right)\ln\left(1-\frac{j-1}{r}\right)\right\},$$

$$B_{j2} = \frac{1}{\theta\sqrt{r}}, \quad j = 1,\ldots,r.$$

Using the matrix in (9.10) and the above elements of the matrix **B** with θ replaced by the MLE $\hat{\theta}_n$ in (9.11), the NRR test $Y1_n^2(\hat{\theta}_n)$ (see Eq. (3.8)) can be used. While using Microsoft Excel, the calculations based on double precision is recommended.

9.5 ELEMENTS OF MATRICES B, C, K, AND V TO TEST THE LOGISTIC DISTRIBUTION

Let $b_0 = -\infty, b_j = \bar{\theta}_{1n} - \sqrt{3}\bar{\theta}_{2n}\ln\left((r-j)/j\right)/\pi, j = 1,2,\ldots,r-1$, and $b_r = +\infty$, be borders of r equiprobable random grouping intervals. Then, the probabilities of falling into each interval are $p_i = 1/r, i = 1,\ldots,r$.

The elements of the $r \times 2$ matrix **B**, for $j = 1,\ldots,r$, are as follows:

$$B_{j1} = \frac{\pi}{\sqrt{3p_j}\theta_2}\left[\frac{\exp\left(-\frac{\pi(b_{j-1}-\theta_1)}{\sqrt{3}\theta_2}\right)}{\left\{1+\exp\left(-\frac{\pi(b_{j-1}-\theta_1)}{\sqrt{3}\theta_2}\right)\right\}^2} - \frac{\exp\left(-\frac{\pi(b_j-\theta_1)}{\sqrt{3}\theta_2}\right)}{\left\{1+\exp\left(-\frac{\pi(b_j-\theta_1)}{\sqrt{3}\theta_2}\right)\right\}^2}\right],$$

$$B_{j2} = \frac{\pi}{\sqrt{3p_j}\theta_2^2}\left[\frac{(b_{j-1}-\theta_1)\exp\left(-\frac{\pi(b_{j-1}-\theta_1)}{\sqrt{3}\theta_2}\right)}{\left\{1+\exp\left(-\frac{\pi(b_{j-1}-\theta_1)}{\sqrt{3}\theta_2}\right)\right\}^2} - \frac{(b_j-\theta_1)\exp\left(-\frac{\pi(b_j-\theta_1)}{\sqrt{3}\theta_2}\right)}{\left\{1+\exp\left(-\frac{\pi(b_j-\theta_1)}{\sqrt{3}\theta_2}\right)\right\}^2}\right].$$

Next, the elements of the $r \times 2$ matrix **C** are as follows:

$$C_{11} = -\frac{(b_1-\theta_1)}{\sqrt{p_1}\left\{1+\exp\left(\frac{\pi(b_1-\theta_1)}{\sqrt{3}\theta_2}\right)\right\}} - \frac{\sqrt{3}\theta_2}{\sqrt{p_1}\pi}\ln\left\{1+\exp\left(-\frac{\pi(b_1-\theta_1)}{\sqrt{3}\theta_2}\right)\right\},$$

$$C_{j1} = \frac{(b_{j-1}-\theta_1)}{\sqrt{p_j}\left\{1+\exp\left(\frac{\pi(b_{j-1}-\theta_1)}{\sqrt{3}\theta_2}\right)\right\}} - \frac{(b_j-\theta_1)}{\sqrt{p_j}\left\{1+\exp\left(\frac{\pi(b_j-\theta_1)}{\sqrt{3}\theta_2}\right)\right\}} + \frac{\sqrt{3}\theta_2}{\pi\sqrt{p_j}}\ln\frac{\left\{1+\exp\left(-\frac{\pi(b_{j-1}-\theta_1)}{\sqrt{3}\theta_2}\right)\right\}}{\left\{1+\exp\left(-\frac{\pi(b_j-\theta_1)}{\sqrt{3}\theta_2}\right)\right\}}, \quad j = 2,\ldots,r-1,$$

$$C_{r1} = \frac{(b_{r-1}-\theta_1)}{\sqrt{p_r}\left\{1+\exp\left(\frac{\pi(b_{r-1}-\theta_1)}{\sqrt{3}\theta_2}\right)\right\}} + \frac{\sqrt{3}\theta_2}{\sqrt{p_r}\pi}\ln\left\{1+\exp\left(-\frac{\pi(b_{r-1}-\theta_1)}{\sqrt{3}\theta_2}\right)\right\},$$

$$C_{12} = \frac{(b_1^2-\theta_1^2-\theta_2^2)}{\sqrt{p_1}\left\{1+\exp\left(-\frac{\pi(b_1-\theta_1)}{\sqrt{3}\theta_2}\right)\right\}} - \frac{2\sqrt{3}\theta_2 b_1}{\sqrt{p_1}\pi}\ln\left\{1+\exp\left(\frac{\pi(b_1-\theta_1)}{\sqrt{3}\theta_2}\right)\right\} - \frac{6\theta_2^2}{\pi^2\sqrt{p_1}}\mathrm{Li}_2\left(-\exp\left(\frac{\pi(b_1-\theta_1)}{\sqrt{3}\theta_2}\right)\right),$$

$$C_{j2} = -\frac{b_j^2-\theta_1^2}{\sqrt{p_j}\left\{1+\exp\left(\frac{\pi(b_j-\theta_1)}{\sqrt{3}\theta_2}\right)\right\}} + \frac{b_{j-1}^2-\theta_1^2}{\sqrt{p_j}\left\{1+\exp\left(\frac{\pi(b_{j-1}-\theta_1)}{\sqrt{3}\theta_2}\right)\right\}} + \frac{2\sqrt{3}\theta_2}{\pi\sqrt{p_j}}\ln\left[\frac{\left\{1+\exp\left(-\frac{\pi(b_{j-1}-\theta_1)}{\sqrt{3}\theta_2}\right)\right\}^{b_{j-1}}}{\left\{1+\exp\left(-\frac{\pi(b_j-\theta_1)}{\sqrt{3}\theta_2}\right)\right\}^{b_j}}\right] + \frac{6\theta_2^2}{\pi^2\sqrt{p_j}}\mathrm{Li}_2\left(-\exp\left(-\frac{\pi(b_j-\theta_1)}{\sqrt{3}\theta_2}\right)\right) - \frac{6\theta_2^2}{\pi^2\sqrt{p_j}}\mathrm{Li}_2\left(-\exp\left(-\frac{\pi(b_{j-1}-\theta_1)}{\sqrt{3}\theta_2}\right)\right) + \frac{\theta_2^2}{\sqrt{p_j}\left\{1+\exp\left(-\frac{\pi(b_{j-1}-\theta_1)}{\sqrt{3}\theta_2}\right)\right\}} - \frac{\theta_2^2}{\sqrt{p_j}\left\{1+\exp\left(-\frac{\pi(b_j-\theta_1)}{\sqrt{3}\theta_2}\right)\right\}}, \quad j=2,\ldots,r-1,$$

$$C_{r2} = \frac{(b_{r-1}^2-\theta_1^2-\theta_2^2)}{\sqrt{p_r}\left\{1+\exp\left(\frac{\pi(b_{r-1}-\theta_1)}{\sqrt{3}\theta_2}\right)\right\}} - \frac{2\sqrt{3}\theta_2 b_{r-1}}{\pi\sqrt{p_r}}\ln\left\{1+\exp\left(-\frac{\pi(b_{r-1}-\theta_1)}{\sqrt{3}\theta_2}\right)\right\} - \frac{6\theta_2^2}{\pi^2\sqrt{p_r}}\mathrm{Li}_2\left(-\exp\left(\frac{\pi(b_{r-1}-\theta_1)}{\sqrt{3}\theta_2}\right)\right),$$

where $\mathrm{Li}_2(-x)$ is Euler's dilogarithm function that can be computed by the series expansion

$$\mathrm{Li}_2 = \sum_{k=1}^{\infty} \frac{(-x)^k}{k^2} \quad \text{for } x \leqslant (\sqrt{5}-1)/2,$$

and by the expansion

$$\mathrm{Li}_2(-x) = \sum_{k=1}^{\infty} \frac{1}{k^2(1+x)^k} + \frac{1}{2}\ln^2(1+x) - \ln x \ln(1+x) - \frac{\pi^2}{6}$$

for $x > (\sqrt{5}-1)/2$ (Prudnikov et al., 1986, p. 763).

Finally, we have the matrices **K** and **V** as

$$\mathbf{K} = \begin{pmatrix} 1 & 0 \\ 2\theta_1 & 2\theta_2 \end{pmatrix} \quad \text{and} \quad \mathbf{V} = \begin{pmatrix} \theta_2^2 & 2\theta_1\theta_2^2 \\ 2\theta_1\theta_2^2 & 4\theta_1^2\theta_2^2 + \frac{16}{5}\theta_2^4 \end{pmatrix}.$$

9.6 TESTING FOR NORMALITY

System requirements for implementing the software of Sections 9.6–9.10 are Windows XP, Windows 7, MS Office 2003, 2007, 2010.

1. Open file Testing Normality.xls;
2. Enter your sample data in column "I" starting from cell 1;
3. Click the button "Compute," introduce the sample size and the desired number of equiprobable intervals ($4 \leqslant r \leqslant 200$). The recommended number of intervals for the NRR test $Y1_n^2(\hat{\boldsymbol{\theta}}_n)$ in (3.8), under close alternatives (such as the logistic), is $4 \leqslant r \leqslant n/5$. The recommended number of intervals for the test $S_n^2(\hat{\boldsymbol{\theta}}_n)$ in (3.24) is $r = n/5$ (see Section 4.4.1). Note that the power of $S_n^2(\hat{\boldsymbol{\theta}}_n)$ can be more than that of the NRR test;
4. Click OK;
5. Numerical values of $Y1_n^2(\hat{\boldsymbol{\theta}}_n)$ and $S_n^2(\hat{\boldsymbol{\theta}}_n)$ are in cells F2 and G2, respectively. Cells F3 and G3 contain the corresponding percentage points at level 0.05. The P-values of $Y1_n^2(\hat{\boldsymbol{\theta}}_n)$ and $S_n^2(\hat{\boldsymbol{\theta}}_n)$ are in cells F4 and G4, respectively.

9.7 TESTING FOR EXPONENTIALITY

9.7.1 Test of Greenwood and Nikulin (see Section 3.6.1)

1. Open file Testing Exp GrNik.xls;
2. Enter your sample data in column "I" starting from cell 1;

3. Click the button "Compute," introduce the sample size and desired number of equiprobable intervals. The recommended number of equiprobable intervals is $2 \leqslant r \leqslant 6$;
4. Click OK;
5. The numerical value of Y_n^2 (see Eq. (3.44)) is in cell F2. The percentage point at level 0.05 and the P-value are in cells F3 and F4, respectively.

9.7.2 Nikulin-Rao-Robson test (see Eq. (3.8) and Section 9.4)

1. Open file Testing NRR 2-param EXP.xls;
2. Enter your sample data in column "I" starting from cell 1;
3. Click the button "Compute," introduce the sample size and desired number of equiprobable intervals. The recommended number of equiprobable intervals is $4 \leqslant r \leqslant 6$;
4. Click OK;
5. The numerical value of $Y1_n^2(\hat{\boldsymbol{\theta}}_n)$ is in cell F2. The percentage point at level 0.05 and the P-value are in cells F3 and F4, respectively.

9.8 TESTING FOR THE LOGISTIC

1. Open file Testing Logistic.xls;
2. Enter your sample data in column "I" starting from cell 1;
3. Click the button "Compute," introduce the sample size and desired number of equiprobable intervals ($4 \leqslant r \leqslant 200$). The recommended number of equiprobable intervals, for close alternatives (such as normal), is $n/20 < r \leqslant n/10$;
4. Click OK;
5. Numerical values of $Y2_n^2(\bar{\boldsymbol{\theta}}_n)$ in (4.9) and $S1_n^2(\bar{\boldsymbol{\theta}}_n)$ in (4.13) are in cells E2 and F2, respectively. Cells E3 and F3 contain the corresponding percentage points at level 0.05. The P-values of $Y2_n^2(\bar{\boldsymbol{\theta}}_n)$ and $S1_n^2(\bar{\boldsymbol{\theta}}_n)$ are in cells E4 and F4, respectively.

9.9 TESTING FOR THE THREE-PARAMETER WEIBULL

1. Open file Testing Weibull3.xls;
2. Enter your sample data in column "I" starting from cell 1;
3. Click the button "Compute," introduce the sample size and desired number of equiprobable intervals ($5 \leqslant r \leqslant 200$). The recommended number of equiprobable intervals for the Exponentiated Weibull and Power Generalized Weibull alternatives is $r = n/5$;
4. Click OK;

5. Numerical values of $Y2_n^2(\bar{\boldsymbol{\theta}}_n)$ in (4.9) and $S1_n^2(\bar{\boldsymbol{\theta}}_n)$ in (4.14) (see also Section 9.2) are in cells F2 and G2, respectively. Note that the power of $S1_n^2(\bar{\boldsymbol{\theta}}_n)$ is usually higher than that of $Y2_n^2(\bar{\boldsymbol{\theta}}_n)$. Cells F3 and G3 contain the corresponding percentage points at level 0.05. The P-values of $Y2_n^2(\bar{\boldsymbol{\theta}}_n)$ and $S1_n^2(\bar{\boldsymbol{\theta}}_n)$ are in cells F4 and G4, respectively.

9.10 TESTING FOR THE POWER GENERALIZED WEIBULL

1. Open file Test for PGW (Left-tailed).xls;
2. Enter your sample data in column "I" starting from cell 1;
3. Click the button "Run," introduce the sample size and desired number of equiprobable intervals ($5 \leqslant r \leqslant 200$). The recommended number of equiprobable intervals for the Exponentiated Weibull, Generalized Weibull, and Three-Parameter Weibull alternatives is $r = n/5$;
4. Click OK. Note that the power of $S_n^2(\hat{\boldsymbol{\theta}}_n)$ in (3.50) is usually higher than that of $Y1_n^2(\hat{\boldsymbol{\theta}}_n)$ in (3.48);
5. Numerical values of $Y1_n^2(\hat{\boldsymbol{\theta}}_n)$ and $S_n^2(\hat{\boldsymbol{\theta}}_n)$ (see Eqs. (3.48), (3.50) and Section 9.3) are in cells F2 and G2, respectively. Cells F3 and G3 contain the corresponding percentage points at level 0.05. The P-values of $Y1_n^2(\hat{\boldsymbol{\theta}}_n)$ and $S_n^2(\hat{\boldsymbol{\theta}}_n)$ are in cells F4 and G4, respectively.

9.11 TESTING FOR TWO-DIMENSIONAL CIRCULAR NORMALITY

1. Open file Testing Circular Normality.xls;
2. Enter your sample data in columns "I" and "J" starting from cell 1;
3. Click the button "Compute," introduce the sample size and the desired number of equiprobable intervals. The recommended number of intervals for the two-dimensional logistic alternative is $5 \leqslant r \leqslant 10$, while the recommended number of intervals for the two-dimensional normal alternative is 3;
4. Click OK;
5. Numerical values of $Y1_n^2(\hat{\boldsymbol{\theta}}_n)$ and $S_n^2(\hat{\boldsymbol{\theta}}_n)$ (see Section 3.5.3) are in cells F2 and G2, respectively. Cells F3 and G3 contain the corresponding percentage points at level 0.05. The P-values of $Y1_n^2(\hat{\boldsymbol{\theta}}_n)$ and $S_n^2(\hat{\boldsymbol{\theta}}_n)$ are in cells F4 and G4, respectively.

Bibliography

Abdel-Aty, S.H., 1954. Approximate formulae for the percentage points and the probability integral of the non-central χ^2 distribution. Biometrika 41, 538–540.

Aguirre, N., Nikulin, M., 1994a. Chi-squared goodness-of-fit tests for the family of logistic distributions. Kybernetika 30, 214–222.

Aguirre, N., Nikulin, M., 1994b. Chi-squared goodness-of-fit tests for the family of logistic distributions. Qüestiió 18, 317–335.

Ahmad, I.A., Alwasel, I.A., 1999. A goodness-of-fit test for exponentiality based on the memoryless property. Journal of the Royal Statistical Society, Series B 61, 681–689.

Akritas, M.G., 1988. Pearson-type goodness-of-fit tests: The Univariate case. Journal of the American Statistical Association 83, 222–230.

Akritas, M.G., Torbeyns, A.F., 1997. Pearson-type goodness-of-fit tests for regression. The Canadian Journal of Statistics 25, 359–374.

Alloyarova, R., Nikulin, M., Pya, N., Voinov, V., 2007. The power-generalized Weibull probability distribution and its use in survival analysis. Communications in Dependability and Quality Management 10, 5–15.

Ampadu, C., 2008. On the powers of some new chi-square type statistics. Far East Journal of Theoretical Statistics 26, 59–72.

Anderson, T.W., Darling, D.A., 1954. A test of goodness of fit. Journal of the American Statistical Association 49, 765–769.

Anderson, G., 1994. Simple tests of distributional form. Journal of Econometrics 62, 265–276.

Anderson, G., 1996. Nonparametric tests of stochastic dominance in income distributions. Econometrica 64, 1183–1193.

Andrews, D.W.K., 1988. Chi-square diagnostic tests for econometric models: Theory. Econometrica 56, 1419–1453.

Ascher, S., 1990. A survey of tests for exponentiality. Communications in Statistics - Theory and Methods 19, 1811–1825.

Bagdonavičius, V., Clerjaud, L., Nikulin, M., 2006. Power generalized Weibull in accelerated life testing, Preprint 0602, Statistique Mathématique et ses Applications, I.F.R. 99, Université Victor Segalen Bordeaux 2, p. 28.

Bagdonavičius, V., Kruopis, J., Nikulin, M., 2011a. Nonparametric Tests for Complete Data, John Wiley & Sons, Hoboken, New Jersey.

Bagdonavičius, V., Kruopis, J., Nikulin, M., 2011b. Nonparametric Tests for Censored Data, John Wiley & Sons, Hoboken, New Jersey.

Bagdonavičius, V., Nikulin M., 2002. Accelerated Life Models. Chapman and Hall, Boca Raton, Florida.

Bagdonavičius, V., Nikulin, M., 2011. Chi-squared goodness of fit test for right censored data. The International Journal of Applied Mathematics and Statistics (IJAMAS), 24, 30–50.

Bagdonavičius, V., Nikulin, M., Tahir, R., 2010. Chi-square goodness-of-fit test for right censored data. In: Chateauneuf, A. (Ed.), Proceedings of ALT'2010, ALT'2010, May 19–21, 2010, Clermont-Ferrand, France, pp. 223–230.

Baglivo J., Olivier D., Pagano M., 1992. Methods of exact goodness-of-fit tests. Journal of the American Statistical Association 87, 464–446.

Baird, D., 1983. The Fisher/Pearson chi-squared controversy: a turning point for inductive inference. British Journal for the Philosophy of Science 34, 105–118.

Bajgier, S.M., Aggarwal, L.K., 1987. A stepwise algorithm for selecting category boundaries for the chi-squared goodness-of-fit test. Communications in Statistics–Theory and Methods 16, 2061–2081.

Balakrishnan, N., 1983. Empirical power study of a multi-sample test of exponential based on spacings. Journal of Statistical Computation and Simulation 18, 265–271.

Balakrishnan, N., 1990. Approximate maximum likelihood estimation for a generalized logistic distribution. Journal of Statistical Planning and Inference 26, 221–236.
Balakrishnan, N. (Ed.), 1992. Handbook of the Logistic Distribution. Marcel Dekker, New York.
Balakrishnan, N., Basu, A.P. (Eds.), 1995. The Exponential Distribution: Theory, Methods and Applications, Taylor and Francis, Newark, New Jersey.
Balakrishnan, N., Brito, M.R., Quiroz, A.J., 2007. A vectorial notion of skewness and its use in testing for multivariate symmetry. Communications in Statistics-Theory Methods 36, 1757–1767.
Balakrishnan, N., Cohen, A.C., 1990. Order Statistics and Inference: Estimation Methods, Academic Press, Boston.
Balakrishnan, N., Kateri, M., 2008. On the maximum likelihood estimation of parameters of Weibull distribution based on complete and censored data. Statistics & Probability Letters 78, 2971–2975.
Balakrishnan, N., Lai, C.D., 2009. Continuous Bivariate Distributions, second ed. Springer, New York.
Balakrishnan, N., Nevzorov, V.B., 2003. A Primer on Statistical Distributions. John Wiley & Sons, Hoboken, New Jersey.
Balakrishnan, N., Scarpa, B., 2012. Multivariate measures of skewness for the skew-normal distribution. Journal of Multivariate Analysis 104, 73–87.
Balakrishnan, N., Zhu, X., 2012. On the existence and uniqueness of the maximum likelihood estimates of parameters of Birnbaum-Saunders distribution based on Type-I, Type-II and hybrid censored samples. Statistics (to appear).
Barlow, R.E., Proshan, F., 1991. Life distribution models and incomplete data, In: Krishnaiah, P.R., Rao, C.R. (Eds.), Handbook of Statistics, vol. 7. North Holland, Amsterdam, pp. 225–250
Bemis, K.G., Bhapkar, V.P., 1983. On BAN estimators for chi squared test criteria. The Annals of Statistics 11, 183–196.
Berkson, J., 1940. A note on the chi-square test, the Poisson and the binomial. Journal of the American Statistical Association 35, 362–367.
Best, D.J., Rayner, J.C.W., 1981. Are two classes enough for goodness of fit test? Statistica Neerlandica 3, 157–163.
Best, D.J., Rayner, J.C.W., 2006. Improved testing for the binomial distribution using chi-squared components with data-dependent cells. Journal of Statistical Computation and Simulation 76, 75–81.
Bhalerao, N.R., Gurland, J., Tripathi, R.C., 1980. A method of increasing power of a test for the negative binomial and Neyman Type A distributions. Journal of the American Statistical Association 75, 934–938.
Billingsley, P., 1968. Convergence of Probability Measures. John Wiley & Sons, New York.
Birch, M.W., 1964. A new proof of the Pearson-Fisher theorem. The Annals of Mathematical Statistics 35, 817–824.
Birnbaum, Z.W., Saunders, S.C., 1969a. A new family of life distributions. Journal of Applied Probability 6, 637–652.
Birnbaum, Z.W., Saunders, S.C., 1969b. Estimation for a family of life distributions with applications to fatigue. Journal of Applied Probability 6, 328–347.
Boero, G., Smith, J., Wallis, K.F., 2004a. Decompositions of Pearson's chi-squared test. Journal of Econometrics 123, 189–193.
Boero G., Smith J., Wallis, K.F., 2004b. The sensitivity of chi-squared goodness-of-fit tests to the partitioning of data. Econometric Reviews 23, 341–370.
Bol'shev, L.N., 1963. Asymptotical Pearson's transformations. Theory of Probability and Its Applications 8, 129–155.
Bol'shev, L.N., 1965. A characterization of the Poisson probability distribution and its statistical applications. Theory of Probability and Its Applications 10, 488–499.
Bol'shev, L.N., Mirvaliev, M., 1978. Chi-square goodness-of-fit test for the Poisson, binomial, and negative binomial distributions. Theory of Probability and its Applications 23, 481–494 (in Russian).
Boulerice, B., Ducharme, G.R., 1995. A note on smooth tests of goodness-of-fit for location-scale families. Biometrika 82, 437–438.
Brownlee, K.A., 1960. Statistical Theory and Methodology in Science and Engineering. John Wiley & Sons, New York.
Brown, B.M., 1975. A method for combining non-independent one-sided tests of significance. Biometrics 31, 987–992.

Brown, B.M., Hettmansperger, T.P., 1996. Normal scores, normal plots, and tests for normality. Journal of the American Statistical Association 91, 1668–1675.

Cabana, A., Cabana, E.M., 1997. Transformed empirical processes and modified Kolmogorov-Smirnov tests for multivariate distributions. The Annals of Statistics 25, 2388–2409.

Callen, E., Leidecker Jr., H., 1971. A mean life on the Supreme Court. American Bar Association Journal 57, 1188–1192.

Castillo, J., Puig, P., 1999. The best test of exponentiality against singly truncated normal alternatives. Journal of the American Statistical Association 94, 529–532.

Chernoff, H., Lehmann, E.L., 1954. The use of maximum likelihood estimates in tests for goodness of fit. The Annals of Mathematical Statistics 25, 579–589.

Chibisov, D.M., 1971. Some chi-squared type criteria for continuous distribution. Theory of Probability and its Applications 16, 4–20.

Chichagov, V.V., 2006. Unbiased estimates and chi-squared statistic for one-parameter exponential family. In: Statistical Methods of Estimation and Hypotheses Testing, vol. 19. Perm State University, Perm, Russia, pp. 78–89.

Clark, J., 1997. Business Statistics, Barron's Educational Series, Inc.

Cochran, W.G., 1952. The χ^2 test of goodness of fit. Annals of Mathematical Statistics 23, 315–345.

Cochran, W.G., 1954. Some methods for strengthening the common χ^2 tests. Biometrics 10, 417–451.

Cohen, A., Sackrowitz, H.B., 1975. Unbiasedness of the chi-square, likelihood ratio, and other goodness of fit tests for the equal cell case. The Annals of Statistics 3, 959–964.

Cohen, J.D., 1988. Noncentral chi-square: some observations on recurrence. The American Statistician 42, 120–122.

Cordeiro, G.M., Ferrari, S.L.D.P., 1991. A modified score test statistic having chi-squared distribution to order n^{-1}. Biometrika 78, 573–582.

Cox, D.R., Hinkley, D.V., 1974. Theoretical Statistics. John Wiley & Sons, New York.

Cramer, H., 1946. Mathematical Methods of Statistics. Princeton University Press, Princeton, New Jersey.

Cressie, N., Read, T.A.C., 1984. Multinomial goodness-of-fit tests. Journal of the Royal Statistical Society – Series B 46, 440–464.

Cressie, N., Read, T.A.C., 1989. Pearson's X^2 and the loglikelihood ratio statistic G^2: a comparative review. International Statistical Review 57, 19–43.

D'Agostino, R.B., Belanger, A., D'Agostino, R.B., Jr. 1990. A suggestion for using powerful and informative tests of normality. The American Statistician 44, 316–321.

D'Agostino, R.B., Pearson, E.S., 1973. Test for departure from normality. Empirical results for the distributions of b_2 and $\sqrt{b_1}$. Biometrika 60, 613–622.

Dahiya, R.C., Gurland, J., 1972a. Pearson chi-squared test of fit with random intervals. Biometrika 59, 147–153.

Dahiya, R.C., Gurland, J., 1972b. Goodness of fit tests for the gamma and exponential distributions. Technometrics 14, 791–801.

Dahiya, R.C., Gurland, J., 1973. How many classes in Pearson chi-squared test? Journal of the American Statistical Association 68, 707–712.

Damico, J., 2004. A new one-sample test for goodness-of-fit. Communications in Statistics–Theory and Methods 33, 181–193.

Davies, R.B., 2002. Hypothesis testing when a nuisance parameter is present only under alternative: linear model case. Biometrika 89, 484–489.

Deng, X., Wan, S., Zhang, B., 2009. An improved goodness-of-fit test for logistic regression models based on case-control data by random partition. Communications in Statistics – Simulation and Computation 38, 233–243.

Ding, Ch.G., 1992. Computing the non-central χ^2 distribution function. Applied Statistics 41, 478–482.

Doornik, J.A., Hansen, H., 1994. An omnibus test for univariate and multivariate normality. Oxford Bulletin of Economics and Statistics 70, 927–939.

Drost, F.C., 1988. Asymptotics for generalized chi-square goodness-of-fit tests. Center for Mathematics and Computer Sciences CWI Tract, 48, Stichting Mathematisch Centrum, Amsterdam.

Drost, F.C., 1989. Generalized chi-squared goodness-of-fit tests for location-scale models when the number of classes tends to infinity. The Annals of Statistics 17, 1285–1300.

Dudley, R., 1976. Probabilities and metrics-convergence of laws on metric spaces with a review of statistical testing, Lecture Notes, Vol. 45, Aarhus University, Denmark.
Dudley, R., 1979. On chi-square tests of composite hypotheses. In: Probability Theory, Banach Center Publications, Warsaw, 5, 75–87.
Dzhaparidze, K.O., 1983. On iterative procedures of asymptotic inference. Statistica Neerlandica 37, 181–188.
Dzhaparidze, K.O., Nikulin, M.S., 1974. On a modification of the standard statistic of Pearson. Theory of Probability and its Applications 19, 851–853.
Dzhaparidze, K.O., Nikulin, M.S., 1982. Probability distributions of the Kolmogorov and omega-square statistics for continuous distributions with shift and scale parameters. Journal of Soviet Mathemetics 20, 2147–2164.
Dzhaparidze, K.O., Nikulin, M.S., 1992. On evaluation of statistics of chi-square type tests. Problems of the Theory of Probability Distributions, vol. 12. Nauka, St. Petersburg, pp. 59–90.
Efron, B., 1988. Logistic regression, survival analysis, and the Kaplan-Meier curve. Journal of the American Statistical Association 83, 414–425.
Efron, B., 1998. R.A. Fisher in the 21st century. Statistical Science 13, 95–122.
Engelhardt, M., Bain, L.J., 1975. Tests of two-parameter exponentiality against three-parameter Weibull alternatives. Technometrics 17, 353–356.
Eubank, R.L., 1997. Testing goodness of fit with multinomial data. Journal of the American Statistical Association 92, 1084–1093.
Feller, W., 1948. On probability problems in the theory of counters. In: Courant Anniversary Volume. Interscience Publishers, New York, pp. 105–115.
Feller, W., 1964. An Introduction to Probability Theory and Its Applications, vol. 1. Mir Publishers, Moscow (in Russian).
Fisher, R.A., 1924. The condition under which χ^2 measures the discrepancy between observation and hypothesis. Journal of the Royal Statistical Society 87, 442–450.
Fisher, R.A., 1925a. Partition of χ^2 into its components. In: Statistical Methods for Research Workers, Oliver and Boyd, Edinburg, Scotland.
Fisher, R.A., 1925b. Theory of statistical estimation. Proceedings of the Cambridge Philosophical Society 22, 700–725.
Fisher, R.A., 1928. On a property connecting the χ^2 measure of discrepancy with the method of maximum likelihood. Atti de Congresso Internazionale di Mathematici, Bologna, 6, 94–100.
Folks, J.L., Chhikara, R.S., 1978. The inverse Gaussian distribution and its statistical applications – a review. Journal of the Royal Statistical Society, Series B 40, 263–289.
Follmann, D., 1996. A simple multivariate test for one-sided alternatives. Journal of the American Statistical Association 91, 854–861.
Gbur, E.E., 1981. On the index of dispersion. Communications in Statistics – Simulation and Computation **10**, 531–535.
Gerville-Reache, L., Nikulin, M., Tahir, R., 2013. On reliability approach and statistical inference in demography. Journal of Reliability and statistical studies (to appear).
Gilula, Z., Krieger, A.M., 1983. The decomposability and monotonicity of Pearson's chi-square for collapsed contingency tables with applications. Journal of the American Statistical Association 78, 176–180.
Gleser, L.J., Moore, D.S., 1983. The effect of dependence on chi-squared and empirical distribution test of fit. The Annals of Statistics 11, 1100–1108.
Graneri, J., 2003. χ^2-type goodness of fit test based on transformed empirical processes for location and scale families. Communications in Statistics – Theory and Methods 32, 2073–2087.
Greenwood, P.E., Nikulin, M.S., 1996. A Guide to Chi-squared Testing. John Wiley & Sons, New York.
Grizzle, J.E., 1961. A new method of testing hypotheses and estimating parameters for the logistic model. Biometrics 17, 372–385.
Guenther, W.C., 1977. Power and sample size for approximate chi-square tests. The American Statistician 31, 83–85.
Gulati, S., Neus, J., 2003. Goodness of fit statistics for the exponential distribution when the data are grouped. Communications in Statistics - Theory and Methods 32, 681–700.

Gulati, S., Shapiro, S., 2008. Goodness-of-fit tests for Pareto distribution. In: Vonta, F., Nikulin, MM., Limnios, N., Huber-Carol, C. (Eds.), Statistical Methods for Biomedical and Technical Systems. Birkhäuser, Boston, pp. 259–274.
Gumbel, E.J., 1954. Applications of the circular normal distribution. Journal of the American Statistical Association 49, 267–297.
Gunes, H., Dietz, D.C., Auclair, P.F., Moore, A.H., 1997. Modified goodness-of-fit tests for the inverse Gaussian distribution. Computational Statistics and Data Analysis 24, 63–77.
Haberman, S.J., 1988. A warning on the use of chi-squared statistics with frequency tables with small expected cell counts. Journal of the American Statistical Association 83, 555–560.
Habib, M.G., Thomas, D.R., 1986. Chi-square goodness-of-fit tests for randomly censored data. The Annals of Statistics 14, 759–765.
Hadi, A.S., Wells, M.T., 1990. A note on generalized Wald's method. Metrika 37, 309–315.
Hall, P., 1985. Tailor-made tests of goodness of fit. Journal of the Royal Statistical Society, Series B 47, 123–131.
Harris, R.R., Kanji, G.K., 1983. On the use of minimum chi-square estimation. The Statistician 32, 379–394.
Harter, H.L. 1991. Weibull, log-Weibull and gamma order statistics. In: Krishnaiah, P.R., Rao, C.R. (Eds.), Handbook of Statistics, vol. 7. North Holland, Amsterdam, pp. 433-466.
Harter, H.L., Balakrishnan, N., 1996. CRC Handbook of Tables for the Use of Order Statistics in Estimation. CRC Press, Boca Raton, Florida.
Harter, H.L., Moore, A.H., 1967. Maximum likelihood estimation, from censored samples, of the parameters of a logistic distribution. Journal of the American Statistical Association 62, 675–684.
Heckman, J.J., 1984. The goodness of fit statistic for models with parameters estimated from microdata. Econometrica 52, 1543–1547.
Henze, N., 2002. Invariant tests for multivariate normality: a critical review. Statistical Papers 43, 467–506.
Henze, N., Meintanis, S.G., 2002. Goodness-of-fit tests based on new characterization of the exponential distribution. Communications in Statistics – Theory and Methods 31, 1479–1497.
Henze, N., Wagner, T., 1997. A new approach to the BHEP tests for multivariate normality. Journal of Multivariate Analysis 62, 1–23.
Hjort, N.L., 1990. Goodness of fit tests in models for life history data based on cumulative hazard rates. The Annals of Statistics 18, 1221–1258.
Hodges, J.L., Le Cam, L., 1960. The poisson approximation to the binomial distribution. Annals of Mathematical Statistics 31, 737-740.
Hollander, M., Pena, E.A., 1992. A chi-squared goodness-of-fit test for randomly censored data. Journal of the American Statistical Association 87, 458–463.
Holst, L., 1972. Asymptotic normality and efficiency for certain goodness-of-fit tests. Biometrika 59, 137–145.
Hsuan, T.A., Robson, D.S., 1976. The goodness-of-fit tests with moment type estimators. Communications in Statistics - Theory and Methods 5, 1509–1519.
Hutchinson, T.P., 1979. The validity of the chi-squared test when expected frequencies are small: A list of recent research references. Communications in Statistics – Theory and Methods 8, 327–335.
Hwang, J.T., Casella, G., Robert, C., Wells, M.T., Farrell, R.H., 1992. Estimation of accuracy in testing. The Annals of Statistics 20, 490–509.
Ivchenko, G.I., Medvedev, U.I., 1980. Divisible statistics and hypothesis testing for grouped data. Theory of Probability and Its Applications 15, 549–560.
Johnson, N.L, Kemp, A.W., Kotz, S., 2005. Univariate Discrete Distributions, third ed. John Wiley & Sons, Hoboken, New Jersey.
Johnson, N.L., Kotz, S., Balakrishnan, N., 1994. Continuous Univariate Distributions, vol. 1, second ed. John Wiley & Sons, New York.
Johnson, N.L., Kotz, S., Balakrishnan, N., 1995. Continuous Univariate Distributions, second ed., vol. 2. John Wiley & Sons, New York.
Johnson, D., 1997. The triangular distribution as a proxy for the beta distribution in risk analysis. Journal of the Royal Statistical Society Series D, 46, 387–398.
Johnson, V.E., 2004. A Bayesian test for goodness-of-fit. The Annals of Statistics 32, 2361–2384.
Jung, S.-H., Ahn, C., Donner, A., 2001. Evaluation of an adjusted chi-square statistic as applied to observational studies involving clustered binary data. Statistics in Medicine 20, 2149–2161.

Jung, S-H., Kang, S-H., Ahn, C., 2003. Chi-square test for $R \times C$ contingency tables with clustered data. Journal of Biopharmaceutical Statistics 13, 241–251.

Kallenberg, W.C.M., Oosterhoff, J., Schriever, B.F., 1985. The number of classes in chi-squared goodness-of-fit tests. Journal of the American Statistical Association 80, 959–968.

Karagrigoriou, A., Mattheou, K., 2010. On distributional properties and goodness-of-fit tests for generalized measures of divergence. Communications in Statistics – Theory and Methods 39, 472–482.

Khatri, C.G., 1968. Some results for the singular normal multivariate regression. Sankhyā, Series A 30, 267–280.

King, G., 1987. Presidential appointments to the Supreme Court. American Politics Quarterly 15, 373–386.

King, G., 1988. Statistical models for political science event counts: bias in conventional procedures and evidence for the exponential Poisson regression model. American Journal of Political Science 32, 838–863.

Kocherlakota, S., Kocherlakota, K., 1986. Goodness of fit tests for discrete distributions. Communications in Statistics – Simulation and Computation 15, 815–829.

Koehler, K.J., Gan, F.F., 1990. Chi-squared goodness-of-fit tests: cell selection and power. Communications in Statistics – Simulation and Computation 19, 1265–1278.

Kolchin, V.Ph., 1968. A class of limit theorems for conditional distributions. Lit. Mat. Sbornik 8, 53–63.

Kotz, S., Balakrishnan, N., Johnson, N.L., 2000. Continuous Multivariate Distributions—Vol. 1, second edition, John Wiley & Sons, New York.

Kowalski, C.J., 1970. The performance of some rough tests for bivariate normality before and after coordinate transformations to normality. Technometrics 12, 517–544.

Koziol, J.A., Perlman, M.D., 1978. Combining independent chi-squared tests. Journal of the American Statistical Association 73, 753–763.

Kundu, D., Kannan, N., Balakrishnan, N., 2008. On the hazard function of the Birnbaum-Saunders distribution and associated inference. Computational Statistics & Data Analysis 50, 2692–2702.

Lancaster, H.O., 1951. Complex contingency tables treated by the partition of χ^2. Journal of the Royal Statistical Society 13, 242–249.

Larntz, K., 1978. Small-sample comparisons of exact levels for chi-squared goodness-of-fit statistics. Journal of the American Statistical Association 73, 253–263.

Lawal, H.B., 1980. Tables of percentage points of Pearson's goodness-of-fit statistic for use with small expectations. Applied Statistics 29, 292–298.

Le Cam, L., Mahan, C., Singh, A., 1983. An extension of the theorem of H. Chernoff and E.L. Lehmann, In: Recent Advances in Statistics, Academic Press, Orlando, Florida, pp. 303–332.

Lehmann, E.L., 1959. Testing Statistical Hypotheses. John Wiley & Sons, New York.

Leiva, V., Riquelme, M., Balakrishnan, N., Sanhueza, A., 2008. Lifetime analysis based on the generalized Birnbaum-Saunders distribution. Computational Statistics & Data Analysis 52, 2079–2097.

Lemeshko, B.Yu., 1998. Asymptotically optimal data grouping in goodness of fit criteria. Industrial Laboratory 64, 56–64 (in Russian).

Lemeshko, B.Yu., Chimitova, E.V., 2000. χ^2 criteria's power maximization. Doklady SB AS HS 2, 53–61 (in Russian).

Lemeshko, B. Yu., Chimitova, E.V., 2002. χ^2 goodness-of-fit criteria and related mistakes. Measuring Techniques 6, 5–11 (in Russian).

Lemeshko, B.Yu., Chimitova, E.V., 2003. On a choice of the number of classes for goodness-of-fit tests. Industrial Laboratory. Diagnostic of Materials 69, 61–67 (in Russian).

Lemeshko, B.Y., Lemeshko, S.B., Akushkina, K.A., Nikulin, M.S., Saaidia, N., 2010. Inverse Gaussian model and its applications in reliability and survival analysis. In: Rykov, V.V., Balakrishnan, N., Nikulin, M.S. (Eds.), Mathematical and Statistical Models and Methods in Reliability. Birkhäuser, Boston, pp. 293–315.

Lemeshko, B.Yu., Lemeshko, S.B., Postovalov, S.N., 2007. The power of goodness of fit tests for close alternatives. Measurement Techniques 50, 132–141.

Lemeshko, B.Yu., Lemeshko, S.B., Postovalov, S.N., 2008. Comparative power analysis of goodness of fit tests for close alternatives. II. Testing compound hypotheses. Siberian Journal of Industrial Mathematics 11, 78–93 (in Russian).

Lemeshko, B.Yu., Lemeshko, S.B., Postovalov, E.V., Chimitova, E.V., 2011. Simulation and Study of Probability Regularities. Computer Approach. NTSU Publisher, Novosibirsk, Siberia.
Lemeshko, B.Yu., Postovalov, S.N., 1997. Applied aspects when using goodness-of-fit tests for compound hypotheses. Reliability and Quality Control 11, 3–17 (in Russian).
Lemeshko, B.Yu., Postovalov, S.N., 1998. On independence of limit distributions of chi-squared Pearson's and likelihood ratio tests on the way of data grouping. Industrial Laboratory 64, 56–63 (in Russian).
Lemeshko, B.Yu., Postovalov S.N., Chimitiva E.V., 2001. On the distribution and power of Nikulin's chi-squared test. Industrial Laboratory, Diagnostic of Materials 67, 52–58 (in Russian).
Li, G., Doss, H., 1993. Generalized Pearson-Fisher chi-square goodness-of-fit tests, with applications to models with life history data. The Annals of Statistics 21, 772–797.
Lindley, D.V., 1996. A brief history of statistics in the last 100 years. The Mathematical Gazette 80, 92–100.
Linnik, Yu. V., 1958. Method of Least Squares and Basics of the Theory of Observations' Reduction, FM, Moscow (in Russian).
Littell, R.C., Folks, J.L., 1971. Asymptotic optimality of Fisher's method of combining independent tests. Journal of the American Statistical Association 66, 802–806.
Lockhart, R.A., Stephens M.A., 1994. Estimation and test of fit for the three-parameter Weibull distribution. Journal of the Royal Statistical Society Series B 56, 491–500.
Looney, S.W., 1995. How to use tests for univariate normality to assess multivariate normality. The American Statistician 49, 64–70.
Lorenzen, G., 1992. A reformulation of Pearson's chi-square statistic and some extensions. Statistics and Probability Letters 14, 327–331.
Loukas, S., Kemp, C.D., 1986. On the chi-square goodness-of-fit statistic for bivariate discrete distributions. The Statistician 35, 525–529.
Malkovich, J.F., and Afifi, A.A., 1973. On the test for multivariate normality. Journal of the American Statistical Association 68, 176–179.
Mann, H.B., Wald, A., 1942. On the choice of the number of class intervals in the application of the chi-squared test. Annals of Mathematical Statistics 13, 306–317.
Marden, J.I., 1982. Combining independent noncentral chi squared or F tests. The Annals of Statistics 10, 266–277.
Mardia, K,V., 1970. Measures of multivariate skewness and kurtosis with applications. Biometrika 57, 519–530.
Mason, D.M., Schuenemeyer, J.H., 1983. A modified Kolmogorov-Smirnov test sensitive to tail alternatives. The Annals of Statistics 11, 933–946.
Mathew, T., Sinha, B.K., Zhou, L., 1993. Some statistical procedures for combining independent tests. Journal of the American Statistical Association 88, 912–919.
McCulloch, C.E., 1980. Symmetric matrix derivatives with applications, FSU Statistics Report M565, Department of Statistics, Florida State University, Tallahassee, Florida.
McCulloch, C.E., 1985. Relationships among some chi-squared goodness of fit statistics. Communications in Statistics – Theory and Methods 14, 593–603.
McEwen, P., Parresol, B.R., 1991. Moment expressions and summary statistics for the complete and truncated Weibull distribution. Communications in Statistics – Theory and Methods 20, 1361–1372.
McLaren, C.E., Legler, J.M., Brittenham, G.M., 1994. The generalized goodness-of-fit test. The Statistician 43, 247–258.
Mecklin, C.G., Mundfrom, D.J., 2004. An appraisal and bibliography of tests for multivariate normality. International Statistical Reviews 72, 123–138.
Members of the Supreme Court (1789 to Present). http:/www.supremecourtus.gov/about/members.pdf (last accessed November 27, 2008).
Mirvaliev, M., 2000. Invariant generalized chi-squared type statistics in problems of homogeneity. O'Zbekiston Respublikasi Fanlar Akademiyasining Ma'Ruzalari 2, 6–10.
Mirvaliev, M., 2001. An investigation of generalized chi-squared type statistics. Doctoral Dissertation, Academy of Sciences of the Republic of Uzbekistan, Tashkent.
Molinari, L., 1977. Distribution of the chi-squared test in non-standard situations. Biometrika 64, 115–121.
Moore, D.S., 1971. A chi-square statistic with random cell boundaries, Annals of Mathematical Statistics 42, 147–156.

Moore, D.S., 1977. Generalized inverses, Wald's method and the construction of chi-squared tests of fit. Journal of the American Statistical Association 72, 131–137.

Moore, D.S., 1982. The effect of dependence on chi-squared tests of fit. The Annals of Statistics 10, 1163–1171.

Moore, D.S., 1984. Measures of lack of fit from tests of chi-squared type. Journal of Statistical Planning and Inference 10, 151–166.

Moore, D.S., 1986. Tests of chi-squared type. In: D'Agostino, R.B., Stephens, M.A. (Eds.), Goodness-of-Fit Techniques. Marcel Dekker, New York, pp. 63–95.

Moore, D.S., Spruill, M.C., 1975. Unified large-sample theory of general chi-squared statistics for tests of fit. The Annals of Statistics 3, 599–616.

Moore, D.S., Stubblebine, J.B., 1981. Chi-square tests for multivariate normality with application to common stock prices. Communications in Statistics - Theory and Methods 10, 713–738.

Mudholkar, G.S., Srivastava, D.K., Freimer M., 1995. The exponentiated Weibull family: a reanalysis of the bus-motor-failure data. Technometrics 37, 436–445.

Mudholkar, G.S., Srivastava, D.K., Kollia, G.D., 1996. A generalization of the Weibull distribution with application to the analysis of survival data. Journal of the American Statistical Association 91, 1575–1583.

Nair, V.N., 1987. Chi-squared-type tests for ordered alternatives in contingency tables. Journal of the American Statistical Association 82, 283–291.

Nair, V.N., 1988. Testing in industrial experiments with ordered categorical data. Technometrics 28, 283–311.

Ng, H.K.T., Kundu, D., Balakrishnan, N., 2003. Modified moment estimation for the two-parameter Birnbaum-Saunders distribution. Computational Statistical & Data Analysis 43, 283–298.

Nikulin, M.S., 1973a. Chi-square test for normality. In: Proceedings of the International Vilnius Conference on Probability Theory and Mathematical Statistics, vol. 2, pp. 119–122.

Nikulin, M.S., 1973b. Chi-square test for continuous distributions with shift and scale parameters. Theory of Probability and its Applications 18, 559–568.

Nikulin, M.S., 1973c. Chi-square test for continuous distributions, Theory of Probability and its Applications 18, 638–639.

Nikulin, M., Balakrishnan, N., Tahir, R., Saaidia, N., 2011. Modified chi-squared goodness-of-fit test for Birnbaum-Saunders distribution. In: Nikulin, M., Balakrishnan, N., Lemeshko, B., (Eds.), Proceedings of the International Workshop on Applied Methods of Statistical Analysis, Simulations and Statistical Inference, AMSA'2011, September 20-22, 2011, pp. 87–99, NSTU Publisher, Novosibirsk, Russia.

Nikulin, M., Haghighi, F., 2006. A chi-squared test for the generalized power Weibull family for the head-and-neck cancer censored data. Journal of Mathematical Sciences 133, 1333–1341.

Nikulin, M.S., Saaidia, N., 2009. Inverse Gaussian family and its applications in reliability: study by simulation. In: Ermakov, S.M., Melas, V.B., Perelyshev, A.N. (Eds.), Proceedings of the 6th St. Petersburg Workshop on Simulation, June 28–July 4, St. Petersburg, Russia, Vol. II, pp. 657–663.

Nikulin, M., Solev, V., 1999. Chi-squared goodness of fit test for doubly censored data, applied in survival analysis and reliability. In: Limnios, N., Ionescu, D. (Eds.), Probabilistic and Statistical Models in Reliability. Birkhäuser, Boston, pp. 101–112.

Nikulin, M.S., Voinov, V.G., 1989. A chi-square goodness-of-fit test for exponential distributions of the first order. Lecture Notes in Mathematics, vol. 1412. Springer-Verlag, New York, pp. 239–258.

Nikulin, M.S., Voinov, V.G., 2006. Chi-squared testing. In: Kotz, S., Balakrishnan, N., Read, C.B.,Vidakovic, B. (Eds.), Encyclopedia of Statistical Sciences, vol. 1, second ed. John Wiley & Sons, Hoboken, New Jersey, pp. 912–921.

Oliver, R.M., 1961. A traffic counting distribution. Operations Research 9, 802–810.

Oluyede, B.O., 1994. A modified chi-square test of independence against a class of ordered alternatives in an $r \times c$ contingency table. The Canadian Journal of Statistics 22, 75–87.

Orlov, A.I., 1997. On goodness-of-fit criteria for a parametrical family. Industrial Laboratory 63, 49–50.

Osius, G., 1985. Goodness-of-fit tests for binary data with (possible) small expectations but large degrees of freedom. Statistics and Decisions, Supplement Issue 2, 213–224.

Paardekooper, H.C.H., Steens, H.B.A., Van der Hoek, G., 1989. A note on properties of iterative procedures of asymptotic inference. Statistica Neerlandica 43, 245–253.

Pardo, L., 2006. Statistical Inference based on Divergence Measures. Chapman & Hall/CRC Press, Boca Raton, Florida.
Park, C.J., 1973. The distribution of frequency counts of the geometric distribution. Sankhyā 35, 106–111.
Peña, E.A., 1998a. Smooth goodness-of-fit tests for the baseline hazard in Cox's proportional hazard model. Journal of the American Statistical Association 93, 673–692.
Peña, E.A., 1998b. Smooth goodness-of-fit tests for composite hypothesis in hazard based models. The Annals of Statistics 26, 1935–1971.
Pearson, K., 1900. On the criterion that a given system of deviations from the probable in the case of a correlated system of variables is such that it can be reasonably supposed to have arisen from random sampling, Philosophical Magazine 50, 157–175.
Pierce, D.A., Kopecky, K.J., 1979. Testing goodness-of-fit for the distribution of errors in regression models Biometrika 66, 1–5.
Plackett, R.L., 1983. Karl Pearson and chi-squared test. International Statistical Review 51, 59–72.
Press, S.J., 1972. Applied Multivariate Analysis, Rinehart and Winston, New York.
Prudnikov, A.P., Brychkov, Y.A., Marichev, O.I., 1981. Integrals and Series, Elementary Functions. Nauka, Moscow.
Prudnikov, A.P., Brychkov, Y.A., Marichev, O.I., 1986. Integrals and Series, Additional Chapters. Nauka, Moscow.
Rao, C.R., 1965. Linear Statistical Inference and its Applications, John Wiley & Sons, New York.
Rao, C.R., 2002. Karl Pearson, chi-square test and the dawn of statistical inference. In: Huber-Carol, C., Balakrishnan, N., Nikulin, M.S., Mesbah, M., (Eds.), Goodness-of-fit Tests and Model Validity, Birkhäuser, Boston, pp. 9–24
Rao, C.R., Mitra, S.K., 1971. Generalized Inverses of Matrices and Its Applications. John Wiley & Sons, New York.
Rao, K.C., Robson, D.S., 1974. A chi-squared statistic for goodness-of-fit tests within the exponential family. Communications in Statistics 3, 1139–1153.
Rayner, J.C.W., Best D.J., 1986. Neyman-type smooth tests for location-scale families. Biometrika 74, 437–446.
Rayner, G.D., 2002. Components of the Pearson-Fisher chi-squared statistic. Journal of Applied Mathematics and Decision Science 6, 241–254.
Read, T.R.C., 1984. Small-sample comparisons for the power divergence goodness-of-fit statistics. Journal of the American Statistical Association 79, 929–935.
Rice, W.R., 1990. A consensus combined p-value test and the family-wide significance of component tests. Biometrics 46, 303–308.
Ritchey, R.J., 1986. An application of the chi-squared goodness-of-fit test to discrete common stock returns. Journal of Business and Economic Statistics 4, 243–254.
Royston, J.P., 1983. Some techniques for assessing multivariate normality based on the Shapiro-Wilk W. Applied Statistics 32, 121–133.
Roy, A.R., 1956. On χ^2 statistics with variable intervals, Technical Report N1, Stanford University, Department of Statistics, Stanford, California..
Rutherford, E., Chadwick, J., Ellis C.D., 1930. Radiations from Radioactive Substances. Cambridge University Press, England.
Saaidia, N., Tahir, R., 2012. A modified chi-squared goodness-of-fit test for the inverse Gaussian distribution and its applications in reliability. In: Statistical Models and Methods for Reliability and Survival Analysis and their Validation, S2MRSA-Bordeaux, July 4–6, 2012.
Sanhueza, A., Leiva, V., Balakrishnan, N., 2008. The generalized Birnbaum-Saunders distribution and its theory, methodology and application. Communications in Statistics – Theory and Methods 37, 645–670.
Sarkar, S.K., Chang, C., 1997. The Simes method for multiple hypothesis testing with positively dependent test statistics. Journal of the American Statistical Association 92, 1601–1608.
Schrödinger, E., 1915. Zur Theorie der Fall- und Steigversuche an Teilchen mit Brownscher Bewegung. Physikalische Zeitschrift 16, 289–295.
Selby, B., 1965. The index of dispersion as a test statistic. Biometrika 52, 627–629.
Seshadri, V., 1993. The Inverse Gaussian Distribution: A Case Study in Exponential Families. Clarendon Press, Oxford.
Sevast'yanov, B.A., Chistyakov, V.P., 1964. Asymptotical normality in the classical pellets problem. Theory of Probability and Its Applications 9, 223–237.

Shapiro, S.S., Wilk, M.B., 1965. An analysis of variance test for normality. Biometrika 52, 591–611.

Sinclair, C.D., Spurr, B.D., 1988. Approximations to the distribution function of the Anderson-Darling test statistic. Journal of the American Statistical Association 83, 1190–1191.

Singh, A.C., 1986. Categorical data analysis for simple random samples. In: Proceedings of the Survey Research Methods Section of ASA, pp. 659–664.

Singh, A.C., 1987. On the optimality and a generalization of Rao-Robson's statistic. Communications in Statistics - Theory and Methods 16, 3255–3273.

Smith, R.L., Naylor., J.C., 1987. A comparison of maximum likelihood and Bayesian estimators for the three-parameter Weibull distribution. Applied Statistics 36, 358–369.

Spinelli, J.J., Stephens, M.A., 1987. Test for exponentiality when origin and scale parameters are unknown. Technometrics 29, 471–476.

Spinelli, J.J., Stephens, M.A., 1997. Cramer-von Mises tests of fit for the Poisson distribution. The Canadian Journal of Statistics 25, 257–268.

Spruill, M.C., 1976. A comparison of chi-square goodness-of-fit tests based on approximate Bahadur slope. The Annals of Statistics 2, 237–284.

Srivastava, M.S., Hui, T.K., 1987. On assessing multivariate normality based on Shapiro-Wilk W statistic. Statistics & Probability Letters 5, 15–18.

Stigler, S.M., 1977. Do robust estimators work with real data? The Annals of Statistics 5, 1055–1098.

Stigler, S.M., 2008. Karl Pearson's theoretical errors and the advances they inspired. Statistical Science 23, 261–271.

Tarone, R.E., 1979. Testing the goodness of fit of the binomial distribution. Biometrika 66, 585–590.

Thomas, D.R., Pierce, D.D., 1979. Neyman's smooth goodness-of-fit test when the hypothesis is composite. Journal of the American Statistical Association 74, 441–445.

Tserenbat, O., 1990. On one modified chi-squared statistic for multivariate normality testing, Izvestiya. Academy of Science of Uzbekistan, Physical and Mathematical Series, Tashkent, FAN Uzbekistan N4, pp. 40–47.

Tumanyan, S.Kh., 1956. Asymptotic distribution of χ^2 criterion when the number of observations and classes increase simultaneously. Theory of Probability and its Applications 1, 131–145.

Tumanyan, S.Kh., 1958. On power of the test against "close" alternatives, Izvestiya of the Armenian Academy of Science. Phys. Math. Series 11, 6–17 (in Russian).

Tweedie, M.C.K., 1957. Statistical properties of Inverse Gaussian Distributions. I. Annals of Mathematical Statistics 28, 362–377.

Ulmer, S.S., 1982. Supreme Court appointments as a Poisson distribution. American Journal of Political Science 26, 113–116.

Van Zwet, W.R., Oosterhoff, J., 1967. On the combination of independent test statistics. Annals of Mathematical Statistics 38, 659–680.

Van Der Vaart, A.W., 1998. Asymptotic Statistics, Cambridge University Press, Cambridge, England.

Voinov, V., 2006. On optimality of the Rao-Robson-Nikulin test. Industrial Laboratory 72, 65–70 (in Russian).

Voinov, V., 2010. A decomposition of Pearson-Fisher and Dzhaparidze-Nikulin statistics and some ideas for a more powerful test construction. Communications in Statistics – Theory and Methods 39, 667–677.

Voinov V., Alloyarova R., Pya N., 2008a. Recent achievements in modified chi-squared goodness-of-fit testing. In: Vonta, F., Nikulin, M., Limnios, N., Huber-Carol, C. (Eds.), Statistical Models and Methods for Biomedical and Technical Systems. Birkhäuser, Boston, pp. 245–262.

Voinov, V., Alloyarova, R., Pya, N., 2008b. A modified chi-squared goodness-of-fit test for the three-parameter Weibull distribution and its applications in reliability. In: Huber, C., Limnios, N., Mesbah, M., Nikulin, M., (Eds.), Mathematical Methods in Survival Analysis, Reliability and Quality of Life, John Wiley & Sons, London, pp. 193–206.

Voinov, V., Grebenyk, A., 1989. Combining dependent criteria and statistical filtration of patterns. MMPR-IV (Mathematical Methods of Pattern Recognition), Abstracts, Part 3, Sec. 2, Riga.

Voinov, V.G., Naumov, A., Pya, N.Y., 2003. Some recent advances in chi-squared testing. In: Proceedings of the International Conference on Advances in Statistical Inferential Methods, Theory and Applications, Almaty, Kazakhstan, pp. 233–247.

Voinov, V., Nikulin, M.S., 1984. A remark on the paper of Bol'shev and Mirvaliev on the exponential model of emission of α-particles. Theory of Probability and Its Applications 29, 174–175.

Voinov, V., Nikulin, M.S., 1993. Unbiased Estimators and Their Applications. Univariate Case, vol. 1. Kluwer Academic Publishers, Dordrecht, The Netherlands.

Voinov, V., Nikulin, M.S., 1994. Chi-square goodness-of-fit test for one and multidimensional discrete distributions. Journal of Mathematical Science 68, 438–451.

Voinov, V., Nikulin, M.S., 1996. Unbiased Estimators and Their Applications. Multivariate Case, vol. 2. Kluwer Academic Publishers, Dordrecht, The Netherlands.

Voinov, V., Nikulin, M.S., 2011. Chi-square goodness-of-fit tests: drawbacks and improvements. In: International Encyclopedia of Statistical Science. Springer, New York, pp. 246–250.

Voinov V., Pya, N., 2004. On the power of modified chi-squared goodness-of-fit tests for the family of logistic and normal distributions. In: Proceedings of the Seventh Iranian Statistical Conference. Allameh Tabatabaie University, pp. 385–403.

Voinov, V., Nikulin, M.S., Pya, N., 2007. Independent χ_1^2 distributed in the limit components of some chi-squared tests. In: Skiadas, C.H. (Ed.), Recent Advances in Stochastic Modeling and Data Analysis. World Scientific Publishers, New Jersey, pp. 243–250.

Voinov, V., Pya, N., 2010. A note on vector-valued goodness-of-fit tests. Communications in Statistics – Theory and Methods 39, 452–459.

Voinov, V., Pya, N., Alloyarova, R., 2009. A comparative study of some modified chi-squared tests. Communications in Statistics – Simulation and Computation 38, 355–367.

Voinov, V., Pya, N., Shapakov, N., Voinov, Y., 2012. Goodness of fit tests for the power generalized Weibull distribution, Communications in Statistics - Simulation and Computation (to appear).

Voinov, V., Pya, N., Voinov, Y., 2011. Goodness of fit tests for the power generalized Weibull probability distribution. In: Proceedings of ASMDA 2011, Applied Stochastic Models and Data Analysis, June 7–10, 2011, Rome, pp. 1395–1401.

Voinov, V., Voinov, E., 2008. Some new modified chi-squared goodness-of-fit tests and their application in particle physics and political science. In: Proceedings of the JSM. Social Statistics Section, American Statistical Association, Alexandria, Virginia, pp. 540–547.

Voinov, V., Voinov, E., 2010. A statistical reanalysis of the classical Rutherford's experiment. Communications in Statistics – Simulation and Computation 39, 157–171.

Voinov, V., Voinov, E., Rakhimova, R., 2010. Poisson versus binomial: Appointment of judges to the U.S. Supreme Court. Australian and New Zealand Journal of Statistics 52, 261–274.

Von Mises, R., 1972. Wahrscheinlichkeit, Statistik, und Wahrheit. Springer-Verlag, Vienna, Austria.

Vu, H.T.V., Maller, R.A., 1996. The likelihood ratio test for Poisson versus binomial distributions. Journal of the American Statistical Association 91, 818–824.

Wald, A., 1943. Tests of statistical hypothesis when the number of observations is large. Transactions of the American Mathematical Society 54, 426–482.

Wallis, W.A., 1936. The Poisson distribution and the Supreme Court. Journal of the American Statistical Association 31, 376–380.

Watson, G.S., 1958. On chi-square goodness-of-fit tests for continuous distributions, Journal of the Royal Statistical Society Series B, 20, 44–61.

Watson, G.S., 1959. Some recent results in goodness-of-fit tests. Biometrics 15, 440–468.

Weiers, R.M., 1991. Introduction to Business Statistics. The Dryden Press, Chicago, Illinois.

Wilk, M.B., Shapiro, S.S., 1968. The joint assessment of normality of several independent samples. Technometrics 10, 825–839.

Williams, C.A., 1950. On the choice of the number and width of classes for the chi-square test of goodness-of-fit. Journal of the American Statistical Association 45, 77–86.

Wilson, J.R., 1989. Chi-square tests for overdispersion with multiparameter estimates. Applied Statistics 38, 441–453.

Yarnold, J.K., 1970. The minimum expectation in goodness of fit tests and the accuracy of approximations for the null distribution. Journal of the American Statistical Association 65, 864–886.

Zacks, S., 1971. The Theory of Statistical Inference. John Wiley & Sons, New York.

Zhakharov, V.K., Sarmanov, O.V., Sevastyanov, B.A., 1969. A sequential chi-squared test. Sbornik Mathematics 79, 444–460.

Zhang, B., 1999. A chi-squared goodness-of-fit test for logistic regression models based on case-control data. Biometrika 86, 531–539.

Zhang, J-T., 2005. Approximate and asymptotic distributions of chi-squared-type mixtures with applications. Journal of the American Statistical Association 100, 273–285.

Zhou, L., Mathew, T., 1993. Combining independent tests in linear models. Journal of the American Statistical Association 88, 650–655.

Index

Printed and bound by CPI Group (UK) Ltd, Croydon, CR0 4YY
08/05/2025
01864852-0001